VOLUME ONE HUNDRED AND FOUR

SEMICONDUCTORS AND SEMIMETALS

Diamond for Quantum Applications Part 2

SERIES EDITORS

CHENNUPATI JAGADISH

Distinguished Professor
Department of Electronic Materials Engineering
Research School of Physics and Engineering
Australian National University
Canberra, ACT2601, Australia

ZETIAN MI

Professor
Department of Electrical Engineering and Computer Science
University of Michigan
1310 Beal Avenue
Ann Arbor, MI 48109
United States of America

VOLUME ONE HUNDRED AND FOUR

SEMICONDUCTORS AND SEMIMETALS

Diamond for Quantum Applications Part 2

Edited by

CHRISTOPH E. NEBEL

Fraunhofer Institute for Applied Solid State Physics, Freiburg, Germany;
Nanomaterials Research Institute, Kanazawa University, Kanazawa, Japan

IGOR AHARONOVICH

Faculty of Science, University of Technology Sydney, Australia

NORIKAZU MIZUOCHI

Kyoto University, Kyoto, Japan

MUTSUKO HATANO

Tokyo Institute of Technology, Tokyo, Japan

ELSEVIER

ACADEMIC PRESS
An imprint of Elsevier

Academic Press is an imprint of Elsevier
50 Hampshire Street, 5th Floor, Cambridge, MA 02139, United States
525 B Street, Suite 1650, San Diego, CA 92101, United States
The Boulevard, Langford Lane, Kidlington, Oxford OX5 1GB, United Kingdom
125 London Wall, London, EC2Y 5AS, United Kingdom

First edition 2021

ISBN: 978-0-323-85024-7
ISSN: 0080-8784

For information on all Academic Press publications
visit our website at https://www.elsevier.com/books-and-journals

Publisher: Zoe Kruze
Acquisitions Editor: Jason Mitchell
Editorial Project Manager: Hilal Johnson
Production Project Manager: Abdulla Sait
Cover Designer: Alan Studholme

Typeset by SPi Global, India

Contents

Contributors *vii*
Preface *xi*

1. Color center formation by deterministic single ion implantation **1**
Tobias Herzig, Tobias Lühmann, Paul Räcke, Clemens Scheuner, Sébastien Pezzagna, and Jan Meijer
1. Introduction 1
2. Principles of ion implantation 3
3. Deterministic implantation of single ions 9
4. Creation and migration of vacancies 17
5. Charge-assisted engineering of color centers 21
6. Outlook 25
Acknowledgments 27
References 27

2. Advanced and in situ transmission electron microscopy of diamond: A review **31**
Wolfgang Jäger
1. Introduction 31
2. TEM for epitaxial CVD synthesis of diamond 35
3. Nanodiamond 58
4. Defects in diamond 65
5. TEM for characterization of physical properties 75
6. TEM for process and device technology 83
7. Conclusions and outlook 89
Acknowledgments 90
References 90

3. Fundaments of photoelectric readout of spin states in diamond **105**
Emilie Bourgeois, Michal Gulka, Daniel Wirtitsch, Petr Siyushev, Huijie Zheng, Jaroslav Hruby, Arne Wickenbrock, Dmitry Budker, Adam Gali, Michael Trupke, Fedor Jelezko, and Milos Nesladek
1. Introduction: Photoelectric detection techniques for quantum physics 106
2. Photoelectric readout of NV spin 108
3. PDMR for magnetic field sensing applications 134

4. Conclusions 141
Acknowledgments 141
References 141

4. Integrated quantum photonic circuits made from diamond 149
Francesco Lenzini, Nico Gruhler, Nicolai Walter, and Wolfram H.P. Pernice

1. Introduction 149
2. Fabrication of diamond-based integrated photonic circuits 152
3. Single photon emitters for waveguide devices 155
4. Active components 161
5. Waveguide integrated single-photon detectors 163
6. Conclusion 167
References 167

5. Diamond membranes for photonic devices 173
Kerem Bray, Blake Regan, and Igor Aharonovich

1. Introduction 173
2. Diamond devices 176
3. Diamond membrane fabrication 178
4. Fabrication of optical resonators from diamond membranes 186
5. Electroluminescence of diamond 191
6. Single crystal diamond membranes for nanoelectronics 197
7. Diamond electroluminescence 210
References 211

6. Diamond quantum nanophotonics and optomechanics 219
Cleaven Chia, Bartholomeus Machielse, Benjamin Pingault, Michelle Chalupnik, Graham Joe, Eliza Cornell, Sophie Weiyi Ding, Stefan Bogdanović, Kazuhiro Kuruma, Afaq Habib Piracha, Smarak Maity, Thomas M. Babinec, Srujan Meesala, and Marko Lončar

1. Introduction 220
2. Photonic devices 222
3. Mechanical devices 230
4. Optomechanical devices 239
5. Conclusions and outlook 243
Acknowledgments 244
References 244

Index *253*

Contributors

Igor Aharonovich
School of Mathematical and Physical Sciences, University of Technology Sydney, NSW, Australia

Thomas M. Babinec
John A. Paulson School of Engineering and Applied Sciences, Harvard University, Cambridge, MA, United States

Stefan Bogdanović
John A. Paulson School of Engineering and Applied Sciences, Harvard University, Cambridge, MA; X, The Moonshot Factory, Mountain View, CA, United States

Emilie Bourgeois
Hasselt University, Hasselt; IMOMEC Division, IMEC, Leuven, Belgium

Kerem Bray
School of Mathematical and Physical Sciences, University of Technology Sydney, NSW, Australia

Dmitry Budker
Helmholtz Institute Mainz, GSI Helmholtzzentrum für Schwerionenforschung; Johannes Gutenberg University Mainz, Mainz, Germany; Department of Physics, University of California at Berkeley, Berkeley, CA, United States

Michelle Chalupnik
Department of Physics, Harvard University, Cambridge, MA, United States

Cleaven Chia
John A. Paulson School of Engineering and Applied Sciences, Harvard University, Cambridge, MA, United States

Eliza Cornell
John A. Paulson School of Engineering and Applied Sciences, Harvard University, Cambridge, MA, United States

Sophie Weiyi Ding
John A. Paulson School of Engineering and Applied Sciences, Harvard University, Cambridge, MA, United States

Adam Gali
Wigner Research Centre for Physics; Department of Atomic Physics, Budapest University of Technology and Economics, Budapest, Hungary

Nico Gruhler
Institute of Physics, University of Muenster, Muenster, Germany

Michal Gulka
Hasselt University, Hasselt; IMOMEC Division, IMEC, Leuven, Belgium; Faculty of Biomedical Engineering, Czech Technical University in Prague, Kladno, Czech Republic

Tobias Herzig
Felix-Bloch Institute for Solid State Physics, University Leipzig, Leipzig, Germany

Jaroslav Hruby
Hasselt University, Hasselt; IMOMEC Division, IMEC, Leuven, Belgium

Wolfgang Jäger
Institute for Materials Science, Christian-Albrechts-University of Kiel, Kiel, Germany

Fedor Jelezko
Institute for Quantum Optics, Ulm University, Ulm, Germany

Graham Joe
John A. Paulson School of Engineering and Applied Sciences, Harvard University, Cambridge, MA, United States

Kazuhiro Kuruma
John A. Paulson School of Engineering and Applied Sciences, Harvard University, Cambridge, MA, United States

Francesco Lenzini
Institute of Physics, University of Muenster, Muenster, Germany

Marko Lončar
John A. Paulson School of Engineering and Applied Sciences, Harvard University, Cambridge, MA, United States

Tobias Lühmann
Felix-Bloch Institute for Solid State Physics, University Leipzig, Leipzig, Germany

Bartholomeus Machielse
Department of Physics, Harvard University, Cambridge, MA, United States

Smarak Maity
John A. Paulson School of Engineering and Applied Sciences, Harvard University, Cambridge, MA, United States

Srujan Meesala
Thomas J. Watson, Sr., Laboratory of Applied Physics, California Institute of Technology, Pasadena, CA, United States

Jan Meijer
Felix-Bloch Institute for Solid State Physics, University Leipzig, Leipzig, Germany

Milos Nesladek
Hasselt University, Hasselt; IMOMEC Division, IMEC, Leuven, Belgium

Wolfram H.P. Pernice
Institute of Physics, University of Muenster, Muenster, Germany

Sébastien Pezzagna
Felix-Bloch Institute for Solid State Physics, University Leipzig, Leipzig, Germany

Benjamin Pingault
John A. Paulson School of Engineering and Applied Sciences, Harvard University, Cambridge, MA, United States

Afaq Habib Piracha
John A. Paulson School of Engineering and Applied Sciences, Harvard University, Cambridge, MA, United States

Paul Räcke
Leibniz Institute of Surface Engineering (IOM), Leipzig, Germany

Blake Regan
School of Mathematical and Physical Sciences, University of Technology Sydney, NSW, Australia

Clemens Scheuner
Felix-Bloch Institute for Solid State Physics, University Leipzig, Leipzig, Germany

Petr Siyushev
Institute for Quantum Optics, Ulm University, Ulm, Germany

Michael Trupke
Faculty of Physics, University of Vienna; Institute for Atomic and Subatomic Physics, Vienna University of Technology, VCQ, Vienna, Austria

Nicolai Walter
Institute of Physics, University of Muenster, Muenster, Germany

Arne Wickenbrock
Helmholtz Institute Mainz, GSI Helmholtzzentrum für Schwerionenforschung; Johannes Gutenberg University Mainz, Mainz, Germany

Daniel Wirtitsch
Faculty of Physics, University of Vienna; Institute for Atomic and Subatomic Physics, Vienna University of Technology, VCQ, Vienna, Austria

Huijie Zheng
Helmholtz Institute Mainz, GSI Helmholtzzentrum für Schwerionenforschung; Johannes Gutenberg University Mainz, Mainz, Germany

Preface

Quantum technologies have revolutionized the 20th century and accomplished vital devices such as lasers, diodes, and transistors. As we embark on the 21st century, the second quantum revolution emerges, bringing together for the first time a vast potential to realize quantum computers, quantum communication networks, and quantum metrology devices. The second-generation quantum applications emerge with atomic centers embedded into materials like diamond, SiC, and Si, using atomic traps or superconducting qubits (Josephson junctions) for quantum computing and polarized photons for quantum information processing. All together remarkable breakthroughs indicate quantum jumps in technology. The introduced basic features however require also new device architectures and control mechanisms to utilize these quantum effects and to integrate them into state-of-the-art electronic or opto-electronic devices. For instance, quantum metrology based on NV centers in diamond requires electron spin-polarization and dephasing-lifetime control. Both are of prime importance. In addition, the use of electron spins demands proper n-type doping for charge-state stabilization of triplet states independent of its position in the host lattice, which can be deep in the bulk or ultra-shallow in direct vicinity of the surface. These centers react extremely sensitive on external perturbations, allowing sensing with unmatched limits. Quantum metrology, therefore, will pave the way toward new magnetometers, electric field sensors, pressure, and temperature detectors with highest sensitivities in combination with nanometer spatial resolution.

Several material platforms are currently investigated, including color centers in diamond and silicon carbide, phosphorous donors in silicon (Si), rare-earth emitters in oxide crystals, and semiconductor quantum dots. Diamond appears to be the material of choice for a variety of applications using, for example, the nitrogen-vacancy (NV) center that has outstanding spin properties even at room temperature. Quantum devices for biosensing in physiological buffer solutions can be realized for nano-NMR or hyperpolarization of nuclear spins in marker molecules like pyruvate, which are possible. In addition, these centers are single-photon emitters, which may be of great importance in quantum communication systems, working as emitters or repeaters, or the entanglement of NV spins may be used in the future for quantum computing or memory applications. Other color centers are also emerging, including group IV–vacancy combinations that possess better optical properties than the NV center in terms of quantum efficiency, brightness, and linewidth.

In "Diamond for Quantum Applications Part 1 & Part 2" (Semiconductor and Semimetals 103 and 104), we will summarize diamond-related quantum phenomena that attracted significant attention recently. In the first volume (SEMI 103), we start with a theoretical discussion of color centers in diamond followed by several chapters that reveal details about growth, doping, and optimization of quantum grade diamond, including the synthesis and doping of nano-diamonds by a high-pressure high-temperature technique. The interaction of light with single centers in diamond as well as a discussion of spins used for dynamic nuclear polarization will be given.

The second part (SEMI 104) of Diamond for Quantum Applications includes topics like (a) deterministic implantation, (b) high-resolution TEM for the detection of defects in diamond, (c) magnetometry using photoexcitation spectroscopy, (d) integrated quantum photonic circuits with polycrystalline diamond, (e) diamond membranes for photonic devices, (f) diamond nano-photonics and opto-mechanics, and (g) NV-doped diamond for high-pressure experiments (anvils). These chapters reveal the potential of NV technology and demonstrate first applications.

The authors of these chapters are leading experts in their respective fields. They have been asked to summarize their results, embedded into background knowledge, to draw a bigger picture for nonspecialized readers. We hope that the presented papers will be attractive to specialists as well as to newcomers, students, and engineers. The book intends to introduce quantum technology topics based on diamond to a large audience of interested readers without requiring specialized knowledge in quantum physics and technology.

The editors as well as the authors are sure that the introduced topics are of great importance for cutting-edge applications in the 21st century as well as for the transformation of established sensing, communication, and computing technologies into the new world of second-generation quantum-based devices.

Christoph E. Nebel
Igor Aharonovich
Norikazu Mizuochi
Mutsuko Hatano

CHAPTER ONE

Color center formation by deterministic single ion implantation

Tobias Herzig[a], Tobias Lühmann[a], Paul Räcke[b], Clemens Scheuner[a], Sébastien Pezzagna[a], and Jan Meijer[a,*]
[a]Felix-Bloch Institute for Solid State Physics, University Leipzig, Leipzig, Germany
[b]Leibniz Institute of Surface Engineering (IOM), Leipzig, Germany
[*]Corresponding author: e-mail address: jan.meijer@uni-leipzig.de

Contents

1. Introduction 1
2. Principles of ion implantation 3
   2.1 100 kV implanter 6
   2.2 Nanobeam 7
3. Deterministic implantation of single ions 9
   3.1 Single ion detection 10
   3.2 Nanoapertures 12
4. Creation and migration of vacancies 17
   4.1 Creation of vacancies 17
   4.2 Radiation-induced damage in diamond and thermal evolution of intrinsic defects 19
5. Charge-assisted engineering of color centers 21
6. Outlook 25
Acknowledgments 27
References 27

1. Introduction

Ion implantation is commonly used in the semiconductor industry to place foreign atoms at a specific depth and position(Current, 2017). This method is one of the key techniques in the manufacturing of integrated circuits. The method has—in combination with lithography—the advantage of creating small structures down to 7 nm in a high throughput volume procedure (Tanimura et al., 2016). For materials with low diffusion properties, ion

Semiconductors and Semimetals, Volume 104
ISSN 0080-8784
https://doi.org/10.1016/bs.semsem.2020.09.001

implantation is even the only technology that can be used for this purpose (Smith et al., 2019). Although lithography has almost reached a resolution that enables the fabrication of quantum systems in solids, this method does not fulfill their specific requirements (Smith et al., 2017).

One of the essential prerequisites for realizing quantum information (Wrachtrup and Jelezko, 2006) or quantum sensor (Chipaux et al., 2015) technology in a solid state is the targeted and deterministic generation of quantum objects (Fukuda et al., 2018; Nizovtsev et al., 2010). These quantum objects must meet the DiVincenzo criteria (Divincenzo, 1995), i.e., they must have coherent states that can be manipulated in a controllable way (DiVincenzo, 1998). A promising group of these quantum objects are color centers mainly in wide band gap materials like diamond. The prominent example is the negatively charged NV center (Nizovtsev et al., 2005), but there are now alternatives such as the ST1 (John et al., 2017) and SiV in diamond, the G center in silicon or praseodymium center in YAG (Kolesov et al., 2013).

Color centers are based on a very simple structure and generally consist of only one foreign atom and one defect (Jelezko and Wrachtrup, 2006). These objects are therefore very well defined and—like an atom—always own the same properties. This is a major advantage over quantum dots or other more complex quantum systems (Beha et al., 2012). Despite the simplicity of these systems, the production of functional structures is difficult, since it must be ensured that the color centers have a defined charge state, long coherence times and spectral stability. In order to couple and entangle the quantum objects, it is necessary to control the positioning very precisely (Meijer et al., 2008). It must be ensured that initialization, manipulation and read out of the quantum state of one center does not disturb the quantum properties of the neighbor systems (Dolde et al., 2013). To achieve all these goals, it is necessary to place the quantum objects with a positioning accuracy of a few nanometers in a solid body (Groot-Berning et al., 2019). The deterministic countable implantation of foreign atoms differs from conventional implantation procedures and requires new methods (Meijer et al., 2008).

Two main conditions have to be fulfilled in order to make color centers usable for quantum technology:

1. Deterministic implantation of single countable foreign atoms in a very defined depth and position.
2. Ensuring that each foreign atom is converted to the desired color center and assumes a stable and defined state of charge with all other necessary quantum properties.

While the first point is a technical task as a kind of next generation nanotechnology (Racke et al., 2018), the second point touches a material science question how color centers are formed and why they take a certain charge state (Luhmann et al., 2019). Both points are extremely challenging and have been studied by a large number of research groups for several decades. This chapter summarizes the main findings. In the first part—besides the basic methods of ion implantation—the special features of deterministic single ion implantation are presented. In the second part, the generation of color centers will be discussed and especially the influence of the chemical potential on the generation mechanism. This recent achievement enables extremely high conversation rates of implanted nitrogen into NV^- centers.

2. Principles of ion implantation

Ion implantation techniques have become an integral part of our modern world. Whether for the doping of semiconductors, the treatment of cancer and tumors, or the measurement and transformation of material properties (Baglin, 2012; Jensen et al., 2011; Sugiyama and Sigesato, 2004). Without particle accelerators, many modern applications would not be available today. And low-energy ion implantation gives us the means to realize quantum systems (Li et al., 2019).

The Coulomb interaction and thus the energy loss of kinetic ions in matter are described by the Bethe-Bloch formula. At high particle energies, the electronic stopping power overbalances, which causes inelastic scattering and results only in excitation or ionization of the involved electrons (Yamada, 1989). However, the creation of defects requires direct interaction between the kinetic ions and the nuclei of the crystal atoms. This is dominant at lower energies, when the nuclear stopping caused by elastic collision of crystal nuclei and kinetic ions, is most efficient (Borschel and Ronning, 2011, 2016). Hence, if the energy transfer during the collision process exceeds the displacement energy of the bound lattice atoms, the atoms can be kicked off from their lattice site and further interact with other crystal atoms. This process can be modeled by the program SRIM (Ziegler et al., 2010), which allows a first approximation of implantation parameters like, e.g., the ion distribution or the number of created vacancies by Monte Carlo simulations.

Fig. 1A and B shows an example of SRIM simulations in diamond in comparison with an optical image. Fig. 1C shows the penetration depth and straggling of hydrogen and nitrogen ions in diamond. Straggling is a

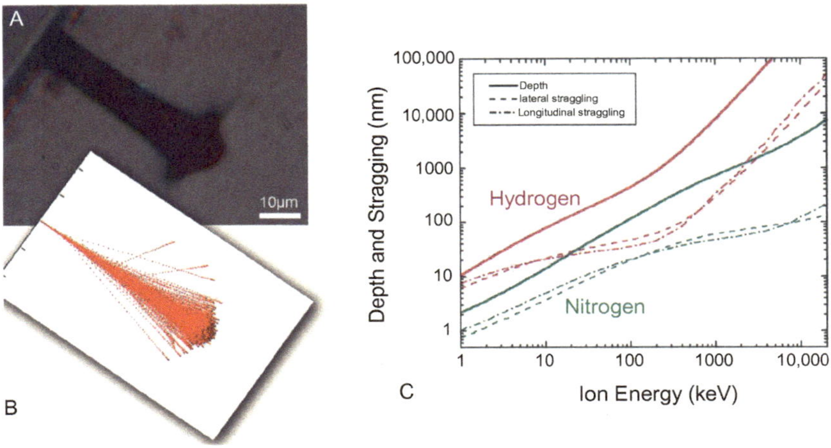

Fig. 1 Penetration depth and the expected statistical deviation from the targeted position using the example of nitrogen and hydrogen ions in diamond. (A) Optical image (side view) of a diamond sample implanted with a 3 MeV focused proton beam at high fluence. (B) SRIM simulation of 3 MeV protons implanted in diamond. (C) Plot of the ion range and straggle (lateral and longitudinal) simulated with SRIM for hydrogen and nitrogen implanted into diamond. *Reproduced with permission from Pezzagna, S., Meijer, J., 2012. High-resolution ion implantation from keV to MeV, INTECH.*

random process thus a high lateral accuracy below 10 nm requires low energetic ions. But SRIM does not take the crystal structure into account and therefore cannot described the "guided" traveling of ions along crystal axes and planes by electric potentials, which is called ion channeling. This effect leads to a higher penetration depth and a smeared defect distribution. Since this is important for precise placement of single ions, more advanced programs like CTRIM are advisable (Posselt, 1994), which considers crystal axes with respect to the ion beam angle. Fig. 2 (left) shows the CTRIM simulations of the ion and vacancy distribution for an implantation of nitrogen ions with a kinetic energy of 4 keV into a (100) diamond, once in channeling direction (0° tilt) and once under 7° tilt. While nearly no channeling effect occurs under 7° tilt, the ion distribution becomes much broader with a doubled implantation depth, when implanted in channeling direction. This indicates that already small deviations from the direction leads to a dechanneling effect. Additionally, the influence of surface roughness was investigated. It was found that experimental implantation profiles are best fitted by introducing a surface roughness by the use of an additional top layer of amorphous carbon in the simulation. Fig. 2 (left) shows the comparison of a

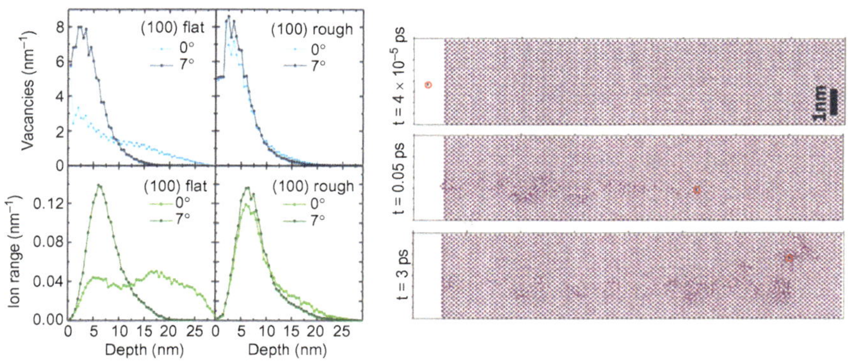

Fig. 2 Simulation methods as shown in (Posselt, 1994). Left: CTRIM simulation of produced vacancies (blue) and ion ranges (green) of 4 keV N ions in (100) diamond. Compared are different incident beam directions (0° vs. 7°) and surface roughnesses (0 nm vs. 1.5 nm, simulated by an additional top layer of amorphous carbon). Right: Molecular dynamics simulations with a 4 keV N ion (red circle) in a (100) diamond lattice (violet). The simulated lattice size was $3.6 \times 3.6 \times 15\,nm^3$. Snapshots of the atoms positions after different times clearly show lattice defect evolvement. Interactions between carbon atoms have been simulated by use of the Brenner potential (Brenner, 1990), whereas carbon-nitrogen interactions have been described by the dimer pair potential (Luhmann et al., 2018).

sample either without roughness or with a roughness of 1.5 nm. There is no distinct difference between the ion distributions for 0° and 7° tilt, thus indicating a strong decrease of the channeling effect due to the surface roughness. However, investigation of the surrounding defect configuration at the end of range of the ion beam within the crystal lattice can be helpful for an understanding of the defect creation. This can be achieved by molecular dynamics simulations (Allen and Wilson, 1989) as shown in Fig. 2 (right). Here, snapshots at different times during an implantation of 4 keV N ions into a (100) diamond lattice of the size $3.6 \times 3.6 \times 15\,nm^3$ are shown. The collision cascade was recorded over a time of 3 ps, which was long enough for the cascade to cool down, making defect migration induced by thermal activation highly improbable. From this simulation, the calculation of the distance from one N atom to its nearest vacancy is possible. Afterward, one can compute the statistical distribution of the number of vacancies and interstitials that are located in a range of 1 nm from the final N atom position (the interstitial formed by the N atom itself was not counted in this statistic). The simulation indicates, that most of the damage results in small disordered clusters, whereas isolated point defect are rather rare. Only in 10% for simulated off-normal impact cases, the N atom was separated by more than 1 nm from all other intrinsic effects, whereas no isolated N atom was found for normal impact cases.

However, new researches show that especially in diamond, a big variety of color centers can be fabricated by low-energy ion implantations (Luhmann et al., 2018) and the yield can be enhanced by, e.g., doping, as discussed later.

In the following, we give an exemplary overview about different implantation techniques to provide quantum technology based on diamond. These tools are, e.g., located at the Leipzig University and used for high-resolution and spatial selective placement of such centers down to the creation of distinguishable single-photon sources by deterministic placement of single ions.

2.1 100 kV implanter

Fig. 3 introduces a universal 100 kV implanter dedicated to the implantation of most of the elements from the periodic table with an energy range of about 5 keV up to 100 keV and an accuracy of a few μm. This is possible by the combination of a source of negatively charged ions (SNICS), a 90° electromagnet and a sample chamber with a particular masking system.

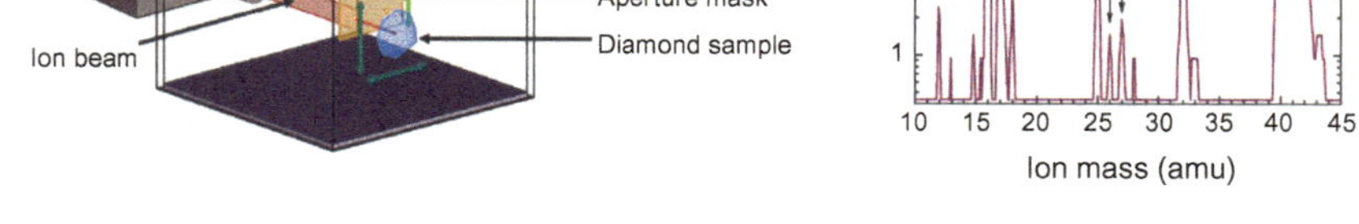

Fig. 3 Left: Main components of the home-made 100 kV ion implanter. Ions are generated within a SNICS sputter source from NEC and extracted by the extractor voltage U_{Ext}, before they are further accelerated up to 100 kV by the high-voltage supply U_{HV}. Different ions can be selected by their mass within the electromagnet. Subsequently, the desired ions are focused and deflected before they enter the sample chamber with the aperture masking system. Right: Typical mass spectrum of a cathode filled with pure Mg powder. Indicated are the different isotopes and clusters with different masses (Luhmann et al., 2018).

The SNICS ion source consists of a cesium oven, which evaporates cesium atoms into the ion source chamber, where they are ionized by the hot surface of the Cs ionizer. The Cs ions are focused and accelerated onto a negatively charged copper cathode, which is filled with powder that contains the desired element (among others). The choice of the cathode and of the focus voltage determines the sputter rate of the powder and therefore the number of created ions. These negatively charged ions are extracted from the source chamber by a positive voltage, before they are further accelerated by the applied high-voltage. Now, having the desired total energy, the ion beam crosses the 90° electromagnet, working as mass spectrometer (see Fig. 3 left) and thus guiding the beam only with ions of the same mass further to an electrostatic einzel lens and octupole for focusing and deflection of the ion beam, before it enters the sample chamber. Here, the ion beam is first tailored by a mirror with a perforation of approx. 1.5 mm in diameter, which also allows the illumination and observation of the sample with a microscope objective during the implantation. Behind the mirror, the ion beam has a diameter of approx. 1.5 mm and can either be aimed to the sample for implantation or further tailored by a home-made and specially designed masking system, which is mounted on a xyz-stage for fast and easy selection of different masks.

These transparent masks are made of a thin membrane (a few μm) of a sheet silicate material (Mica). By ion beam milling (see Section 3.2) one can perforate them and thus transform them to apertures of different shapes and sizes (approx. 10^{-7}–10^{-4} m). This enables the creation of a low-energy ion beam with high resolution. After crossing the mica foil, the ion beam reaches the sample on the sample holder, which is also mounted on a xyz-stage with micrometer precision. After measurement of the ion beam current by an isolated Faraday cup mounted on the sample holder, the ion beam position is located by implanting a PMMA sample and observing the color change of it. Afterward, the implantation of low-energy ions with micrometer precision into the desired sample can start.

2.2 Nanobeam

Ion implantation with nanometer resolution requires high accuracy in all directions. Thus the kinetic energy is limited to 10 keV and the positioning of the ion beam must be controlled with such precision and surface profiles need to be considered. For this purpose, the compact nanobeam setup was developed (Fig. 4).

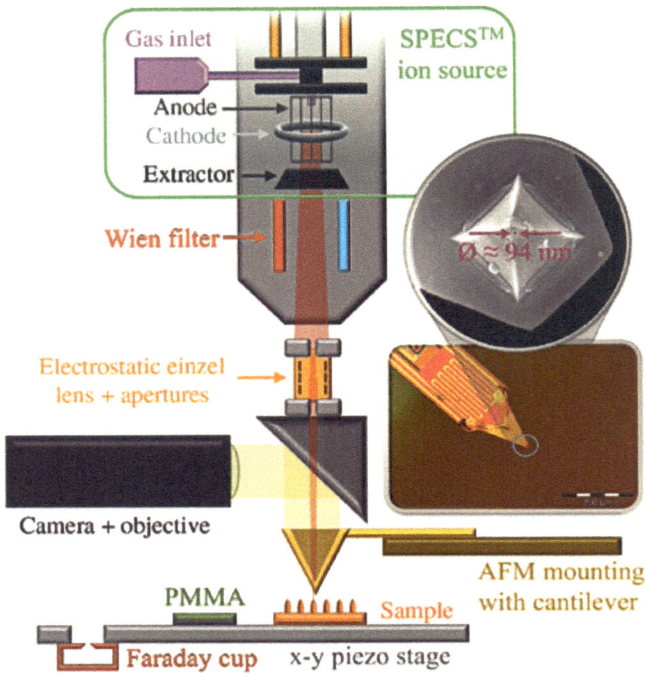

Fig. 4 Principle of the 5 keV nanoimplanter. The combination of a compact low-energy accelerator setup and an AFM setup with a pierced cantilever allows ion implantation with nanometer resolution. The AFM is commercially available from nano analytik GmbH (Smith et al., 2019).

It consist of a commercial SPECS™ gaseous ion source that allows the formation of a beam of ions with energies of up to 10 keV for double charge ions. Electrons from an Ω-shaped iridium ribbon cathode are accelerated into a grid cylinder anode, which is placed within the cathode. The electrons hit gas particles, ionize them and form an electron-ion plasma. The ions are accelerated by a conical extractor electrode and subsequently cross an interconnected Wien filter for selection of desired ions by their energy. The beam is furthermore tailored and focused by a system of apertures and an "einzel" lens. Afterward, it crosses a perforated mirror that enables the observation of a cantilever of an AFM setup that was previously pierced with a diameter of a few tens of nanometer by a focused ion beam (Inset of Fig. 4.). Thus, the desired surface can be scanned by the AFM and subsequently implanted through the hole in the AFM cantilever, enabling targeted implantations in, e.g., nanostructures with nanometer resolution.

Single-defect creation not only requires high spatial resolution, but also controlled placement of single distinguishable ions by deterministic ion implantation and a high production efficiency. Both will increase the formation probability of desired quantum centers and will be discussed in the next sections.

3. Deterministic implantation of single ions

A conventional ion source emits single particles at random instances in time. The average number of ions sent into the direction of a desired implantation spot is a well-controllable quantity. However, as described by the Poisson distribution, the maximal success probability for exactly one single ion to be implanted is only about 37% in the best case scenario (see Fig. 4). If the task is to implant an array of 40 single ions, which is still a rather small number for a qubit register, the probability of creating a flawless array is virtually zero (in fact it is in the order of 0.37^{40} (10^{-18})). Hence, the deterministic implantation of single ions is a prerequisite for the deterministic formation of color centers (Fig. 5).

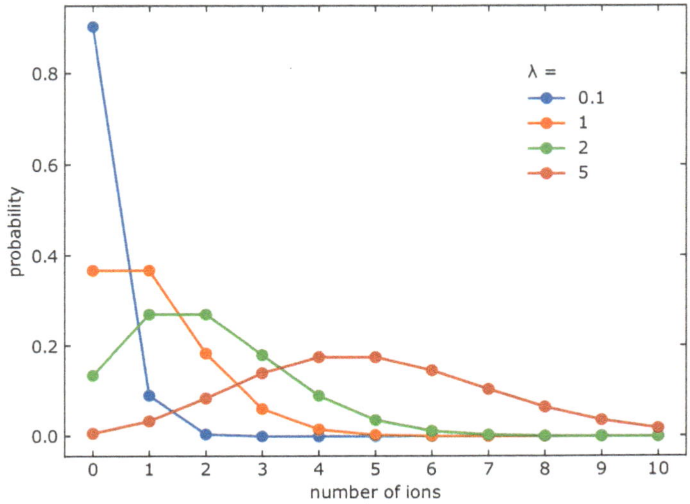

Fig. 5 Probability to implant a certain number of ions given by the Poisson distribution with average value (l), which is adjusted in the implantation process (Räcke, 2020).

3.1 Single ion detection

Approaches to realize the detection of single ions can be categorized according to the order of detection and implantation, which determines the requirements for the detection efficiency. In post-detection, the ion is detected after it has already been implanted, for example, via the electron-hole pairs that arise from the electronic stopping (IBIC) (Jamieson et al., 2017; Pacheco et al., 2017; van Donkelaar et al., 2015). In contrast, pre-detection is carried out while the ion is still traveling along its trajectory to the target. In all approaches, the inevitable noise in the measurement electronics sets the limit for the sensitivity of detecting single ions. At the same time, the ion species, which is used influences the produced detection signal through its mass, charge and velocity.

When aiming to detect the electron-hole pairs upon impact in the sample material, as it is done in the ion beam induced charge (IBIC) technique, the target itself acts as a particle detector. Commonly, a pin-diode structure is fabricated, for example in silicon, where single implanted donors are interesting for quantum applications (Morello et al., 2018). One challenge is that the sample needs to be a good IBIC detector and compatible with the later steps of fabrication for a quantum computer at the same time. Silicon is a typical particle detector material, but recently there has also been some progress in the fabrication of diamond single ion IBIC detectors (Abraham et al., 2016). In general, the number of electron-hole pairs decreases with decreasing projectile kinetic energy. Thus, the signal-to-noise ratio (SNR) becomes critically small for the low kinetic energies, which are desired for spatially precise implantation with low straggling.

In contrast to post-detection methods, pre-detection techniques are independent from the target properties. A Paul trap was used successfully as a deterministic source of single ions, which might be considered as a pre-detection method (Groot-Berning et al., 2019; Meijer et al., 2006; Schnitzler et al., 2009). Ions are trapped, cooled and then ejected from the trap toward the target. This is a considerable experimental effort, but allows for extremely precise, truly deterministic implantations, since the cooled ion inside the trap is detected with high certainty and localized very close to an ion optical point source. The creation of praseodymium color centers in YAG crystals was already demonstrated (Groot-Berning et al., 2019).

Using a conventional non-deterministic ion source, the pre-detection must be performed along the ion trajectory with only minimal disturbances.

One idea to realize this is based on image charge detection (Räcke, 2020; Racke et al., 2018, 2019). According to the Shockley-Ramo theorem, a moving charged particle induces a moving image charge, and thus a current, in nearby electrodes (Ramo, 1939; Shockley, 1938). The sensitivity follows from the ion parameters, the characteristics of the amplification electronics and the capacitance of the electrodes (Räcke, 2020). In principle, a longer detector together with slower ions with higher charge states enhance the SNR. Theoretically, a detector with single ion sensitivity can be constructed, but a set of external requirements, such as ion optical limitations, limits to the total electrode capacitance, desired ion implantation kinetic energies, etc. need to be taken into account. In an ion implanter set-up, a beam blanker would permanently shield the sample from any incoming ions. Only when the pre-detector correctly registers an ion, the path to the target is opened for a single ion implantation. The time-of-flight between the detector and the blanker is in the order of microseconds, which is a realistic order of magnitude for fast switching (Racke et al., 2019).

More generally, deterministic ion implantation using any detection strategy can be described by binary detection theory (see Table 1 and Fig. 6). As sketched in Fig. 6, any given detection observable has a probability density function (PDF), used to compute the probability of attaining a value inside some interval in a measurement. The sketch shows two cases: the PDF for pure noise in a detector (red) and the PDF for the same measurement, but with an additional single ion signal. The resulting probability to get a positive detection result is the integral of the respective PDF on the positive side of a defined threshold. If there was indeed an ion to be detected in this specific measurement instance, then the result is called "true positive (TP)," with a probability PTP. In case in reality there was no ion to be detected present, so the detector only falsely responded to noise, a false positive error is the result. Complementary, false negative errors are possible, as well, which results in a detection rate smaller than 100%. In terms of single ion implantation, false positive errors correspond to missing atoms in an array,

Table 1 Detection description for a binary logic.

Detection \ Reality	Signal + noise	noise
Positive	True positive (TP)	False positive (FP)
Negative	False negative	True negative

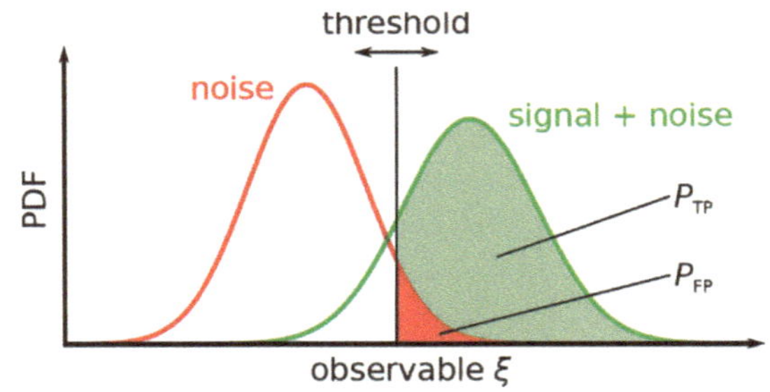

Fig. 6 Probability density function (PDF) for a binary detection logic. The green area defines the true positive (P_{TP}) and the red area the false positive rates (P_{FP}). True and false negative signals below the threshold will not be considered, which is unproblematic in pre-detection methods (Räcke, 2020).

because the implantation procedure registered the false positively detected implantation and goes over to the next implantation spot. This is the same for pre- and post-detection, whereas the consequences of false negative errors differ considerably, comparing the two. In a pre-detection scheme, false-negative errors only reduce the implantation rate, but the target is protected from incoming, undetected ions. Consequently, even if the PDFs of signal and noise exhibit a considerable overlap, a threshold can be chosen to minimize false positive errors, without any unwanted additionally placed atoms (Racke et al., 2019). For a post-detection process, false positive and false negative error probabilities must be small at the same time for a high success probability to implant a faultless array of single atoms (Fig. 7).

Both on the side of pre-detection and post-detection approaches, interesting and promising techniques are currently emerging from the endeavor to create scalable qubit registers by ion implantation.

3.2 Nanoapertures

Since electrostatic and magnetic focusing in the desired resolution are very challenging, the collimation of the ion beams with nanopores can be an excellent alternative for low current applications like single ion implantation. In general, there are two common techniques to create nanopores: ion track etching and FIB (Focused Ion Beam) milling.

For ion track etching swift heavy ions are used to penetrate a crystalline material, e.g., Muscovite (mica). Since the energy of the heavy ions is chosen to be far above the needed minimum for full penetration straggling on the

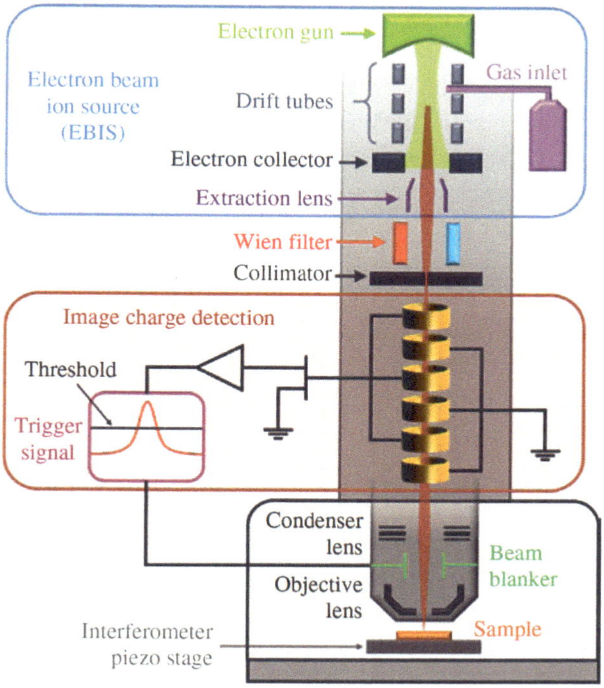

Fig. 7 Concept for deterministic ion implantation: ions, coming from an EBIS, are separated by a Wien filter and collimator, before entering the image charge detector. The image charge signals, induced into the detection tubes, are amplified and measured, thus allowing the distinction of single ions. The connected beam blanker is controlled by the trigger signal allowing deflection of ions with unknown or unwanted properties, thus ensuring deterministic implantation of defined single ions (Smith et al., 2019).

ion path can be neglected. This is leading to a straight path of the ion on which the crystal structure is destroyed by local heating due to energy transfer of the electronic stopping. With this, a cylindrical region of damaged and mainly amorphized material is created, with a diameter of a few nanometers depending mainly on the ion energy and species. In a subsequent etching step this damaged material can be removed. In the case of Muscovite this can be done in Hydrofluoric acid (HF). There the etching rate is slowest at oxygen planes of the crystal leading to diamond shaped pores (Sun et al., 2000). Due to the orientation of the pores on the crystal planes they are all aligned all in the same direction and have about the same size. Furthermore the pore walls are very parallel and smooth (Chien et al., 2002).

The size of the nano pores is easily controlled by the etching time (Pezzagna et al., 2011). 33 nm wide pores in 5 μm thick Muscovite sheets were reported (Sun et al., 2000). When an ion is penetrating the Muscovite sheet it is creating an ion track and thus a pore. Therefore, the nanopores are statistically distributed over the area irradiated by the beam of swift heavy ions and the mean distance between the pores can be controlled by the ion fluence. In order to get single pores a Muscovite sheet with very few pores in combination with further masking or a single ion track can be used. A single ion track can be created by using a low flux swift heavy ion beam in combination with a fast blanking unit and an ion detection system, for detecting ions that penetrated the muscovite sheet (Toimil-Molares, 2012). This technique can achieve great aspect rations enabling the collimation of MeV beams down to a few tens of nanometers. The minimum energy that can be used is given by the thickness of the sheet. Knowing the material type and the thickness a simulation with, e.g., SRIM can be performed to estimate the maximum penetration depth of the ions in the material hosting the nanopore. This maximum range should be well below the material thickness. Nanopores in Muscovite were already successfully used as irradiation mask to create a pair of close-by Nitrogen-Vacancy centers in diamond having been entangled at room temperature (Dolde et al., 2014).

The creation of nanopores via FIB milling is a different approach. Focused ion beam devices are widely available and used for material manipulation and microscopy purposes. They consist of a high brightness liquid metal ion source (LIMS) and a focusing unit. Due to the LIMS usually Gallium ions are used, and the choice of ion species is limited. The sample in which the nanopores will be created is irradiated with a highly focused ion beam of several 10 keV. This leads to sputtering, where the incident ion creates collision cascades reaching the sample surface and disposing there enough energy to remove surface atoms. This loss of material can be used to structure the surface and drill nanopores in the sample up to a few tens of nanometers. This technique offers the advantage of creating nanopores of arbitrary shapes in a wide variety of materials. For example, it was shown that FIB milling of a nanopore in an AFM cantilever can be used for high precision single ion implantation (Meijer et al., 2008). But due to the limited aspect ratios of this technique to reach the desired pore diameters, free-standing thin membranes with thicknesses of a few 100 nm must be used as a milling sample. These membranes are capable to stop ions up to a few 100 keV. In Fig. 8 an example of nanopores created by ion track etching and FIB milling is shown (Scheuner et al., 2017).

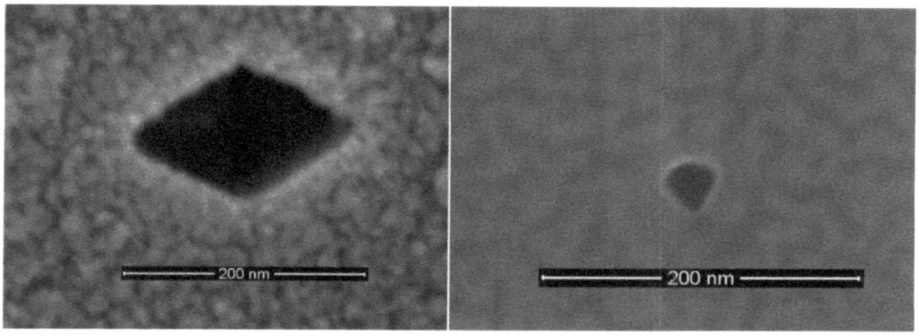

Fig. 8 SEM Micrograph of nanopores. Left: Track etched pore in Muscovite. Right: Fib milled pore in silicon nitride with a diameter of 40 nm. The deviation from a circular form is due to focusing artifacts.

When using nanopores as collimators for ion beams scattering must be considered. Most of the ions transmitted through the pore do not interact with the pore walls but a certain fraction does so, depending on the divergence of the ion beam and the aspect ratio of the pore. Ions getting close to the pore walls can be deflected by surface atoms of the pore walls and change their flight direction and energy. The ions can travel a certain range through the pore wall material, depending on their energy. The zone in which the ions penetrate the pore walls and get back into the ion beam is called transparency zone and strongly depends on the shape and surface roughness of the pore (Adamczewski et al., 1996). Even if the pore is just a few micrometers away from the sample where the ions should be placed, this scattering could lead to a significant broadening of the area met by the ion beam (Raatz et al., 2019). Therefore, it is necessary to place the pores as close as possible to the sample or use additional collimators if the pore is used in an ion optical system away from the sample.

Experiments with muscovite nanopores of extremely high aspect ratios have shown that there is an unexpected surplus of transmitted ions in the MeV regime. This could be ascribed to an effect that is similar to crystal channeling (see before) and leading to a more efficient collimation of the ion beam and to a reduced energy loss of the transmitted ions (Gemmel, 1974; Scheuner et al., 2017). In the case of the muscovite pores, the ions are "guided" by the smooth pore walls created by ion track etching and small misalignments from the pore being parallel to the ion beams do not hinder the ions to be transmitted through the pore.

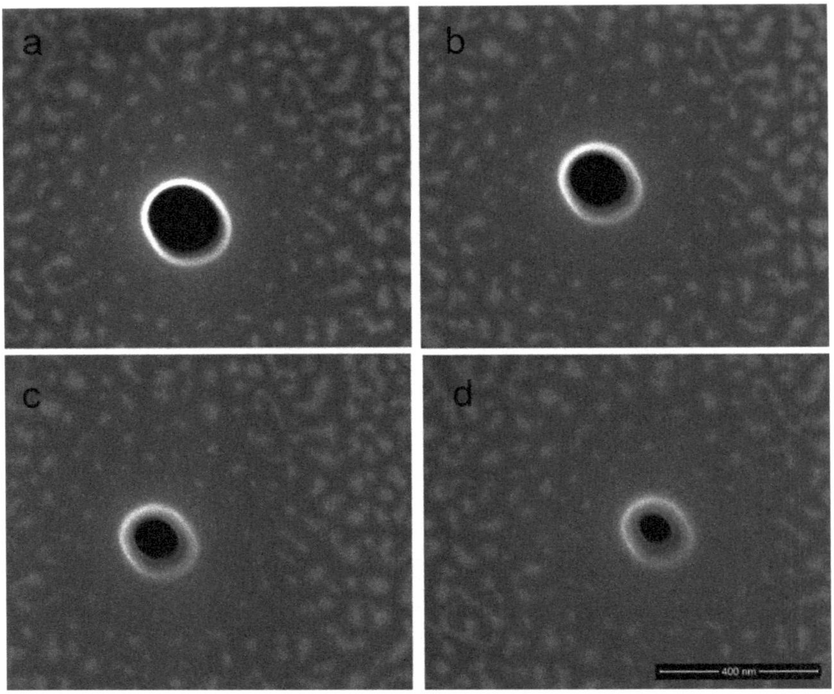

Fig. 9 Closing of a nanopore in silicon nitride under electron irradiation. The irradiation dose increases from (A) (0 nC/μm^2) to (D) (657 nC/μm^2). The open-pore diameter shrinks down from 180 to 80 nm.

While using the nanopores as a collimator in the lower keV energy regime, it was observed in the case of FIB milled pores that the pore diameter changes over the irradiation time. This is ascribed to the activation of freely diffusing surface adatoms and addimers activated by the ion or electron bombardment. If these particles move in the nanopore, they get trapped there due to a free energy gain caused by surface reduction. This mechanism is leading to material transport in the pore that shrinks it (Storm et al., 2003). Fig. 9 displays this effect. As the effect is diffusion related it is also dependent on the temperature. A greater temperature will lead to a faster transport of the material in the pore. However, since material in the pore (as well as on the surface) can be removed by sputtering in the case of ion irradiation, an opposite effect can occur under a certain temperature (Li et al., 2001). As promoted by sputtering this change in the behavior is dependent on irradiation parameters and the pore material. Understanding this effect, it can be used for a better-controlled use of nanopores for ion implantation and furthermore to shrink down the size of nanopores to diameters not available by FIB milling.

4. Creation and migration of vacancies

Once nitrogen has been gently incorporated into the diamond at a specific depth and density, a secondary step of introducing vacancies into the diamond may be needed to create additional NV centers. Various vacancy creation techniques can be used to preserve the pristine CVD-grown diamond lattice and many of them feature additional advantages, such as lateral spatial localization and tunable NV densities (Smith et al., 2019).

4.1 Creation of vacancies

There are different ways to create vacancies in diamond: form a mechanical treatment to Laser-, gamma-, neutron- electron- or ion- irradiation (Smith et al., 2019). On the surface, this displacement energy is only a few eV. Surface defects caused by mechanical processing are therefore often an undesirable phenomenon when polishing and show up by a strong NV luminescence. Also a high temperature treatment of more than 1800 °C leads to NV centers on the surface (Pezzagna et al., 2011). However, such processes cannot be controlled and are therefore undesirable. In addition, one would like to produce the NVs at a controlled depth of a few nm (Spinicelli et al., 2011).

Inside the crystal a transferred energy of about 30 eV is required to displace a carbon atom into an interstitial state (Delgado and Vila, 2011). This energy corresponds to a temperature of 3.5×10^5 K and can only be achieved by means of femto-laser pulses. Through the targeted application of several pulses, it has been possible to control the generation and healing of defects and to create NV centers. The disadvantage is that the nitrogen atom must already be in the crystal (Smith et al., 2019). Also the depth cannot be controlled with the precision required to fabricate a quantum computer. A more direct method is the creation of defects by a direct impact of a carbon atom. The maximum transferable energy depends thereby on the mass ratio of the two impact partners. For electrons, a kinetic energy of 145 keV is required to release a carbon atom on the lattice (McLellan et al., 2016). For He it is a few keV and for carbon atoms only a few eV in a direct collision

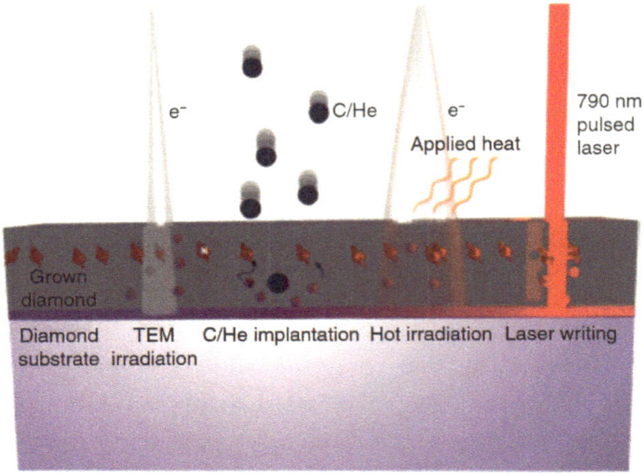

Fig. 10 A schematic depiction of various gentle vacancy creation techniques for CVD-doped nitrogen. Left to right: low energy TEM irradiation is capable of gently generating localized vacancies. Ion implantation at a depth that is different from the nitrogen layer is another way of gently introducing vacancies. In this approach, a subsequent anneal will activate the NV centers as the vacancies diffuse to the nitrogen layer. Simultaneous annealing with vacancy creation is new technique whereby vacancies are introduced slowly during annealing. Laser writing vacancies is a promising technique for use in nitrogen-doped diamond (Smith et al., 2019).

(Huang et al., 2013). As shown in the next chapter, the diffusion range of defects is only a few nm for typical annealing temperature of 1100 °C. The probability of converting implanted nitrogen atoms by subsequent irradiation of carbon or helium is therefore very small. Experiments with this procedure step were not very helpful and increase the conversion rate only by a few percent. However, the implantation of molecules such as CN can be an advantage to create shallow NV centers (Spinicelli et al., 2011). In this case the distance of the implanted N atom and the defects caused by the additional C ion are very small. In fact, this results in an improved performance, even if only to a small extent. A variety of creation methods are schematically depicted in Fig. 10. A nitrogen layer is represented by the red spins. From left to right, focused electron irradiation from a transmission electron microscope (TEM) offers an energetically tunable way to write nanoscale patterns of vacancies into a diamond (Smith et al., 2019). Carbon (Ohno et al., 2014) and helium (Huang et al., 2013) implantation are alternative methods for introducing vacancies in a controlled and gentle way. Their energy can be chosen so that they stop at a much large depth than the nitrogen layer, thus minimizing collateral damage at the nitrogen layer. An additional annealing step is then required for vacancy diffusion to form NV centers and to remove other residual defects, such as paramagnetic di-vacancies that may contribute additional sources of decoherence. The best choice is to use the vacancies created by the Nitrogen ion during the implantation procedure (Luhmann et al., 2019).

4.2 Radiation-induced damage in diamond and thermal evolution of intrinsic defects

When diamonds are irradiated with electrons, ions or gamma photons, a number of defects are formed. The type and quantity of the defect depends on the energy, the mass of the projectile, the dose and the diamond type. The most frequent defects are vacancies. The single vacancies can be in different states of charge. The fluorescence of the neutral vacancy is known as GR1 with a ZPL at 741 nm (Zaitsev, 2001). If the vacancy has an additional electron it is absorbent with the ZPL at 394 nm and known as ND1 (Pu et al., 2000). The positive charge state is also possible (Dannefaer and Iakoubovskii, 2008). Another kind of defects are the vacancy-correlated ones. The simplest are the double vacancies TH5 (Leawilson et al., 1995), the Frenkel defect, NV center and (N-V-N) H3 (Collins and Kiflawi, 2009). With a correspondingly high dose and ion mass, large vacancy clusters M1, M2 (Dannefaer and Iakoubovskii, 2008) and aggregates are created. Another kind of defect are the dislocated carbon atoms. They occur in different configurations. As simple interstitial carbon atoms, in the Frenkel defect, as split self-interstitial R2, the double split self-interstitial R1(Iakoubovskii et al., 2003, 2005), the double interstitial 3H center (Iakoubovskii et al., 2005) and Th12 (Naydenov et al., 2009).

The counted centers were not only caused by the irradiation. Their occurrence and quantity is also determined by the thermal treatment of the irradiated diamonds. This means that some centers form, decay or heal at certain temperatures or form even more complex centers. The Frenkel defect partially decays during irradiation and disappears by recombination or dissociation at 400 °C (Iakoubovskii et al., 2005). At 300 °C, 3H centers are formed from R1 centers. The 3H centers are stable above 800 °C.

With the beginning of the mobility of vacancies at 600 °C TH5, M1, M2 and, depending on the nitrogen content, NV centers can be formed with maximum concentration at 800 °C. The yield of these various vacancy-correlated centers are in competition with each other. But they decay at different temperatures. The TH5 defective decays again already at 900 °C, but M1 and M2 are still stable up to 1000 °C (Dannefaer and Iakoubovskii, 2008). The NV centers are stable up to 1150 °C (Luhmann et al., 2018). If the vacancy concentration inside the Diamond is above $1 \times 10^{22}\,cm^{-3}$ and the diamond is thermally treated at 1200 °C, graphite is formed in the diamond matrix (Luhmann et al., 2019). From 1500 °C the nitrogen becomes mobile and if there is enough nitrogen the H3 center is formed.

A major difference in the occurrence and concentration of the different centers is the degree of doping of the diamonds. This defines in a broader sense the type of diamond. The main difference is made between type Ib with many isolated nitrogen atoms, type IIa (with very little almost undetectable nitrogen content) and type IIb (with a high boron content). The type IIa are the purest diamonds. It is the one with the lowest content of Frenkel defects formed during irradiation, because they heal immediately (Iakoubovskii et al., 2005). Above 600 °C TH5 centers are formed in these diamond (Clark, 1956), with maximum efficiencies at 800 °C. Most of the vacancies in this type are neutrally charged, therefore they form more TH5. In these diamonds also R2, R1 and finally at 300 °C 3H defects can form. In Type IIb diamonds, the formation of TH5 is enhanced and in addition, larger vacancy aggregates M1 and M2 are formed (Dannefaer and Iakoubovskii, 2008). In type Ib it behaves differently. On the one hand, there are much more Frenkel defects after irradiation which are additionally negatively charged, and on the other hand the majority of vacancies are negatively charged as well. At 400 °C the Frenkel defects dissociate and there is an even further increase in negative vacancies (Iakoubovskii et al., 2005). At this temperature the nitrogen atoms can trap the carbon interstitial atoms, but above 400 °C they release them again. If larger nitrogen aggregates are present as in type IaA and type IaB, the carbon interstitial atoms are trapped tightly. At higher temperatures, due to the charged vacancies, a significantly lower formation of TH5 is observed (Leawilson et al., 1995). The negative vacancies disappear at 750 °C. On the one hand they recombine with carbon interstitial atoms and on the other hand they are trapped by the nitrogen atoms and form NV centers. This process is dominant and dependent on the single nitrogen content.

5. Charge-assisted engineering of color centers

As shown above, it exists a large variety of point defects in an implanted material: the native ones, the implanted atoms and the defects induced by the implantation. The native ones are related to the quality of the synthetic starting material. Most of the time, nitrogen and boron concentrations are given by the supplier. Hydrogen is however never specified but plays an important role, especially by its larger diffusion coefficient and its ability to build complexes which can passivate the color centers (Lesik et al., 2016; Luhmann et al., 2018; Stacey et al., 2012). Depending on the growth technique, these impurities can be found in different concentrations. The implanted species are the easiest to control. Of course, in the case of deterministic implantation (Groot-Berning et al., 2019), the number of atoms introduced in the material can be counted exactly. Nevertheless, even in the case of a stochastic implantation, the uncertainty on the number of ions is given by Poisson statistics, for which the standard deviation equals $\sqrt{\overline{n}}$ where $\overline{n}$ is the average number of implanted ions. Beside the number of ions, their mass (isotopes) is easily controlled and the spatial resolution can be less than 20 nm in lateral and a few nm in the depth direction (Pezzagna et al., 2010a). The implantation-induced defects consist of vacancies, carbon interstitials and combinations of these.

The implanted diamonds are heated to heal out the implantation defects, to place the implanted impurities in substitutional sites and to form the wished color centers. Now, there is a complex interplay between all these point defects present in the material. Interestingly, the observations are often very reproducible. For example, the creation efficiency of NV centers (ratio of formed NVs to the number of implanted nitrogen) in pure diamond is a good indication of the energy at which the nitrogen was implanted, because this behavior is extremely reproducible and well known (Pezzagna et al., 2010b). Another example, in nitrogen rich diamonds, the ratio of NV centers to NVN centers (H3) is a good indication of the temperature at which a diamond was treated (Felton et al., 2009). Now for quantum computer applications, most of the uncertainty relies in the "invisible" species, which can strongly influence the formation of color centers and operation of the qubits, especially at the single level when only a few ions are implanted. This can be seen as a competition between the different forms. In the search for scalability, one aims to reach the ideal case for which one would create only the targeted species and avoid the other ones to form. In this purpose, one can play with three parameters: temperature, pressure and annealing time (and dynamics as well). This is a rather time-consuming and sample-consuming problem because these conditions apply necessarily to one sample at a time and can lead to deviations owing to inhomogeneity from sample to sample.

On the other hand, a fourth degree of freedom was proposed recently: the charge state of the different species. Indeed, the formation energies and diffusion can vary strongly. As shown in Section 4.2, the implantation-induced vacancies tend to form complexes for larger ion masses, whereas more isolated vacancies are found for lighter ions (Luhmann et al., 2018). One assumes in this case neutral vacancies, which is what is occurring in ultrapure diamond (unintentionally doped). This can be modified when the diamond is doped with donors or acceptor impurities.

In this purpose, two studies were dedicated to NV centers and the complementary search for vacancies. The first one (de Oliveira et al., 2017) used a sacrificial boron p-type doped layer (grown by CVD) to charge vacancies positively. The vacancy charging prior to the annealing was not shown but this led to an enhancement of the NV creation yield (for a nitrogen implantation energy of a few keV) from 1% to 2% and most interestingly a twofold to tenfold increase of the coherence time of the NV electron spin. This effect was attributed to reduction of paramagnetic di-vacancies V_2 defects. The second study used n-type doping, to charge vacancies negatively to also avoid di-vacancy formation and likely to take advantage of positively charged nitrogen during the thermal annealing (Fig. 11C). This led to strong enhancement of the NV center formation using the Coulomb forces (Luhmann et al., 2019). The charging of the implantation-induced vacancies was proven by optical means. As seen in Fig. 11A, the implantation of carbon into diamond of different type (n-type, p-type or intrinsic) leads to very different charge states. It can be seen that in donor-doped diamond, neutral vacancies are not observed, whereas they are present in the p-type one and in a mixed state in the intrinsic region. Note that no real Fermi level can be defined in diamond and that the charge state of point defects is determined the presence (or not) of neighboring electrically active impurities. An additional and interesting effect can be remarked: the damage threshold of diamond is pushed toward higher implantation fluences in the donor-doped region, highlighting the repulsion effect between negatively charged vacancies. Upon nitrogen implantation (which has a mass close to carbon) the vacancy

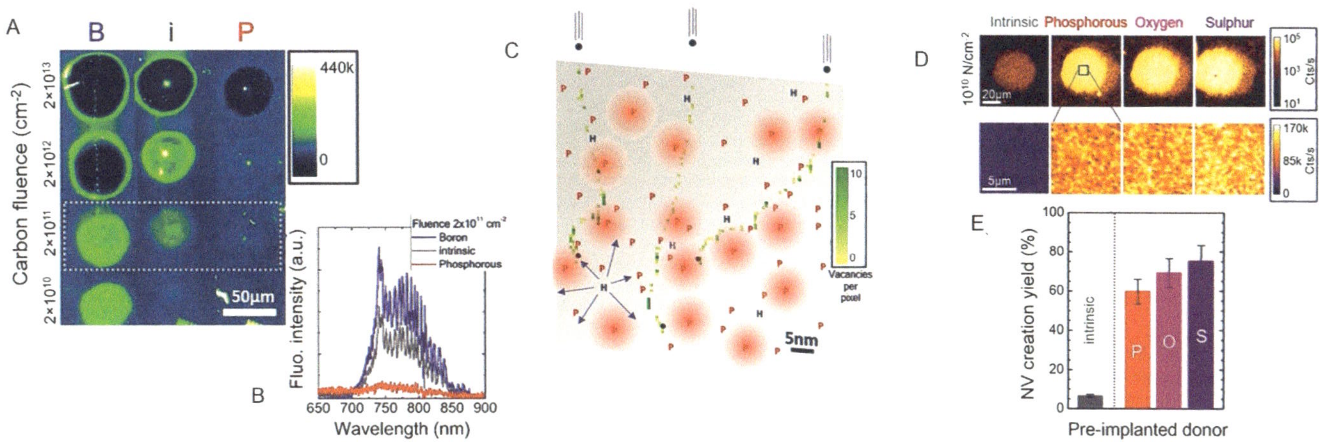

Fig. 11 Coulomb-driven defect engineering using doping. (A) Confocal fluorescence scan of an ultrapure diamond implanted with carbon and not annealed to image the implantation-induced vacancies. (B) accounts for boron doping ($2 \times 10^{18}\,cm^{-3}$), i for intrinsic and P for phosphorous doping ($2 \times 10^{18}\,cm^{-3}$). Laser excitation 532 nm and detection with longpass 700 nm. Only the neutral charge state of vacancies contributes to the fluorescence, with a characteristic ZPL at 741 nm. In the phosphorous region, no fluorescence is observed, indicating that vacancies are negatively charged. (B) Fluorescence spectra recorded in the 3 different regions for the carbon fluence of $2 \times 10^{11}\,cm^{-2}$. Each carbon atom produces about 100 vacancies. (C) Scheme of the situation with the most abundant species in the case of a phosphorous-doped layer. (D) Confocal fluorescence scans of diamond implanted with nitrogen in areas previously prepared with different dopants, and annealed at 800 °C to produce NV centers. Laser excitation 532 nm, detection in the range 660–735 nm. (E) NV creation yield as a function of the pre-implanted donor. A tenfold enhancement from 7% to 75% is achieved (Luhmann et al., 2019).

distribution is very similar. However, the nitrogen vacancy center formation is strongly modified. At the energy of 40 keV, the creation yield is of about 7% typically and very reproducible. When donor doping is employed, a tenfold improve can be reached, up to 75% was observed (Fig. 11D and 11E). This effect is not only working with NV centers. It was demonstrated that it can be applied as well to SnV and MgV centers, revealing that it is a more general method.

These two pioneering works are at the early stages of complex and rich engineering possibilities, since the other parameters which are temperature, pressure and time can be as well combined to find the optimal conditions, dedicated to each particular situation. As an example, varying the temperature for color centers in the same diamond but prepared with different doped regions has shown very different results (Fig. 12). Whereas SnV and MgV centers are formed more and more under a temperature increase, the number of NV centers reduces and this reduction is strongly depending on the doping type. Interestingly, the NV thermal stability was extensively studied and it is known that at this temperature of 1000 °C they do not dissociate (Luhmann et al., 2019). This indicates that the NV centers are here passivated, with hydrogen being the most probable candidate (Lesik et al., 2016; Luhmann et al., 2018; Stacey et al., 2012).

The same kind of charge-assisted engineering can be envisaged by realizing the thermal annealing under light illumination or under application of

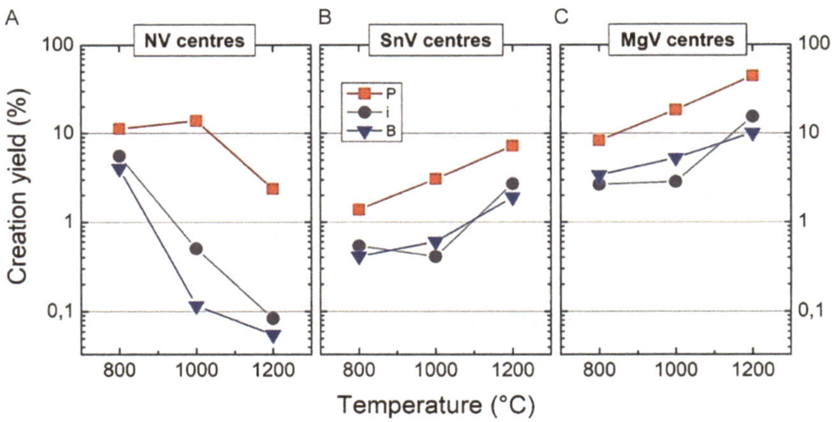

Fig. 12 Temperature evolution of the population of NV, SnV and MgV centers for different doping conditions. The density of each implanted species (N, Sn and Mg) is $3 \times 10^{11}\,cm^{-2}$. Each annealing step was done for 4 h in vacuum (Luhmann et al., 2019).

an electrical bias to the sample. Under light of appropriate wavelength and intensity, a quasithermal equilibrium can be reached which induces a steady-state where the defects are charged differently than in the dark. This offers moreover the possibility to associate light illumination and temperature changes to refine the engineering. Concerning the use of electrical bias, it was shown for example that the diffusion of deuterium is strongly influenced by the bias applied to the sample. The diffusion coefficient can be modified by orders of magnitude and activation energies were estimated to be. This local modification of the chemical potential can be combined with electrical structures such as pin, nin or pip junctions. Then a more refined band bending can be engineered for the different color centers or targetted applications.

6. Outlook

To build a quantum computer based on color centers it is necessary to be able to generate the color centers deterministically with high lateral resolution in a defined depth. For this purpose, foreign atoms and defects have to be implanted into a crystal in order to generate the color centers with a stabilized charge state of almost 100%. Especially the NV center in diamond is in the focus of the investigations, because it has a long coherence time (T_2) even at room temperature and thus allows a quantum computer at natural environment. The question to be solved is therefore: Can foreign atoms and defects be introduced into a solid body in such a targeted manner and is it possible to shape them with high probability to the desired color center. The color center should be close to the surface in order to have an electronic access but deep enough not to be disturbed by the surface.

Indeed, much progress were achieved recently to reach this goal. It was shown that ions can be implanted deterministically with high lateral resolution at a certain depth. Also, by a targeted diamond doping with sulfur it has been possible to transform NV centers with a probability of 75%. These advances represent a "game changer" and allow to use the color centers not only for sensors but also for quantum information. In more recent approaches, NV centers are linked with nuclear spins. Up to 10 ^{13}C nuclear spins were coupled to 1 NV center. The NV center can furthermore be read out electrically and controlled by nanowires in near-field. Fig. 13 shows a scheme of a corresponding setup. In contrast to quantum computers based

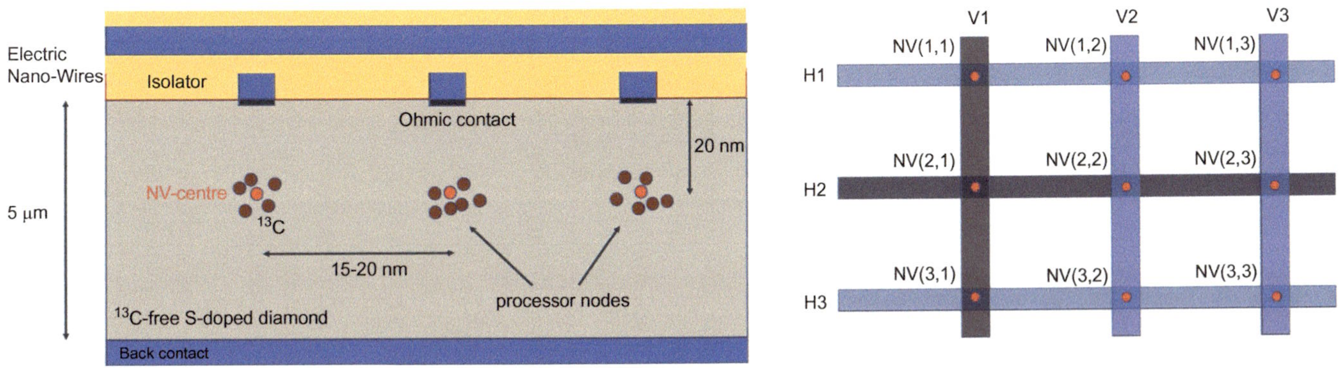

Fig. 13 Concept of a Quantum computer based on a NV-grid surrounded by ^{13}C nuclear spin (A) The sideview show the nanowires in blue above the NV center surrounded by ^{13}C nuclear spins. The crossed wires are isolated and can be used to address a microwave or radiowave pulse but also to electrical readout the NV center. The backgate will be used to recharge the NV center (B) Top view of the grid concept: Behind every cross a NV center will be implanted. E.g., to readout a NV center a applied suction voltage will be used to address a single NV (black lines).

on superconducting qubits in this scheme offers that only 20 wires are necessary to control up to 1000 qubits. This allows an extremely high degree of integration and low-energy consumption. Even if the current concepts will first be tested at low temperature about 10 K, the way to a room temperature quantum computer is possible owing to the unique properties of NV centers at 300 K.

Acknowledgments

We acknowledge the financial support from the European Union (ASTERIQS, MICROSENS), the BMBF: DiaquantFAB as well as the Deutsche Forschungsgemeinschaft (Ulysses). We are further grateful to receive funding by the Leibniz Association (SAW-2015-IOM-1) and the European Union together with the Sächsisches Ministerium für Wissenschaft und Kunst (Project No. 100308873).

References

Abraham, J.B.S., et al., 2016. Fabrication and characterization of a co-planar detector in diamond for low energy single ion implantation. Appl. Phys. Lett. 109 (6), 6.

Adamczewski, J., et al., 1996. Optimization of material and shape for nuclear microprobe apertures. Nucl. Inst. Methods Phys. Res. B 114 (1–2), 172–184.

Allen, M.P., Wilson, M.R., 1989. Computer-simulation of liquid-crystals. J. Comput. Aided Mol. Des. 3 (4), 335–353.

Baglin, J.E.E., 2012. Ion beam nanoscale fabrication and lithography-a review. Appl. Surf. Sci. 258 (9), 4103–4111.

Beha, K., et al., 2012. Diamond nanophotonics. Beilstein J. Nanotechnol. 3, 895–908.

Borschel, C., Ronning, C., 2011. Ion beam irradiation of nanostructures—a 3D Monte Carlo simulation code. Nucl. Inst. Methods Phys. Res. B 269 (19), 2133–2138.

Borschel, C., Ronning, C., 2016. Low energy ion beam modification of nanostructures. In: Ion Beam Modification of Solids. Ion-Solid Interaction and Radiation Damage, vol. 61, pp. 475–500.

Brenner, D.W., 1990. Empirical potential for hydrocarbons for use in simulating the chemical vapor-deposition of diamond films. Phys. Rev. B 42 (15), 9458–9471.

Chien, C.L., et al., 2002. Electrodeposited magnetic nanowires: arrays, field-induced assembly, and surface functionalization. J. Magnetism Magnetic Mater. 249 (1–2), 146–155.

Chipaux, M., et al., 2015. Nitrogen vacancies (NV) centers in diamond for magnetic sensors and quantum sensing. In: Quantum Sensing and Nanophotonic Devices Xii, p. 9370.

Clark, C.D., 1956. The absorption spectra of irradiated diamonds after heat treatment. Proc. R. Soc. Lond 237.

Collins, A.T., Kiflawi, I., 2009. The annealing of radiation damage in type Ia diamond. J. Phys. Condensed Matter 21 (36), 36.

Current, M.I., 2017. Ion implantation of advanced silicon devices: past, present and future. Mater. Sci. Semicond. Process. 62, 13–22.

Dannefaer, S., Iakoubovskii, K., 2008. Defects in electron irradiated boron-doped diamonds investigated by positron annihilation and optical absorption. J. Phys. Condensed Matter 20 (23), 23.

de Oliveira, F.F., et al., 2017. Tailoring spin defects in diamond by lattice charging. Nat. Commun. 8.

Delgado, D., Vila, R., 2011. Statistical molecular dynamics study of displacement energies in diamond. J. Nucl. Mater. 419 (1–3), 32–38.

Divincenzo, D.P., 1995. Quantum computation. Science 270 (5234), 255–261.
DiVincenzo, D.P., 1998. Quantum gates and circuits. Proc. R. Soc. a-Math. Phys. Eng. Sci. 454 (1969), 261–276.
Dolde, F., et al., 2013. Room-temperature entanglement between single defect spins in diamond. Nat. Phys. 9 (3), 139–143.
Dolde, F., et al., 2014. High-fidelity spin entanglement using optimal control. Nat. Commun. 5.
Felton, S., et al., 2009. Electron paramagnetic resonance studies of nitrogen interstitial defects in diamond. J. Phys. Condensed Matter 21 (36), 36.
Fukuda, R., et al., 2018. Lithographically engineered shallow nitrogen-vacancy centers in diamond for external nuclear spin sensing. New J. Phys. 20.
Gemmel, 1974. Channeling and related effects in the motion of charged particles. Rev. Mod. Phys. 46, 129.
Groot-Berning, K., et al., 2019. Deterministic single-ion implantation of rare-earth ions for nanometer-resolution color-center generation. Phys. Rev. Lett. 123 (10), 10.
Huang, Z., et al., 2013. Diamond nitrogen-vacancy centers created by scanning focused helium ion beam and annealing. Appl. Phys. Lett. 103 (8), 8.
Iakoubovskii, K., et al., 2003. Annealing of vacancies and interstitials in diamond. Phys. B Condens. Matter 340, 67–75.
Iakoubovskii, K., Dannefaer, S., Stesmans, A., 2005. Evidence for vacancy-interstitial pairs in Ib-type diamond. Phys. Rev. B 71 (23), 67.
Jamieson, D.N., et al., 2017. Deterministic doping. Mater. Sci. Semicond. Process. 62, 23–30.
Jelezko, F., Wrachtrup, J., 2006. Single defect centres in diamond: a review. Phys. Status Solidi Appl. Mater. Sci. 203 (13), 3207–3225.
Jensen, A.D., Munter, M.W., Debus, J., 2011. Review of clinical experience with ion beam radiotherapy. Br. J. Radiol. 84, S35–S47.
John, R., et al., 2017. Bright optical Centre in diamond with narrow, highly polarised and nearly phonon-free fluorescence at room temperature. New J. Phys. 19.
Kolesov, R., et al., 2013. Mapping spin coherence of a single rare-earth ion in a crystal onto a single photon polarization state. Phys. Rev. Lett. 111 (12), 12.
Leawilson, M.A., Lomer, J.N., Vanwyk, J.A., 1995. Electron-spin-resonance of the R4/W6 defect in irradiated diamond. Philos. Mag. B 72 (1), 81–89.
Lesik, M., et al., 2016. Production of bulk NV Centre arrays by shallow implantation and diamond CVD overgrowth. Phys. Status Solidi A Appl. Mater. Sci. 213 (10), 2594–2600.
Li, J., et al., 2001. Ion-beam sculpting at nanometre length scales. Nature 412 (6843), 166–169.
Li, W.Q., et al., 2019. A review of recent applications of ion beam techniques on nanomaterial surface modification: design of nanostructures and energy harvesting. Small 15 (31), 166.
Luhmann, T., et al., 2018. Screening and engineering of colour centres in diamond. J. Phys. D Appl. Phys. 51 (48), 1.
Luhmann, T., et al., 2019. Coulomb-driven single defect engineering for scalable qubits and spin sensors in diamond. Nat. Commun. 10.
McLellan, C.A., et al., 2016. Patterned formation of highly coherent nitrogen-vacancy centers using a focused Electron irradiation technique. Nano Lett. 16 (4), 2450–2454.
Meijer, J., et al., 2006. Concept of deterministic single ion doping with sub-nm spatial resolution. Appl. Phys. a-Mater. Sci. Process. 83 (2), 321–327.
Meijer, J., et al., 2008. Towards the implanting of ions and positioning of nanoparticles with nm spatial resolution. Appl. Phys. a-Mater. Sci. Process. 91 (4), 567–571.
Morello, A., et al., 2018. Scalable quantum computing with ion-implanted dopant atoms in silicon. In: 2018 IEEE International Electron Devices Meeting (Iedm).

Naydenov, B., et al., 2009. Engineering single photon emitters by ion implantation in diamond. Appl. Phys. Lett. 95 (18), 18.
Nizovtsev, A.P., et al., 2005. A quantum computer based on NV centers in diamond: optically detected nutations of single electron and nuclear spins. Opt. Spectrosc. 99 (2), 233–244.
Nizovtsev, A.P., et al., 2010. Quantum registers based on single NV + n C-13 centers in diamond: II. Spin characteristics of registers and spectra of optically detected magnetic resonance. Opt. Spectrosc. 108 (2), 239–246.
Ohno, K., et al., 2014. Three-dimensional localization of spins in diamond using C-12 implantation. Appl. Phys. Lett. 105 (5), 5.
Pacheco, J.L., et al., 2017. Ion implantation for deterministic single atom devices. Rev. Sci. Instrum. 88 (12), 5.
Pezzagna, S., et al., 2010a. Creation efficiency of nitrogen-vacancy centres in diamond. New J. Phys. 12, 2017.
Pezzagna, S., et al., 2010b. Nanoscale engineering and optical addressing of single spins in diamond. Small 6 (19), 2117–2121.
Pezzagna, S., et al., 2011. Creation of colour centres in diamond by collimated ion-implantation through nano-channels in mica. Phys. Status Solidi A Appl. Mater. Sci. 208 (9), 2017–2022.
Posselt, M., 1994. Crystal-trim and its application to investigations on channeling effects during ion-implantation. Radiation Effects and Defects in Solids 130, 87–119.
Pu, A., et al., 2000. Positron annihilation investigation of vacancies in as-grown and electron-irradiated diamonds. Diamond Relat. Mater. 9 (8), 1450–1463.
Raatz, N., et al., 2019. Investigation of ion channeling and scattering for single-ion implantation with high spatial resolution. Phys. Status Solidi A Appl. Mater. Sci. 216 (21), 21.
Räcke, P., 2020. Image charge detection for deterministic ion implantation. PhD Thesis, Universität Leipzig, Universitätsbibliothek Leipzig.
Racke, P., et al., 2018. Detection of small bunches of ions using image charges. Sci. Rep. 8, 305103.
Racke, P., et al., 2019. Image charge detection statistics relevant for deterministic ion implantation. J. Phys. D Appl. Phys. 52 (30), 305103.
Ramo, S., 1939. Currents induced by electron motion. Proc. IRE 27, 584.
Scheuner, C., et al., 2017. Nanometer collimation enhancement of ion beams using channeling effects in track-etched mica capillaries. Sci. Rep. 7, 1.
Schnitzler, W., et al., 2009. Deterministic ultracold ion source targeting the heisenberg limit. Phys. Rev. Lett. 102 (7), 070501.
Shockley, W., 1938. Currents to conductors induced by a moving point charge. J. Appl. Phys. 9, 635.
Smith, M.A., et al., 2017. High-resolution, high-throughput, CMOS-compatible electron beam patterning. In: Advances in Patterning Mater. Processes Xxxiv, p. 10146.
Smith, J.M., et al., 2019. Colour Centre generation in diamond for quantum technologies. Nanophotonics 8 (11), 1889–1906.
Spinicelli, P., et al., 2011. Engineered arrays of nitrogen-vacancy color centers in diamond based on implantation of CN- molecules through nanoapertures. New J. Phys. 13.
Stacey, A., et al., 2012. Depletion of nitrogen-vacancy color centers in diamond via hydrogen passivation. Appl. Phys. Lett. 100 (7).
Storm, A.J., et al., 2003. Fabrication of solid-state nanopores with single-nanometre precision. Nat. Mater. 2 (8), 537–540.
Sugiyama, M., Sigesato, G., 2004. A review of focused ion beam technology and its applications in transmission electron microscopy. J. Electron Microsc. (Tokyo) 53 (5), 527–536.
Sun, L., Chien, C.L., Searson, P.C., 2000. Fabrication of nanoporous single crystal mica templates for electrochemical deposition of nanowire arrays. J. Mater. Sci. 35 (5), 1097–1103.

Tanimura, H., et al., 2016. 10 nm-Deep n+/p and p plus/n Ge Junctions with High Activation Formed by Ion Implantation and Flash Lamp Annealing (FLA). In: 2016 16th International Workshop on Junction Technology (Iwjt), pp. 77–80.
Toimil-Molares, M.E., 2012. Characterization and properties of micro- and nanowires of controlled size, composition, and geometry fabricated by electrodeposition and ion-track technology. Beilstein J. Nanotechnol. 3, 860–883.
van Donkelaar, J., et al., 2015. Single atom devices by ion implantation. J. Phys. Condensed Matter 27 (15), 15.
Wrachtrup, J., Jelezko, F., 2006. Processing quantum information in diamond. J. Phys. Condensed Matter 18 (21), S807–S824.
Yamada, I., 1989. A review of material processing by low-energy ion-beams at the ion-beam Engineering Laboratory of Kyoto-university. Nucl. Inst. Methods Phys. Res. B 40-1, 557–561.
Zaitsev, A., 2001. Optical Properties of Diamaond. Springer, Berlin.
Ziegler, J.F., Ziegler, M.D., Biersack, J.P., 2010. SRIM—the stopping and range of ions in matter (2010). Nucl. Inst. Methods Phys. Res. B 268 (11 – 12), 1818–1823.

CHAPTER TWO

Advanced and in situ transmission electron microscopy of diamond: A review

Wolfgang Jäger*
Institute for Materials Science, Christian-Albrechts-University of Kiel, Kiel, Germany
*Corresponding author: e-mail address: wj@tf.uni-kiel.de

Contents

1. Introduction 31
2. TEM for epitaxial CVD synthesis of diamond 35
   2.1 Synthesis of diamond: Some fundamentals 35
   2.2 CVD diamond heteroepitaxy on Si and SiC 36
   2.3 CVD diamond heteroepitaxy on Ir (100) and on Ir (111) 49
   2.4 Homoepitaxial CVD synthesis of diamond 55
3. Nanodiamond 58
4. Defects in diamond 65
   4.1 Inclusions and voidites 66
   4.2 Grain boundaries, stacking faults, platelets 69
   4.3 Dislocations 72
   4.4 Point defects and defect complexes 73
5. TEM for characterization of physical properties 75
   5.1 In situ TEM deformation of diamond nanopillars 75
   5.2 TEM in measurements of electronic properties 77
   5.3 (S)TEM, PL and CL spectroscopy in characterizations of optical emission properties 78
   5.4 Thermal conductivity of thin diamond films 82
6. TEM for process and device technology 83
7. Conclusions and outlook 89
Acknowledgments 90
References 90

1. Introduction

Diamond is known to be a unique *material with outstanding physical properties* and has found numerous applications in technology. High pressure high temperature (HPHT) synthesis, plasma-assisted chemical vapor

Semiconductors and Semimetals, Volume 104
ISSN 0080-8784
https://doi.org/10.1016/bs.semsem.2020.08.003

deposition (CDV), the hot-filament technique, or detonation synthesis are prominent methods to fabricate synthetic diamond as material with single crystalline, poly-crystalline, nano-crystalline, or ultra-nano-crystalline structure. Properties of interest for applications of diamond are the high mechanical strength, the high thermal conductivity, the wide-band gap semiconducting and optical properties, the broadband spectral transparency, the high charge carrier mobility, the high dielectric breakdown field, favorable spin properties and Debye temperature, the chemical durability, and the biocompatibility.

High *demand for diamond* exists in industrial sectors, as diverse as oil and gas drilling, mining, wear-resistant tool coatings, abrasives, construction, wood-working, automotive, road maintenance, and aerospace. The progress of diamond fabrication and doping technology enabled electronic applications of diamond as semiconducting material for power and high-temperature electronics, supercapacitors, microelectromechanical systems, sensors, and detectors. The fabrication of single crystal wafer scale diamond of highest quality for device applications is still subject of current research. The various types of nanodiamond are used for a wide variety of applications in electrochemistry, in superconductivity, in field emission, for nanocomposites, and for photocatalytic conversion of carbon dioxide to chemicals and fuels. Fascinating are the perspectives of nanodiamond for biological sensing or for fluorescence imaging in biomedical applications in medicine and brain research. Based on the peculiar properties of single defect centers in diamond, such as the nitrogen-vacancy center, magnetometry on the nanoscale has been realized, and numerous research activities focus on exploring their potential for quantum sensing and quantum information technology.

The vast continued interest in *research of diamond* is reflected by numerous review articles. The volume on "Research, applications and challenges in CVD diamond" (edited by Nemanich et al., 2014) collects the topics of homoepitaxial CVD diamond for electronic applications (Yamasaki et al., 2014), large-area high quality single crystal diamond (Schreck et al., 2014), MEMS/NEMS based on mono-, nano-, and ultrananocrystalline diamond films (Sumant et al., 2014), surface functionalization and nanodiamond particle surfaces for biological applications (Szunerits et al., 2014), diamond electrodes (Einaga et al., 2014), diamond electron emission (Lin et al., 2014), and diamond surface conductivity and properties, devices, and sensors (Pakes et al., 2014). Further reviews focus on the electronic properties of CVD diamond (Nebel, 2003, 2017), nanocrystalline diamond

(Williams, 2011), quantum photonic networks in diamond (Loncar and Faraon, 2013), nitrogen-vacancy centers in diamond close to the surface (Wrachtrup et al., 2013), diamond for electrochemistry at the nanoscale (Yang et al., 2016), nanoscale sensing and imaging in biology using the nitrogen-vacancy center in diamond (Hall et al., 2013), science and engineering of nanodiamond particle surfaces for biological applications (Shenderova and McGuire, 2015), diamond for monolithically integrated optical and optomechanical devices (Rath et al., 2015), on diamond X-ray optics (Shvyd'ko et al., 2017), and on diamond as material for X-ray refractive lenses (Roth et al., 2017). The role of surface functionalization for photoelectrochemical applications in dye-sensitized solar cells or in photocatalytic applications for reduction of carbon monoxide, ammonia synthesis and water splitting is described in a review by Raymakers et al. (2019). Fascinating are the perspectives and developments in research for the application of defect complexes for nanophotonics (Aharonovich et al., 2011; Bradac et al., 2019), for nanosensing (Achard et al., 2020; Batzer et al., 2019; Radtke et al., 2019), and in quantum information technologies (e.g., Pfender et al., 2017; Morishita et al., 2020). More information on the science and technology of diamond can be also found in further overviews, see, for instance, Williams (2011), Arenal (2012), Shenderova and Ciftan Hens (2013), Arnault and Girard (2014), Williams (2014), Krueger (2014), Arnault (2017), Nebel (2017), May et al. (2018), Kato and Teraji (2018), Koizumi et al. (2018), Arnault (2019), Sang et al. (2019), Yang (2019).

Transmission electron microscopy (TEM, the term being synonymously used for microscopy as method and for the microscope instrumentation) and its advanced characterization methods belong to the most important techniques in research for materials and their physical properties. Starting with the construction of a transmission electron microscope in 1932 (Knoll and Ruska, 1932; Ruska, 1987, Nobel lecture), the methods of high resolution electron microscopy have made substantial contributions to key areas of materials science, condensed matter physics, biosciences and nanosciences and to the development of device technologies. The invention of aberration-corrected electron optics (Rose, 1990, 2009, 2012) and the efforts taken to realize the first aberration-corrected advanced instrumentation for high-resolution (HR)TEM (Haider et al., 1998a,b,c) and for scanning TEM (STEM) (Krivanek et al., 1999) led to improvements in the resolution in electron microscopy to its theoretical limits and were a huge step forward by creating novel opportunities for instrumentation and methods and for research and development in materials science, life sciences, and in the associated technologies. The current instrument developments focus on creating novel methodological possibilities for materials characterization and support a substantial industry of manufacturers. The modern field of transmission electron microscopy uses now sophisticated characterization platforms that combine specialized computer-controlled instruments with various types of advanced detector systems and of sample stages, enabling sensitive atomic-level characterization in many areas of nanoscience and nanotechnology as well as in applications of cryo-EM for structural biology and life science.

Frequently applied for *high-resolution characterization* of structure, defects, interfaces and surface phenomena of diamond and other materials on a microscopic and nanoscopic scale are the imaging methods of conventional TEM, HRTEM and STEM. The instruments provide collimated electron beams that either use broad beam illumination of samples in a stationary mode (typically nm to μm in diameter) or a highly focused scanning electron beam (typically nm to sub-nm in diameter) at energies of between 20 keV and 1 MeV. Samples are self-supporting thin films of typically few 10–100 nm in thickness or nano-dimensional powders on electron-transparent carrier films. The spectroscopic methods of electron energy loss spectroscopy (EELS) and energy-dispersive X-ray spectroscopy (XEDS) provide chemical and electronic bonding information and elemental mappings at high spatial resolution. Selected-area electron diffraction from micrometer-size regions or nano-beam electron diffraction provide crystallographic information. Furthermore, environmental TEM instruments and (S)TEM instrumentation with specialized sample holder configurations can be utilized as 'nano-laboratories' for in situ experiments, providing the potential to study the 3D structure and morphology in tomography experiments or the dynamics of atomic or nanoscale phenomena of the physical behavior of materials in response to external stimuli, such as gas environment, liquids, temperature, stress, and electric or magnetic fields. Electron holography can provide quantitative information about local variations in magnetic field and electrostatic potential both within and surrounding the specimen (Dunin-Borkowski et al., 2019). Differential phase contrast imaging has been applied to probe electric fields of materials, including nanoscale electric fields (Brown et al., 2017; Müller et al., 2014; Rose, 1977; Shibata et al., 2012, 2017). Excellent overviews on these and various further methods of (S)TEM and the related techniques and on their vast potential for materials science can be found, for instance, in the articles of handbooks edited by van Tendeloo et al. (2012) and by Hawkes and Spence (2019) and in textbooks (e.g., Reimer and Kohl, 2008). Dedicated low-voltage high-resolution and STEM instruments have been used for the characterization of low dimensional materials, particularly graphene, and for studies of single atoms and molecules (see, e.g., Kaiser and Stöger-Pollach, 2014, Bell et al., 2014, Suenaga, 2014, Colliex et al., 2012). High-resolution methods of TEM have contributed substantially also to structural biology for which the 2017 Nobel prize in Chemistry was awarded (Dubochet et al., 2017). The methods and applications of cryo-electron microscopy for structural biology are reviewed, e.g., by Plitzko and Baumeister (2019).

This chapter concentrates on the *materials science of diamond* with the advanced imaging and spectroscopy methods of high-resolution and in situ (S)TEM. Contributions to different topical fields are reviewed, comprising the epitaxial chemical vapor deposition (CVD) synthesis of diamond, nanodiamond, defects, physical properties, and device technology. The progress in these areas is illustrated by selected examples of published research during the past 25 years. These contributions were pivotal for the development, monitoring and control of diamond synthesis, for understanding of the microstructure and its relationship with physical properties, and for the assessment of device processing and device performance.

2. TEM for epitaxial CVD synthesis of diamond

2.1 Synthesis of diamond: Some fundamentals

The *synthesis of diamond* as particle, thin film and coating material utilizes the methods of high-pressure high-temperature (HPHT) synthesis and low-pressure CVD techniques for which the activation of reactants, leading to deposition of carbon on the substrate, takes place in a plasma, in a flame, or at the surface of a hot filament. Diamond films of high structural quality and with outstanding properties are synthesized ("grown") by homoepitaxial CVD on diamond seed crystals or by heteroepitaxial CVD on various substrates, in the form of polycrystalline, nano-crystalline, ultra-nanocrystalline and "single crystal" diamond films. Microwave-plasma CVD is considered to be advantageous for the synthesis of diamond with desired properties and quality because of its potential for improved parameter control and reproducibility of conditions. Among the various methods for the synthesis of nanodiamond particles of few nanometers in size, the detonation method has shown to be advantageous. Different techniques for the synthesis of diamond are reviewed, for instance, by Lettingdon and Steeds, 1994, Arnault and Girard, 2014, Nemanich et al., 2014, Sang et al., 2019.

Homoepitaxial CVD synthesis of diamond is often performed with HPHT diamonds as substrates. The usable surface size of a few tens of mm^2 limits the dimension of the epitaxial diamond layer. Among the several methods developed in order to enlarge the usable surface for homoepitaxial growth, the mosaic wafer growth is an approach often applied for the fabrication of freestanding large-area single-crystal diamond wafers (Yamada et al., 2012; Yamasaki et al., 2014). In *heteroepitaxial CVD synthesis of diamond,* some obvious considerations for the choice of substrates are the lattice mismatch and the differences in the thermal expansion coefficients between the diamond and substrate crystal lattices, the availability and cost of substrate material, and the compatibility with device technology, for instance, in microelectronics. The heteroepitaxial CVD synthesis of polycrystalline and "single crystal" diamond thin films has been investigated extensively with the objective to understand, optimize and control the processes of nucleation and crystal growth. Goals are the synthesis of poly-crystalline films of desired physical properties or of large-area "single-crystal" diamond layers of highest crystal perfection, i.e., with relatively low defect densities (for reviews see, for instance, Kawarada, 1998, Dischler and Wild, 1998, Schreck and Arnault, 2018, Jiang, 2019). Thus, a solid understanding could be established of the relationship between deposition parameters and structural and physical film properties. Fig. 1 illustrates the heteroepitaxial high-quality large-area CVD synthesis of diamond films on (001)-oriented single-crystal Si, SiC and Ir substrates, respectively, and depicts an example of a high-resolution TEM image of different types of structural defects in CVD diamond. Diamond films synthesized by optimized CVD process technology have become already early candidate materials for numerous applications, for instance, in high power and high temperature electronics, in microtechnology, for sensors and for microelectromechanical systems, or in medicine (Adamschik et al., 2001; Denisenko et al., 2001; Gluche et al., 1998a,b; Güttler et al., 1999; Kawarada, 1998; Kawarada et al., 1994; Lettingdon and Steeds, 1994; Werner and Locher, 1998).

As illustrated in the following, the methods of TEM have contributed to investigations of the deposition process, of the diamond nucleation, of the microstructure development during diamond layer growth, and of defect and doping phenomena and their role for crystal growth phenomena.

2.2 CVD diamond heteroepitaxy on Si and SiC

Microwave plasma assisted CVD synthesis of diamond on (001)-oriented substrates of Si, β-SiC, and Ir turned out to be successful for synthesizing

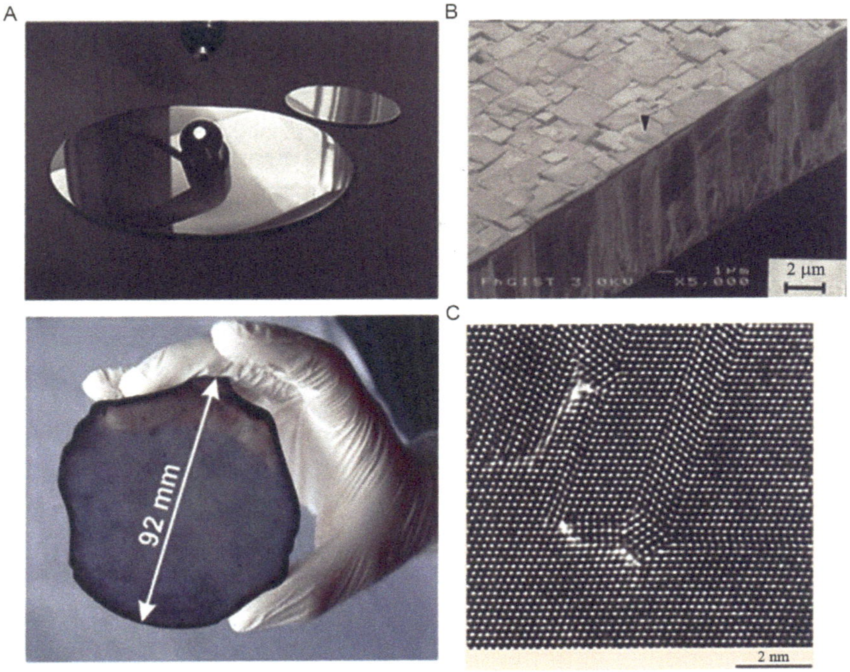

Fig. 1 CVD diamond. (A) (Above) Highly-oriented (HOD) diamond layers about 15 µm thick on Si (001) substrate wafers 1 in. and 2 in. in diameter. The mirror-like optical appearance results from the extremely low surface roughness. (Below) Freestanding unpolished diamond "single crystal" wafer in thickness synthesized by optimized large-area CVD on Ir/YSZ/Si(001). The thickness of the disc is about 1.6 mm, its weight is 155 carat. (B) Highly-oriented CVD diamond film on Si (001) substrate. The scanning electron microscope (SEM) micrograph of the film with columnar grain structure shows perspective views of surface and cross-section. (C) Defects in a CVD deposited diamond film: twins, stacking faults, and nanograins, and local disorder at the intersection of the stacking faults and twins. High-resolution TEM lattice image, [110] zone axis projection. *Panel A (above): Copy of fig. 1 in: Güttler, H., Flöter, A., Kohn, E., Gluche, P., Jiang, X., et al., 1999. Evaluation of semiconducting CVD diamond films for electronic and sensor applications. In: Kempter, K., Haußelt, J. (Eds.), Werkstoffwoche '98, Symp. 1, Materials for Information Technology. Wiley-VCH, Weinheim, ISBN 3-527-29938 6, pp. 189–194. (Below) Copy of fig. 5 in: Schreck, M. et al., (2017). Ion bombardment induced buried lateral growth: the key mechanism for the synthesis of single crystal diamond wafers. Sci. Rep. 7, 44462. Panel (B): Copy of fig. 2.3.10 in: Jiang, X., 2019. Ultra-wide Bandgap Semiconductor Materials Ch. 2.3 Heteroepitaxial growth of diamond. Elsevier Inc. Panel (C): Copy of fig. 1.2 in: Kirkland, A.I., Chang, S.L.-Y., Hutchison, J.L., 2019. Atomic resolution transmission electron microscopy. In: Hawkes, P.W., Spence, J.C.H. (Eds.), Springer Handbook of Microscopy, Springer Handbooks, Springer Nature, Switzerland AG, Ch. 1, pp. 3–47.*

large-area "single crystal" diamond films (for review see, e.g., Güttler et al., 1999, Jäger and Jiang, 2001, Jiang, 2019, Schreck and Arnault, 2018, Werner and Locher, 1998). The surface layers of such films contain large fractions of (001)-oriented crystal grains with lateral dimensions of several 10 μm or more and possess a high degree of crystal orientation texture, with only small-angle lattice misorientations between grains. Narrow deposition parameter windows for the CVD process are chosen in order to control the nucleation and subsequent crystal growth of the diamond films. Important considerations are the base pressure in the deposition chamber, the purity of gas sources applied, and the surface treatment and temperature of the substrate, and, in microwave-assisted deposition, the microwave power. Generally, a first stage of bias-assisted (or bias-enhanced, BEN) nucleation is applied in order to obtain a high density of diamond nuclei. During this stage, the substrate is exposed to a mixture of methane gas diluted with hydrogen at low gas pressures (typically about 50 Torr), while a negative bias voltage (typically up to about 200 V) is applied for a short time (typically up to 30 min). Thus, positively charged ions of the plasma are accelerated toward the substrate and can assist in forming diamond nuclei. Influences of ion bombardment induced defects on nucleation or of renucleation of grains have been revealed by studying the conditions for formation of misoriented crystals (Jiang et al., 1996). An important role for the diamond nucleation plays also the kinetics, depending on parameters like substrate temperature, methane concentration, and process pressure. For optimized BEN-assisted CVD synthesis of diamond on silicon and on silicon carbide, graphite formation is largely suppressed, and nanocrystalline diamond crystals form in large numbers on the substrate surface. For the second stage of film synthesis at temperatures of typically about 800–900°C, also called growth stage, different deposition protocols can be chosen. For instance, "highly-oriented" diamond (HOD) films can be achieved choosing parameters for which overgrowth of grains with large misorientations with respect to the desired crystal orientation and columnar growth are favored. The predominating crystal defects in the top layers of such films are small-angle grain boundaries. Secondary carbon phases at grain boundaries are largely absent (Güttler et al., 1999; Jiang et al., 1998a; Wittorf, 1999).

2.2.1 Early stages of bias-assisted CVD synthesis: microstructure and interface phenomena

For the BEN-assisted CVD synthesis on Si(001) substrates, highly-oriented diamond crystallites can form directly on the Si substrate or on a nanocrystalline β-SiC interlayer (Kohl et al., 1993; Locher et al., 1994). HOD diamond layers can be routinely achieved under conditions leading to the formation of an SiC interlayer. Based on the microscopic investigations by high-resolution transmission electron microscopy and electron diffraction experiments, different processes controlling the early stages of diamond synthesis can be discerned (Wittorf, 1999; Wittorf et al., 1997c, 2000): (i) Si substrate surface roughening, redeposition of Si and formation of {111}-facets, (ii) formation of a nanocrystalline β-SiC interlayer, with Si in the plasma being one of the sources for Si, (iii) {111} facet formation of SiC nanocrystallites resulting from fast growth of {100}-facets, and (iv) nucleation and growth of diamond crystallites on Si and on the {111}-facets of SiC, resulting from implantation of carbon from the plasma by the applied bias. Figs. 2 and 3C illustrate the formation of domains with nearly periodic arrays of nanoscopic surface ripples or Si islands, with predominantly {111}-faceted ridges along the two ⟨110⟩ surface directions, and the range of small misorientations between the crystal lattices of

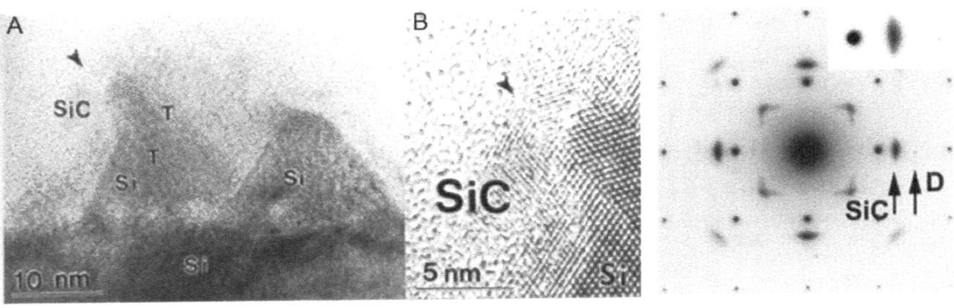

Fig. 2 Early stage of CVD synthesis on Si (001) substrates with β-SiC interlayer formation. (A) Prismatic Si islands, {111} microtwins (T) and β-SiC nanocrystals. The {111}-oriented part of the SiC/Si interface shows epitaxial lattice orientation relationship between the Si nanocrystals and the silicon carbide nanocrystals. (right) Interface between a prismatic Si island and a SiC nanocrystal: interconnected twins in SiC and in Si (arrow). Cross-section HRTEM lattice images. ⟨110⟩ zone axis. (B) (001) zone-axis selected area diffraction pattern showing reflections from Si, from SiC (arrow) and from diamond (D, arrow to sharp intensity spot). The electron diffraction intensity is shown as dark contrast. The insert depicts at higher magnification the broadening of SiC reflections with respect to those of Si and diamond, indicating small misorientations of the lattices of SiC and of diamond nanocrystals. Plane-view TEM specimen. ⟨001⟩ zone axis orientation. *Panel (A): Copy of fig. 1 in: Wittorf, D., Jäger, W., Jia, C. L., Urban, K., Flöter, A., Güttler, H., Zachai, R., 1997. Interfaces of CVD diamond films on silicon(001). Inst. Phys. Conf. Ser. 157, 451. Panel (B): Copy of fig. 7 in: Wittorf, D., Jäger, W., Dieker, C., Flöter, A., Güttler, H., 2000. Electron microscopy of interfaces in chemical vapor deposition diamond films on silicon. Diam. Relat. Mater. 9, 1696–1702.*

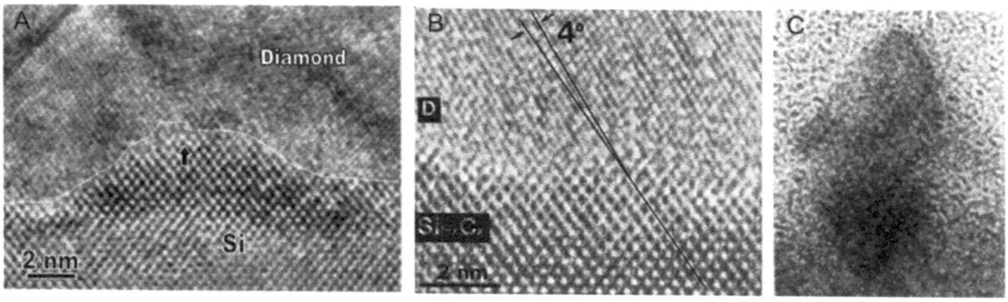

Fig. 3 CVD diamond synthesis on Si (001) substrates, on $Si_{1-x}C_x$ buffer layers, and on β-SiC interlayers. (A) Diamond grain formed on top of a Si substrate hillock. The crystal lattice is misoriented with respect to the Si lattice by about 6° around the [110] axis. A small terrace edge on the (001) facet on top of the hillock (marked by the arrow) induced a twin boundary. Cross-section HRTEM micrograph. ⟨110⟩ zone axis. (B) Diamond grain epitaxially formed on a $Si_{1-x}C_x$ buffer layer ($x_{nom}=0.014$). The diamond lattice rotation about the ⟨110⟩ axis direction is 4°. Cross-section HRTEM micrograph in ⟨110⟩ zone axis projection. (C) Diamond nanocrystal epitaxially nucleated on SiC interlayer. The crystallite shows facets parallel to {111}-lattice planes and twins T. Cross-section HRTEM micrograph. ⟨110⟩ zone axis. The lattices of diamond and SiC appear superimposed onto each other in the lower part of the projected lattice image. *Panel (A): Copy of fig. 3 in: Jiang, X., Jia, C.L., 2000. Direct local epitaxy of diamond on Si(100) and surface-roughening-induced crystal misorientation. Phys. Rev. Lett. 84 (16), 3658–3661. Panel (B): Copy of fig. 5 in: Wittorf, D., Jäger, W., Urban, K., Gutheit, T., Güttler, H., Schulz, G., Zachai, R., 1997. Microstructure and growth of MWCVD diamond on $Si_{1-x}C_x$ buffer layers. Diam. Relat. Mater. 6, 649–653. Panel (C): Copy of fig. 6 in: Wittorf, D., Jäger, W., Dieker, C., Flöter, A., Güttler, H., 2000. Electron microscopy of interfaces in chemical vapor deposition diamond films on silicon. Diam. Relat. Mater. 9, 1696–1702.*

Si, SiC, and diamond. The quasi-periodic Si surface ripple structure with {111} facets offers favorable sites for nucleation of SiC nanocrystals and of diamond crystallites (Wittorf et al., 1997c, 2000) (Figs. 2A and 3C). The use of $Si_{1-x}C_x$ alloy buffer layers ($x<0.035$) as substrate, aiming at reducing the lattice mismatch during heteroepitaxy, may lead to undesirable strain relaxation phenomena of the buffer layers, such as twin formation for higher C concentrations or formation of β-SiC precipitates. Fig. 3B shows an example for the formation of an epitaxial diamond grain directly on a $Si_{1-x}C_x$ alloy buffer layer (Wittorf et al., 1997a).

Diamond nanocrystals form also during bias-assisted CVD synthesis on interlayers of amorphous carbon (Morris and Pehrsson, 1991) or directly on Si surface hillocks and along small Si (001) terraces (Jia et al., 1995; Jiang and Jia, 2000) (Fig. 3). For CVD synthesis on amorphous interlayers, the morphology of grains is generally different as compared to CVD synthesis on crystalline surfaces (see, e.g., Tzou et al., 1994). The heteroepitaxial growth of smooth and continuous diamond thin films on silicon substrates via high quality silicon carbide buffer layers and the heteroepitaxial growth of highly oriented diamond on cubic silicon carbide are reported by Kawarada et al. (1995, 1997). The formation of hillocks containing β-SiC, diamond and amorphous components and the preferred nucleation on hillocks (Wurzinger et al., 1996) as well as the growth of individual diamond crystals from nuclei in a β-SiC layer (Stammler et al., 1997) have been also reported for bias-assisted CVD synthesis conditions. Successful synthesis of highly oriented {111} diamond films on 3C-SiC/Si {111} substrates has been achieved also by pulsed bias-assisted nucleation and epitaxial grain selection by oxidative etching (Suto et al., 2017).

Various possible influences on nucleation and the early stages of BEN-assisted heteroepitaxial CVD synthesis of diamond have been discussed (for reviews see, e.g., Kawarada, 1998, Jäger and Jiang, 2001, Jiang, 2019). Substrate surface roughening by etching and redeposition of Si are the results of the exposure to the H-rich plasma. Surface defects, surface sites of particular morphology, such as hillocks, strained surface inclusions, or surfaces with faceted nanoscopic surface ripples are favorable surface sites for nucleation of nanocrystals. Shallow carbon implantations (sometimes referred to as subimplantation) influence the nucleation of crystals, being dependent on flux and direction of impinging ions, as could be shown by deposition experiments on structured Si surfaces and on Si surfaces containing pores. Due to the ion bombardment associated with the BEN process, the lattices of nanocrystals can be slightly misoriented with respect to the substrate orientation. A mechanism for diamond nucleation from energetic ion species was suggested, involving spontaneous nucleation of a diamond cluster in a dense, amorphous carbon hydrogenated matrix and its subsequent growth. The surface diffusion of C—H species may influence the formation of immobile nuclei under conditions of only moderate substrate surface roughening (Jiang et al., 1994). The area density of diamond nanocrystals on the substrate surface is likely to be determined also by "depletion zones" around nuclei, providing an increased barrier for subsequent formation of a critical nucleus. The mechanism of forming such zones was ascribed to surface adatom diffusion, with the nuclei acting as a network of sinks (Jiang et al., 1994). Theoretical considerations suggest that depletion zones occur as result of an increased nucleation barrier due to surface strain (Ascarelli et al., 1997). The total density of small diamond crystals achieved for CVD synthesis on Si substrates under BEN conditions can be as high as

10^{11} cm^{-2} (Lifshitz et al., 2002). Strategies for the CVD synthesis of diamond on crystalline Si and SiC substrates and of ultrathin diamond films by diamond nucleation and seeding techniques are described in overviews, for instance, by Arnault and Girard (2014), Yaita et al. (2018), Jiang (2019).

2.2.2 Microstructure phenomena and morphology during layer growth

During the early stage of the optimized CVD synthesis, the diamond nanocrystals possess mostly cube-octahedral morphology with {001} and {111}-type surface facets. From distributions of such nanocrystallites, diamond films have been synthesized that consist of well-oriented diamond grains, with a high degree of orientation texture and only small misorientations between their crystal lattices (Wild et al., 1994). The fabrication of such films utilizes synthesis routes that favor coarsening of the grain structure by overgrowth of misoriented grains (Jiang et al., 1993, 1998a,b) and, for small grain misorientations of 2° or below, by coalescence of adjacent grains, as well as increasingly columnar growth for higher film thickness (Fig. 4).

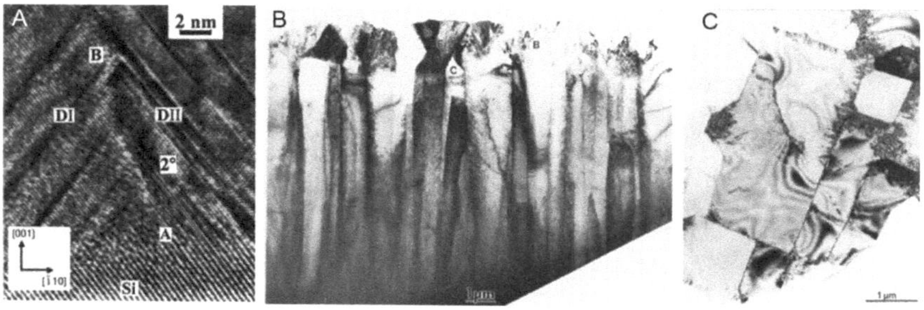

Fig. 4 Microstructure phenomena of heteroepitaxial CVD diamond films on Si (001). (A) Coalescence of small grains (at B) near Si substrate surface (marked A) and a small-angle grain boundary (marked A-B) between grains DI and DII with small lattice misorientation of 2°. Cross-section TEM lattice imaging, ⟨110⟩ zone axis projection. (B) Columnar grain structure and change of grain boundary orientations due to overgrowth of misoriented grains in the diamond top layer (e.g., of grain B by grain A or of grain C by grain E) when changing to CVD synthesis conditions with boron doping. Cross-section TEM bright-field imaging near ⟨110⟩ zone axis. (C) Large boron-doped grains in the top surface region of the diamond film, resulting from coalescence of columnar grains. The predominating grain boundary orientations are parallel to (110) lattice planes. Crystal defects are present only near grain boundaries (dark contrast). Plane-view TEM bright-field imaging near (001) zone axis. *Copies of figs. 4, 6, 3 in: Jiang, X., Schiffmann, K., Klages, C.-P., Wittorf, D., Jia, C. L., Urban, K., et al., 1998. Coalescence and overgrowth of diamond grains for improved heteroepitaxy on Si(001). J. Appl. Phys 83, 2511.*

The resulting films of several μm in thickness contain nearly epitaxially oriented surface grains with pronounced facets parallel to {111} and {100} lattice planes. The formation of {001} surface facets at layer thickness above about 10 μm results from grain growth with preferred growth of {111} facets (Jiang et al., 1998a,b).

Crystal growth modes can be initiated that favor fast growing facets (Jiang et al., 1998a,b; Kawarada et al., 1997) or, by adding boron to the gas phase, inhomogeneous boron incorporation (Wurzinger et al., 1997) and lateral grain growth (Jiang et al., 1998a,b). Applying boron doping initiates a step-flow growth mode by enhancing the surface diffusivity of C—H compounds, leading to the described growth phenomena and to the changes of grain boundary orientations as observed by TEM (Fig. 4B). Thus, the size and orientation distributions of grains are influenced during film synthesis. CVD parameter conditions of high methane concentration, low substrate temperature and additions of small amounts (several tens of ppm) of molecular nitrogen in the gas phase are able to favor ⟨001⟩ textured crystal growth (Jiang and Jia, 1996; Wild et al., 1994). For such conditions, the growth velocities of the {001} grain surfaces are higher than those for {111} surfaces, leading to the selective growth of ⟨001⟩-oriented grains (sometimes also referred to as "evolutionary selection") (Wild et al., 1994). Ion bombardment during the continuous deposition is able to also influence the crystal growth (Jiang et al., 1996, 1998a,b). The columnar growth of diamond grains which is achieved by applying such conditions, enables to synthesize diamond layers of thickness in the μm range with well-oriented large surface grains at small misorientations between the top layer grains (Fig. 4B and C) (Jiang et al., 1998a,b). The published best values of the average polar and azimuthal spreads of grain orientations for diamond deposition on Si (as deduced from X-ray rocking curves) are 2° (Jiang, 2019; Jiang et al., 1998a,b).

2.2.3 Grain boundaries and defects in highly-oriented diamond films

Various types of grain boundaries have been identified in CVD diamond films: small-angle grain boundaries (Jiang et al., 1998a,b; Sivazlian et al., 1994; Wittorf et al., 1997b), large-angle grain boundaries (Wittorf et al., 1997b) and twin boundaries (Shechtman et al., 1993a,b; Wittorf et al., 1997b). Highly-oriented HOD diamond films contain predominantly small-angle grain boundaries along ⟨110⟩ crystal directions in regions near to the film surface (Jäger et al., 1999; Jiang et al., 1998a,b, 2000; Wittorf et al., 1997b) (Figs. 4, 5). Small-angle grain boundaries with grain misorientations of less than 10° generally possess, on a microscopic scale, asymmetric and

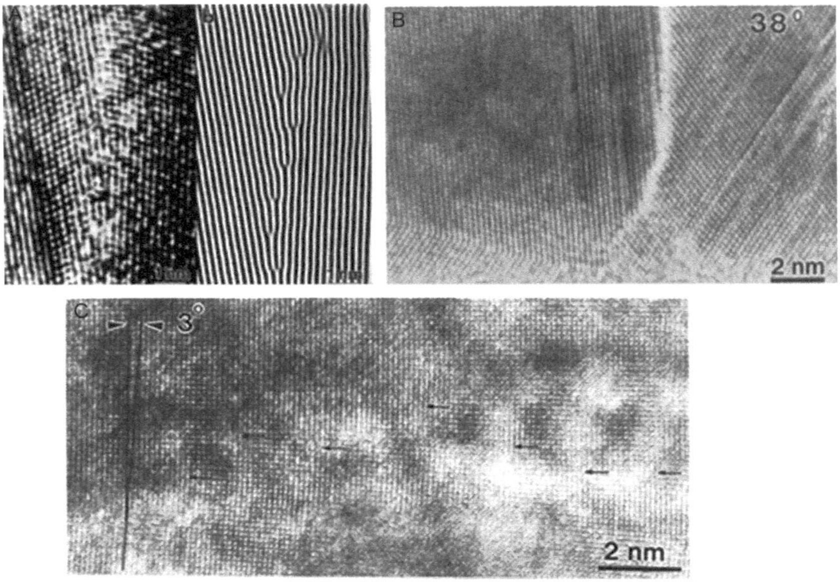

Fig. 5 Grain boundaries in CVD diamond films on Si (001) substrates. (A) Small-angle grain boundary with an asymmetric geometry. The rotation between both lattices about the [110]-zone axis is about 11°. Sections of the boundary are slightly inclined with respect to the image normal leading to local lattice superposition and contrast enhancement. The respective Fourier-filtered image obtained using (-1-11)-reflections for image synthesis illustrates the locations of grain boundary dislocations. (B) Large-angle grain boundary between diamond grains. The lattices are rotated by 38° about the [110]-zone axis. High-resolution TEM lattice image in [110]-zone axis projection. (C) Small-angle grain boundary between surface grains of a highly-oriented HOD CVD diamond film in a top-view ⟨001⟩ zone axis projection. The {220}-diamond lattice planes with interplanar distances of about 0.13 nm are well resolved for both lattices. The misorientation between the adjacent grains is about 3°. An array of dislocations with an average distance of 1.9 nm is visible (arrows). The detailed analysis reveals that these dislocations have a projected Burgers vectors b with a modulus a/4[110]. This atomic scale image shows that these boundaries do not contain any pores or secondary phases. High-resolution TEM lattice image. ⟨001⟩ zone axis projection. *Copies of figs. 1, 2, 4 in: Wittorf, D., Jia, C.L., Jäger, W., Grushko, B., Urban, K., Jiang, X., et al., 1997. Grain boundaries in diamond films on Si(100). Proc. 1996 MRS Fall Meeting, Symposium 'Atomic Resolution Microscopy of Surfaces and Interfaces', Mat. Res. Soc. Symp. Proc. Vol. 466, 27–32.*

symmetric parts (Fig. 5A and C). Their nanoscale structure can be described by the classical dislocation model (Wittorf et al., 1997b). For optimized CVD synthesis conditions, these grain boundaries generally do not contain any significant amounts of inclusions of additional secondary carbon phases

(Wittorf, 1999; Wittorf et al., 1997b), thus confirming that the crystal lattices of the grains are in direct contact with each other. Any other types of inclusions, like those observed in natural diamond (e.g., Chen and Van Tendeloo, 1999) are also absent (Wittorf et al., 1997b). The near-surface diamond grains still contain defects near grain boundaries, however, the interior of grains is largely free of defects (Fig. 4C). These defects are predominantly dislocations, microtwins and stacking faults (Shechtman et al., 1993a,b; Wittorf et al., 2000) which have formed during layer synthesis, for instance from incorporation of impurities, or as result of local mechanical strain relaxation between the columnar diamond grains (Jiang et al., 1998a,b; Wittorf, 1999; Wittorf et al., 1997b). For CVD diamond films with large fractions of (001)-oriented near-surface grains, large-angle grain boundaries are generally observed only near to the substrate interfaces, accommodating misorientations between grains of up to 40° (Fig. 5B). These boundaries contain often large fractions of {111} microfacets, their presence reflecting the stability of {111} facets during CVD film growth (Wittorf et al., 1997b). Frequently, coherent twins are observed in the vicinity of the grain boundaries. Further defects in CVD diamond film grains are N-induced defect centers (Nebel et al., 1997). The grain boundary orientations, as is the grain growth, can be influenced by boron-assisted CVD growth conditions, thus enabling to improve the heteroepitaxy (Jiang et al., 1998a,b). Fig. 4B illustrates the change of grain boundary orientations due to overgrowth of misoriented grains for a film with columnar grain structure at simultaneous direct coupling between doped and nominally undoped grain regions. The origin is associated with a step-flow growth mode on (001) planes, changing the ratio of growth rates of {111} facets and {100} facets. This mechanism leads to the formation of a film of cubic grain morphology with small-angle grain boundaries.

2.2.4 Diamond-silicon and diamond-silicon carbide interfaces

High-resolution TEM analyses of the interface structures have contributed to understanding the lattice mismatch accommodation and the causes for grain misorientation. Direct lattice plane coupling is observed for the interface between the diamond lattice and silicon or a nanocrystalline β-SiC interlayer (Wittorf et al., 1997a, 2000). The diamond-silicon interface is characterized by a ratio of 3/2 of the numbers of diamond-to-Si lattice planes. For the diamond-SiC interface, this ratio is 5/4. In three dimensions both configurations can be described with a modified coincidence site lattice (CSL) model (Balluffi et al., 1982) generalized to describe strained interfaces, the near-coincidence site lattice (NCSL) model. This model describes energetically favored crystallographic orientation relationships with lattice sites of the NCSL common to both lattices adjacent to the interface, allowing for small elastic lattice distortions. The coincidence lattice points form a cubic face centered unit cell for both types of interfaces. The crystallographic relation between the diamond and silicon lattice has been described by a NCSL model first by Jia et al. (1995). Investigations of the interfaces between diamond and nanocrystalline SiC, between diamond and a SiC substrate, and between SiC and the Si substrate have shown that the NCSL model can be successfully used also to describe the crystallographic relationships between diamond and silicon carbide (Wittorf, 1999; Wittorf et al., 2000). Small angle deviations of the crystal lattices of diamond and silicon and of diamond and silicon carbide are connected with an excess of diamond {111} planes terminating at the interface, being nearly periodically arranged (Fig. 6). The period

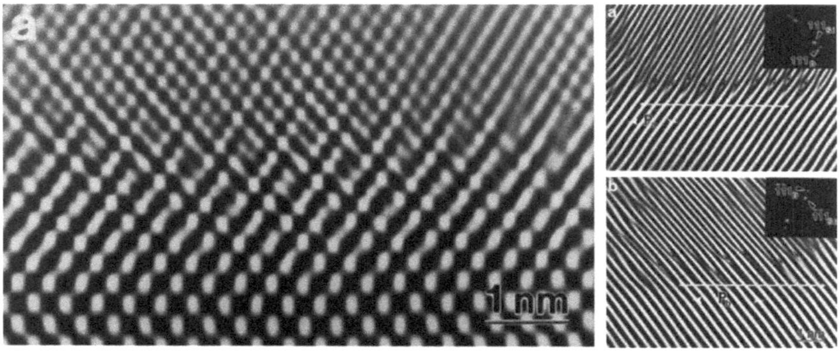

Fig. 6 Interface between CVD diamond and a Si (001) substrate surface. (Left) Fourier-filtered and digitally processed high-resolution image obtained from a cross-section HRTEM lattice image in [1-10] zone axis projection of the interface between diamond and a $Si_{1-x}C_x$ (001) substrate for small substitutional C content ($x=0.007$). The image is synthesized using the +-(111) reflections. The lattice misorientation between Si (below) and diamond (above) amounts to 4°. (Right) High-resolution images depicting arrays of excess {111} planes terminating at the interface. The images are digitally processed using two opposite sets of {111} diffraction spots of power spectra obtained from the experimental image. The excess planes are responsible for the mismatch accommodation and the tilt. *(Left): Copy of fig. 3 in: Jäger, W., Wittorf, D., Jia, C.L., Urban, K., 1999. High-resolution electron microscopy of grain boundaries and interfaces in heteroepitaxial diamond films. In: Vincenzini, P. (Ed.), Proc. 9th Cimtec—World Forum on New Materials, Topical Symposium Diamond Films, Techna Srl, Faenza. Advances in Science and Technology 21, 137–144. (Center, right): Copy of fig. 4.26 in: Wittorf, D., 1999. Dissertation D 82 RWTH Aachen, Research Centre Jülich Report, Germany, JÜL-3637.*

lengths can vary for (111) and (11-1) planes. Such interfaces can be described with a model similar to the dislocation model for small-angle tilt grain boundaries, with the terminating planes accounting for the tilt misorientation between the lattices (Wittorf, 1999, Wittorf et al., 2000).

Typical for the epitaxial diamond-Si (001) interface is a matching of, on the average, three (111)-diamond lattice planes with two (111)-Si lattice planes ("3/2-registry"). The remaining lattice mismatch is determined to be 1.5% at room temperature (or 1.4% at the deposition temperature of around 800 °C). Assuming an isotropic contraction of the Si lattice of 1.5% allows to construct a NCSL lattice with a face-centered cubic structure for this type of interface (Jia et al., 1995; Wittorf et al., 1997a, 2000).

Effective strain-relaxation of diamond(001)-Si(001) interfaces can be achieved also with an epitaxial interlayer of SiC formed on the Si substrate with surface islands during the early deposition stages (Fig. 2) (Wittorf et al., 1997c, 2000). The diamond-Si interfaces and the SiC-Si interfaces are facetted parallel to {111} lattice planes and show regular arrays of terminating {111} lattice planes (Fig. 7). Spatially averaged along the interfaces, a "5/4-registry" of five (-111)-lattice planes of diamond matching with four (-111)-lattice planes of SiC lattice planes has been deduced for these interfaces (Wittorf, 1999; Wittorf et al., 1997c, 2000). This arrangement compensates almost completely the nominal misfit of 22% between diamond and cubic β-SiC. The remaining lattice mismatch is calculated to be 2.3%. Assuming isotropic expansion of the SiC lattice by 2.3% allows to construct a NCSL lattice with a face-centered cubic structure for this type of interface (Wittorf, 1999, Wittorf et al., 2000).

The Si-SiC interfaces of the Si{111} facets are characterized, on the average, by a "5/4"—registry of 5 (-111)-planes of SiC and 4 planes of Si (Fig. 7). Hence, the effective "25/16-registry" of 25 (111) planes for diamond and 16 planes for silicon for the interface with SiC interlayers is again close to the 3/2 registry of the diamond-silicon interface. Comparing the two cubic near coincidence site lattices shows that the lattice mismatch remains smaller for the diamond-Si interface as compared to the diamond-SiC interface. Faceting parallel to {111} lattice planes is a general phenomenon and is observed for Si islands of the surface ripple structure during the early stages of synthesis, for the diamond-SiC interfaces and for the large angle grain boundaries in diamond. It results from higher growth rates of {100} facets as compared to the rate of growth of {111} facets during the early stages of CVD synthesis.

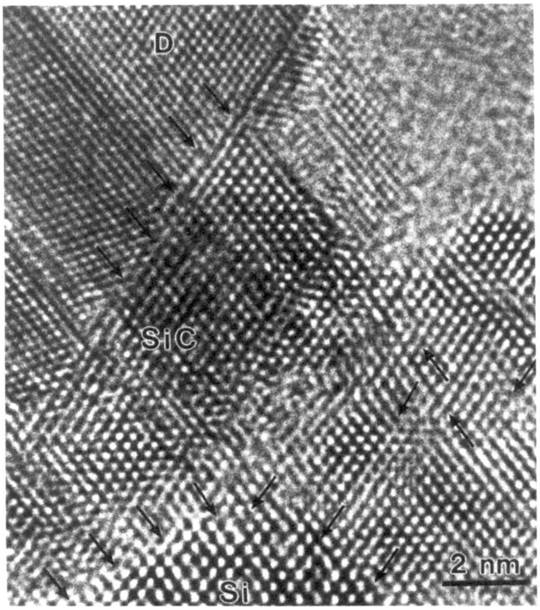

Fig. 7 Interface between a diamond nanocrystal (top left) and a SiC interlayer nanocrystal (center) and the interfaces between the SiC interlayer and the top region of a prismatic Si island of the Si (001) substrate (below). The interfaces are epitaxial, connecting the lattices with each other. Terminating {111}-lattice planes are indicated by arrows. Cross-section HREM micrograph, ⟨110⟩ zone axis. *Copy of fig. 8 in: Wittorf, D., Jäger, W., Dieker, C., Flöter, A., Güttler, H., 2000. Electron microscopy of interfaces in chemical vapour deposition diamond films on silicon. Diam. Relat. Mater. 9, 1696–1702.*

A different orientation relationship has been reported for the structure of the interface between diamond and SiC composites by Park et al. (2007). It is characterized by the diamond (111) plane being parallel to the cubic SiC (111) plane, while the ⟨110⟩ direction of one crystal being aligned with the ⟨112⟩ direction of the other. This orientation relation can give rise to a smaller degree of lattice mismatch between the two materials while preserving the tetrahedral coordination across the interface.

Diamond films synthesized under optimized conditions may also contain near-interface regions with small amorphous inclusions of nanometer size between the regions of direct lattice plane coupling. High-resolution analyses of the C—K near-edge fine structure in electron energy loss spectroscopy enabled to identify the bonding structure of carbon and the relative fractions of sp^2-bonded and sp^3-bonded carbon (Wittorf et al., 2000).

Such analyses show that the inclusions consist predominantly of amorphous C. Amorphous phases, structurally modified surfaces and defects, such as microtwins, are expected to contribute locally to mismatch accommodation and elastic layer strain relaxation (Wittorf, 1999, Wittorf et al., 2000). The thermal interface resistance for diamond films with a high fraction of amorphous secondary phases is increased as compared to films with a high fraction of interface regions with direct lattice plan coupling (Verhoeven et al., 1997, Fig. 18).

2.3 CVD diamond heteroepitaxy on Ir (100) and on Ir (111)

The heteroepitaxial CVD diamond synthesis of diamond on single crystalline metallic iridium belongs to the most promising routes for synthesizing high-quality large-area diamond with a lower density of crystal defects as compared to deposition on Si substrates. The synthesis utilizes bias-assisted CVD deposition on Ir single crystals or on Ir single-crystal films fabricated on $SrTiO_3$ (001)/Si(001) or on yttria-stabilized zirconia (YSZ)/Si(001) substrates (Ohtsuka et al., 1996; Schreck and Arnault, 2018). The quality of the diamond synthesis by bias-assisted CVD depends obviously on many factors, among those the reactor configuration, the substrate quality, and the parameter choice for nucleation and growth conditions. Synthesis on Ir is advantageous due to its comparatively smaller lattice constant difference, thus reducing the layer stress and defect formation during the early stages of deposition, and due to a low solubility for carbon. Layer growth under appropriate process conditions can further decrease the defect density at larger thickness. Many investigations by TEM focused on understanding the mechanisms of nucleation and growth of crystallites and layers for bias-assisted CVD synthesis (Fig. 8). Diamond films synthesized under optimized conditions contain well-oriented columnar grains and only low densities of dislocation arrays in near-surface regions (Fig. 9). For such "single crystal" CVD diamond films on Ir, measured average polar and azimuthal grain orientation distributions determined from X-ray rocking curves were less than 1° (Bauer et al., 2005; Ohtsuka et al., 1996; Schreck et al., 2001). Based on high-quality diamond crystals from synthesis on iridium, MESFET devices could be successfully fabricated. The concepts of CVD diamond synthesis on Ir/metal-oxide/Si substrates are reviewed in detail by Schreck and Arnault (2018).

2.3.1 Diamond synthesis on Ir (001): Early stages

For *CVD synthesis on Ir (001)substrates*, an early study of the interface structure of diamond crystallites reports the presence of misfit dislocations,

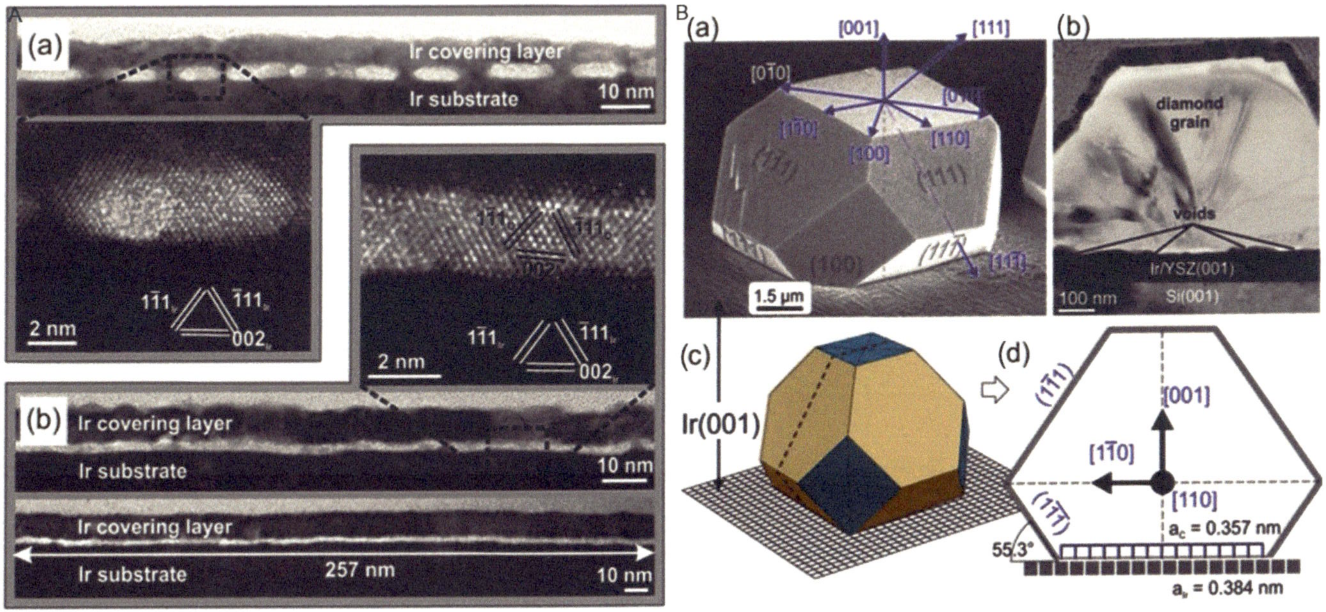

Fig. 8 Early stages of CVD diamond synthesis on Ir (001). (A) Isolated grains (a) and single crystalline layer about 260 nm wide of diamond grains (b) formed after termination of a BEN treatment (40–45 min) and film growth for a period of 5 s. HRTEM imaging in ⟨110⟩ zone axis projection. For the cross-section TEM sample preparation, a cover layer ~10 nm thick was deposited. The CVD diamond conditions applied were virtually identical: $T_{Substrat} = 730–750\,°C$, 7% CH_4/H_2, $U_{Bias} = -260\,V$. (B) Individual diamond crystallite formed during the initial stage of CVD of diamond on Ir (001). (a) Scanning electron microscope (SEM) secondary electron micrograph of a truncated diamond octahedron. Crystallographic notations describe facet planes and plane normal, respectively. (b) Cross-section bright-field TEM image showing voids at the interface between the diamond island and Ir (c) isometric projection of the truncated octahedron (d) cross section along the (110) plane (lattice parameters a). *Panel (A): Copy of fig. 2 in: Schreck, M., Gsell, S., Brescia, R., Fischer, M., 2017. Ion bombardment induced buried lateral growth: the key mechanism for the synthesis of single crystal diamond wafers. Sci. Rep. 7, 44462. Panel (B) Copy of fig. 1 in: Lebedev, V., Yoshikawa, T., Giese, C., Kirste, L., Žukauskaitė, A., Graff, A., et al., 2019. Formation of icosahedron twins during initial stages of heteroepitaxial diamond nucleation and growth. J. Appl. Phys. 125, 075305.*

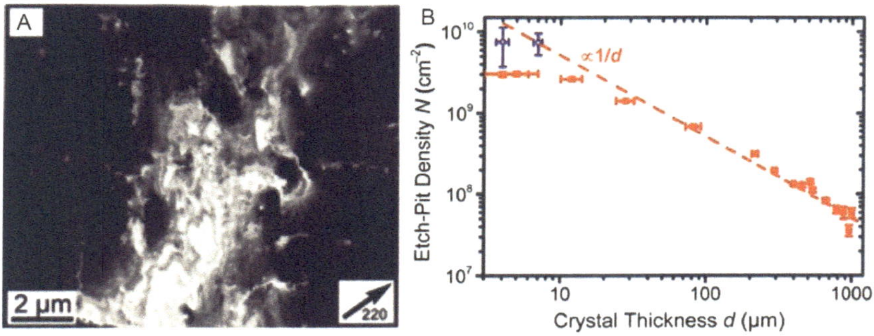

Fig. 9 Microstructure phenomena in thick CVD diamond films on Ir (001). (A) Arrays of individual dislocations near a grain boundary in the near-surface region of a diamond film 34 µm in thickness. Individual dislocations (line contrast) and the slight local misorientation between the lattices across the boundary appear in bright contrast. Dark-field TEM micrograph, (220) reflection. (B) Dislocation density derived from TEM measurements (blue open circles) and from the analysis of etch pits (red squares) as function of the diamond crystal thickness d. The dashed red line shows a 1/d fit to the experimental data measured for the layer thickness range from 20 to 1000 µm. *Panel (A): Copy of fig. 4 in: Schreck, M., Hörmann, F., Roll, H., Lindner, J.K.N., Stritzker, B., 2001. Diamond nucleation on iridium buffer layers and subsequent textured growth: A route for the realization of single-crystal diamond films. Appl. Phys. Lett. 78 (2), 192. Panel (B): Copy of fig. 2 in: Stehl, C., Fischer, M., Gsell, S., Berdermann, E., Rahman, M.S., Traeger, M., et al., 2013. Efficiency of dislocation density reduction during heteroepitaxial growth of diamond for detector applications. Appl. Phys. Lett. 103, 151905.*

suggesting strong chemical bonding between diamond and Ir (Sawabe et al., 2000). Furthermore, nucleation conditions for epitaxial diamond deposition can be chosen that do not lead to significant roughening of the Ir film surface. As result, the (001) interfaces with Ir remain flat, and the crystalline lattices are directly connected with each other, not showing any intermediate layer. The misorientations of diamond crystallites directly grown on the iridium are smaller than 1° (Hörmann et al., 2002). The earliest appearance of diamond crystallites was found to occur after applying crystal growth conditions (subsequently to bias-assisted conditions) for very short times ranging from 5 s to 1 min. For instance, applying growth conditions for 5 s results in crystallites of 2 nm in thickness and about 10 nm in lateral extension, and an amorphous carbon precursor phase that is formed by the bias-assisted ion bombardment has largely disappeared. From these observations it is concluded that a 1–2 nm-thick amorphous carbon layer formed on the substrate surface is only stable under biasing conditions (Brescia et al., 2008). Possibly resulting from an initially amorphous hydrogenated carbon matrix, the formation of a surface layer on Ir (001) consisting of local domains of diamond nuclei and of graphitic C structures is reported (Schreck et al., 2009).

Referring to the role of ion bombardment induced buried lateral growth as the key mechanism for the synthesis of single crystal diamond wafers, a possible *mechanism for heteroepitaxial diamond nucleation on iridium* surfaces is proposed in a recent overview (Schreck and Arnault, 2018; Schreck et al., 2017). Ion bombardment under BEN conditions is proposed to be key to induce formation and lateral growth of epitaxial diamond within a ~1 nm thick buried carbon layer (Fig. 8A). The model proposes that the ion bombardment induces formation and lateral spread of epitaxial diamond within a ~1 nm thick a-C:H precursor matrix layer at the interface to the Ir substrate, starting from a primary event of epitaxial crystallite nucleation and expanding laterally over distances of several microns. During this epitaxial lateral growth, numerous isolated secondary nuclei are thought to be generated continuously. Crystallite growth in perpendicular direction is limited which is ascribed to the effects of defect formation and back etching by atomic hydrogen. It is reported that diamond single crystals with ~90 mm in diameter and a weight of 155 carat are grown from such a carbon film, initially consisting of about 2×10^{13} individual grains.

2.3.2 Diamond synthesis on Ir (001): Layer growth

Under optimized *conditions for film growth*, CVD synthesis protocols are applied that result in size distributions of diamond crystals in favor of forming larger grains and initiate overgrowth of slightly 'misoriented' crystals. Initially isolated grains merge to form a closed diamond films that consist of grain structures with narrow orientation distributions, indicating a high degree of epitaxial orientation relationships. At a thickness of around 600 nm, the film morphology consists of largely interconnected grains separated by grain boundaries. The measured average scatter in grain orientations (mosaic spread) as measured from XRD (full-width half-maximum, FWHM) line widths measurements is about 1° and therefore smaller compared to diamond films of similar thickness on Si substrates. At about 20–30 μm, large well-oriented grains and a network of small angle grain boundaries start to predominate the microstructure, and the layer surface smoothens. Such "defective single-crystal" diamond films contain grain boundaries that can be described as an interconnected network of dislocation bands several μm in length (Fig. 9A).

The *reduction of dislocation density*, the development of micro-strain and the role of dislocations during film growth were investigated for a thickness up to 1 mm (Stehl et al., 2013). The dislocation density is found to be inversely proportional to the thickness of the diamond layer (Fig. 9B). The Raman line width as a measure of microstrain shows a decrease to values close to the value of perfect single crystals. The thickness dependence of the measured average dislocation density is thought to be related to dislocation reactions that effectively reduce the dislocation density. Assuming that a scaling law describes this thickness dependence also for larger film thickness, the synthesis of thicker crystals may result in even lower dislocation density (Stehl et al., 2013). For comparison, the dislocation density decreased to about $5–10 \times 10^8$ cm^{-2} for a thickness of around 30 μm. For a single-crystalline diamond film on Ir about 1.6 mm thick and about 90 mm in diameter (Stehl et al., 2013), the dislocation density is about $4 \times 10^7\,cm^{-2}$. In comparison, this value is considerably lower than values observed for standard natural Type IIa crystals (10^8–$10^9\,cm^{-2}$). For films of 35 μm thickness, a mosaic spread value of about 0.2° is reported (Schreck and Arnault, 2018).

2.3.3 Diamond synthesis on Ir (111)

For *CVD synthesis on Ir (111)* substrates, the diamond nucleation is similar as for depositions on Ir (001). Lattice misorientations of 0.2° in tilt and of 0.28° in twist are reported for twin-free heteroepitaxial diamond on Ir/YSZ/Si(111) (Fischer et al., 2008). Investigations of the propagation of threading dislocations in heteroepitaxial diamond films on Ir (111) and their role in the formation of intrinsic stress revealed a correlation between the varying film stress state and the dislocation propagation direction. This result is attributed to the combined effects of dislocation climb and of step-flow growth on the dislocation propagation direction. From TEM observations it is concluded that anisotropic film stress is correlated with anisotropies of dislocation line tilting (Gallheber et al., 2017). Detailed investigations of the developments of azimuthal mosaic spread, dislocation density, and film stress show that the optimized quality of crystal growth is comparable with the quality of the best heteroepitaxial layers on Ir (001) (Gallheber et al., 2018).

For CVD synthesis of diamond on {111} surfaces, the perfect alignment of NV centers along a single crystallographic direction and the preferential orientation of NV center axes are reported, being of interest for development of quantum metrology and quantum information applications of diamond (Michl et al., 2014; Miyazaki et al., 2014). Perfectly oriented

NV centers in single-crystal (111) diamond are of interest, for instance, as highly sensitive magnetometers for bio-imaging with magnetic nanoparticles as a label and the diamond substrate as a magnetic field imager (Hatano et al., 2018; Ozawa et al., 2018). Single crystal (111) diamond is also advantageous for the fabrication of diamond lateral p–n junction diodes (Sato et al., 2015), or may find applications for quantum nanophotonic devices (Regan et al., 2019).

2.3.4 Toward improving wafer-scale single-crystal heteroepitaxy

Further developments are to be mentioned here briefly, aiming at improving wafer scale single-crystal diamond heteroepitaxy for fabrication of wafer-scale thin films and free-standing substrates:

Patterned nucleation growth. By *patterned nucleation growth (PNG)* in diamond syntheses, epitaxial diamond crystals grow laterally across non-nucleated Ir substrate surfaces and coalesce with each other. This approach, being reported also for the epitaxial lateral overgrowth of diamond on Ir(001)/MgO(001) substrates (Ando et al., 2012), is considered to be promising for fabricating free-standing heteroepitaxial diamond (001) films for industrial applications. The patterns of nucleation regions contribute to the stress state, thus influencing the early stages of diamond growth on Ir (001) and the elastic deformation of free-standing wafers (Yoshikawa et al., 2015).

Epitaxial lateral overgrowth (ELO) has been investigated and compared for both uniformly nucleated surfaces and for prepatterned Ir (001) surface with the intention of controlling strain and coalescence of the diamond grains. Using electron microscopy and atomic force microscopy as well as Raman spectroscopy, together with thermodynamic and kinetic concepts, detailed studies focus on the initial stages of growth of heteroepitaxial diamond films and on the crystallization mechanisms. The formation of μm-scale multiply twinned diamond particles (MTP) with icosahedron twins (Fig. 8B), being detrimental for the synthesis of high-quality thin diamond films on iridium, is found to be controlled by topologic, thermodynamic, and kinetic factors provided during plasma-assisted CVD on iridium. As driving forces for the formation of the icosahedron and the icosahedron twin structures are the low {111} stacking fault energy, high supersaturation conditions, and limited surface diffusion are identified. The formation of MTP particles can be efficiently suppressed during the precoalescence growth phase. It is proposed that icosahedron MTPs might find applications, e.g., as cavities in photonic devices based on diamond color centers, as biosensors, or as carriers in hydrogen storage systems (Lebedev et al., 2019).

Optimization strategies for CVD synthesis. In the development of *optimized homoepitaxial CVD synthesis of single crystal diamond* on seed crystals, the investigations of the effects of pressure, temperature and holder geometry on growth rate and crystal morphology enabled to establish conditions for which no twins or cracks are observed, and films of high purity and with no luminescent defects are obtained (Widmann et al., 2016). *Designs of substrate holder configurations* are developed for improving the homogeneity of bias-enhanced nucleation on larger substrate areas and thus provide improved bias-assisted nucleation conditions for wafer-scale diamond heteroepitaxy (Yoshikawa et al., 2019).

2.4 Homoepitaxial CVD synthesis of diamond

CVD synthesis of diamond for homoepitaxial crystal growth is often performed using HPHT diamonds as substrate. Their surface area, typically a few tens of mm^2, is limiting the size of epitaxial diamond layers. Several methods were developed in order to enlarge the effective surface for homoepitaxial growth. An approach often applied for the fabrication of freestanding large-area single-crystal diamond wafers is the so called mosaic wafer growth (Yamada et al., 2012). For industrial production and applications, the high-speed crystal growth of single-crystal diamond by microwave plasma-assisted chemical vapor deposition is an important issue (Tabakoya et al., 2019). Considerable progress has been made in the development of this technique that also requires uniform deposition rate for large-area synthesis processes and associated reactor technologies. The reduction of the number of crystal defects in the epitaxial diamond is one of the remaining process problems. Defects form either during the crystal growth process or originate from defects that are present in the original seed crystal. The mechanism of diamond growth is still not well understood, one of the reasons being the large number of experimental parameters. Several overviews document the progress of fabrication of single crystal diamond wafer fabrication, the homoepitaxial CVD synthesis of diamond films of different crystalline orientation, and problems related to defects and doping (see, e.g., Achard and Tallaire, 2018, Arnault and Girard, 2014, Asmussen et al., 2008, Jin et al., 2019, Mokuno et al., 2018, Ohmagari et al., 2019, Teraji, 2018, Teraji et al., 2015, Tokuda, 2019, Yamasaki et al., 2014).

The following examples illustrate some applications of TEM for process optimization and for understanding defect and doping phenomena as well as for methodological developments of improved defect analyses.

Step-free diamond surfaces. Step-free diamond surfaces are promising for improving the performance of devices based on diamond heterostructures. Their fabrication generally uses either selective growth with a mask process or a mask-less mesa process, utilizing a lateral CVD growth mode without terrace nucleation. Fig. 10A shows as example a diamond {111} surface after homoepitaxial lateral growth of diamond by MPCVD on an array of $50 \times 50\,\mu m^2$ mesas fabricated on single-crystal diamond substrates through a conventional lithographic technique. In the light optical observation, the mesa regions appear either featureless or contain at least one hillock, induced by a screw dislocation (Fig. 10B). The mesa regions without features possess an atomically step-free surface and contain no dislocations (Fig. 10C) (Tokuda, 2019).

Metal-assisted termination for dislocation control. The development of large single-crystal diamond wafers with decreased numbers of defects is required in view of applications in high-performance diamond electronics. A method of so called *metal-assisted termination* is proposed in which the propagation of dislocations is suppressed by local stresses around metal impurities of large atomic radius introduced by in situ doping. This approach is considered effective for both mosaic and heteroepitaxial substrates (Ohmagari et al., 2019).

Nature of hillocks on homoepitaxial diamond. Analyses of dome-shaped hillocks formed on B-doped CVD homoepitaxial diamond films by optical microscopies, scanning electron microscopy and TEM reveal that the hillocks are crystalline and possess epitaxial orientations with the film and the substrate. They contain dislocations, most of which extend to the boundary with the film, and twins with {111} twin planes (Tsubouchi et al., 2019).

Doping-related dislocation formation. The *formation of dislocations in boron-doped* (111) and (001) homoepitaxial diamond layers and the existence of a critical B-doping level for generation of dislocations were observed to depend on growth parameters and growth direction, the critical boron concentrations being in the range between 6×10^{20} at/cm^3 and 3×10^{21} at/cm^3. A mechanism involving substitutional boron pairs is proposed for the generation of dislocations in epilayers (Alegre et al., 2014). The *orientation dependence of B-doping induced dislocation formation* has been investigated for different growth plane orientations of hillocks for heavily B-doped CVD diamond. It is suggested that local strain effects connected with boron complexes at the growth surfaces may induce the generation of extended defects during growth (Lloret et al., 2015). For nonplanar CVD homoepitaxial diamond growth, boron-doping proximity effects on

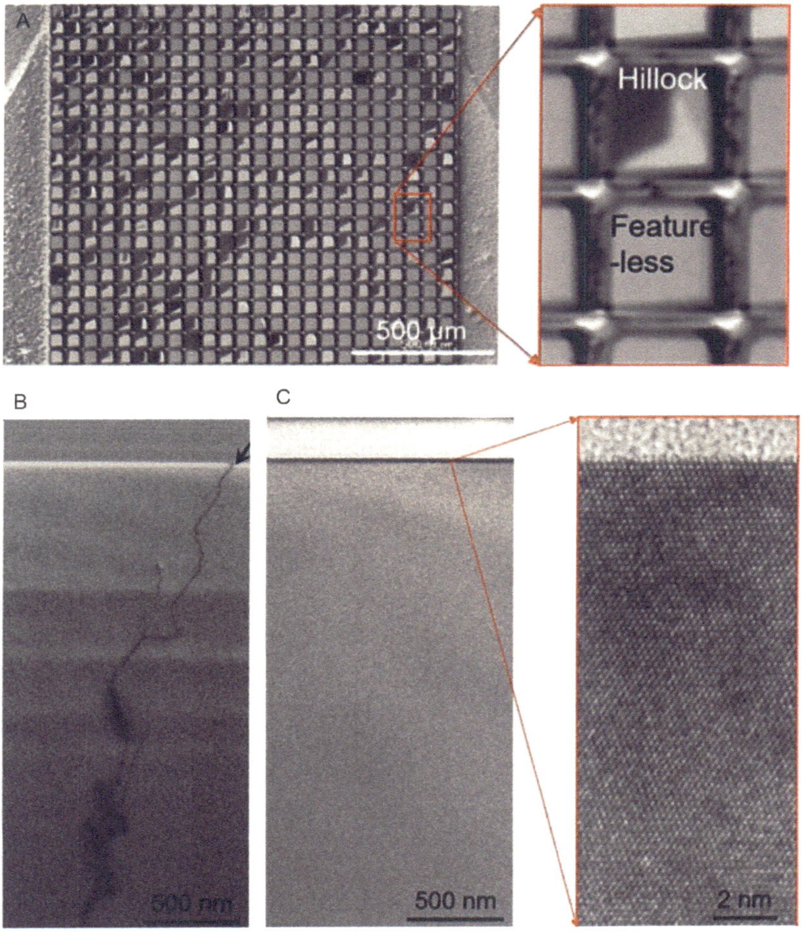

Fig. 10 Homoepitaxial diamond {111} films on single-crystal diamond substrates with $50 \times 50\,\mu m^2$ mesas. (A) Light-optical images depicting mesa regions that appear either featureless or contain at least one hillock, induced by a screw dislocation. (B) Cross-section bright-field TEM image obtained from a mesa region with hillock. A screw dislocation (dark line contrast) extends through the diamond layer to the surface (arrow). The contrast region on top is due to a protecting cover layer. (C) Bright-field TEM and HRTEM lattice images of a cross-section obtained from a featureless mesa region. The surface is atomically sharp. The contrast regions on top are due to a protecting cover layer. *Copies of figs. 1.18, 1.19, 1.21 from Tokuda, N., 2019. Homoepitaxial Diamond Growth by Plasma-Enhanced Chemical Vapor Deposition. In: Yang, N. (Ed.), Topics in Applied Physics 121. Novel Aspects of Diamond—From Growth to Applications, second ed. Springer, Ch. 1.*

dislocation generation are described by Lloret et al. (2018). The evaluation of boron doping and distribution of defects from cross-sections of homoepitaxial diamond layers is described by Araújo et al. (2011).

Correlation between threading dislocation and surface etch pits. Cross-section TEM investigations confirm the correlation between a single grown-in threading dislocation and an etch pit on the surface of CVD homoepitaxial single crystal diamond. The threading dislocation line is observed to be along a near-[001] growth direction from the apex of the pit to the deeper region. Quantitative TEM contrast analyses reveal short segments of undissociated perfect 60°-dislocations, 30°-dislocations and screw-dislocations with b = 1/2[110], 1/2[011] and 1/2[101] lying on the {111} glide plane (Tsubouchi and Mokuno, 2016). TEM contrast analyses of grown-in threading dislocations bundles in CVD homoepitaxial diamond plates yield similar results as reported for the study of a single dislocation (Tsubouchi and Mokuno, 2017).

Dislocation imaging in B-doped diamond by SEM. The dislocation imaging using secondary electron imaging at 20 kV in a field emission scanning electron microscope has been investigated for low boron doped diamond epilayers. A specific contrast behavior of secondary electron imaging is found to be correlated with a dislocation network. Boron impurities seem to play a major role in the contrast formation which is possibly related to enrichment of boron at dislocations and trapping of secondary electrons. Combined measurements applying high-resolution STEM-EELS and secondary emission are suggested for clarifying the role of boron in the contrast formation (Barbay et al., 2019).

3. Nanodiamond

Nanodiamond particles are synthesized from hydrocarbons at high pressures, by the detonation of explosives, or by grinding of diamond. *High static pressure (HPHT) technology* provides favorable conditions for the controlled synthesis of nanodiamonds with sizes as small as 1–2 nm. The *detonation nanodiamond (DND)* is formed in a detonation shockwave where pressures and temperatures are achieved which correspond to the stability region of diamond in the phase diagram. From detonation synthesis, small diamond particles around 4–5 nm in size can be obtained. In view of the prospects for applications from quantum physics to biomedicine, an efficient method for the controllable size reduction of DND particles to the molecular-size level has been developed (Stehlik et al., 2016).

Size-controllable synthesis of ultrasmall diamonds from halogenated adamantanes at high static pressure and their optical properties are described by Ekimov et al. (2019) and Ekimova et al. (2020). Spherical nanodiamond particles can be formed in carbon onion structures by *electron irradiation*, with the carbon onions acting as nanoscopic pressure cells for diamond formation (Banhart and Ajayant, 1996). The low-pressure transformation of graphite to diamond under irradiation is reported by Lyutovich and Banhart (1999). Nanodiamond particles with sizes of 20 nm and above are known to behave like bulk diamond. High quality NCD diamond layers have been synthesized from nanodiamond seeds buried under a Cr layer on a Si substrate (Degutis et al., 2016). The formation of nanocrystalline diamond films has been investigated for co-deposition of high aspect ratio diamond plates and octahedral diamond grains at low deposition temperatures ($<410\,°C$) (Drijkoningen et al., 2017).

Nanocrystalline diamond is considered for a *wide variety of applications*, for instance, in electrochemistry, in superconductivity, in field emission, for nanocomposites, in biological sensing, for biomedical applications, for quantum technology, for magnetometry on the nanoscale, and more. Nanodiamond particles with sizes of 20 nm and above are used, for instance, in abrasive industries. Films of crystalline grains with sizes around 5 nm, called ultra-nanocrystalline diamond (UNCD), contain a considerable amount of amorphous grain boundaries, similar to diamond-like carbon (DLC) which has no crystalline structure. Spatially resolved electron energy loss spectroscopy enabled to understand the n-type electrical conduction of UNCD films of diamond core nanostructures and sp^2-bonded C sheaths that incorporated N (Arenal et al., 2009). Nanocrystalline diamond films (NCD) with grains of typically below 100 nm generally contain regions of less sp^2 hybridization, thus being, e.g., more transparent than UNCD films. Novel applications of nanodiamond may emerge by achieving p-type and n-type conduction through doping of NCD films using boron and phosphorus (Williams, 2014). In view of applications in electrochemistry, thermal management, seeding, or the study of fundamental properties, such as superconductivity, boron-doped diamond nanoparticles with a crystallite size of 10–60 nm have been produced from polycrystalline diamond films using a multistep milling procedure (Heyer et al., 2014). Analyses by HRTEM and EELS revealed the substitutional lattice occupation and the tetragonal coordination of boron. The substitutional and homogeneous incorporation of phosphorus (of the order of $10^{19}\,cm^{-3}$) in nanocrystalline CVD diamond thin films has been reported by Janssen et al. (2014).

Topics of *nanodiamond research* are synthesis, fundamental structure-property relations for physical and chemical properties, and research for applications of nanodiamond as particle material or as nanodiamond films. The interest in the research of nanodiamond is strongly motivated by the particular properties. Many diamond particle and film properties for crystallite sizes below 10 nm deviate substantially from that of bulk diamond. The high fraction of surface area enhances disorder, sp^2 bonding, hydrogen content and scattering of electrons and phonons, and leads to more complex density of states within the bandgap, to a reduction of Young's modulus, or to increased optical absorption (Shenderova and Gruen, 2012; Shenderova and McGuire, 2015; Shenderova et al., 2019; Williams, 2011).

The *surface properties* and availability in colloidal suspensions, the good biocompatibility and low toxicity, and the luminescence properties of specific impurity defects make nanodiamond attractive for biological sensing applications or for biomedical applications as nanocarriers for drug delivery. The *optical luminescence properties of color centers*, of interest for biological sensing and imaging and for quantum optics, depend on their proximity to the particle surface. Surfaces of diamond nanoparticle and nanowire materials can be functionalized with biological and therapeutic molecules. For specific applications the surfaces of nanodiamond particles must be "engineered." For developing these functions, a fundamental understanding is required of the surface structure, surface chemistry, effects of surface functionalization and of the colloidal properties but also of the "bulk" structure of ND particles. Topical overviews can be found in several books and journals (Arenal, 2012; Arnault, 2017, 2019; Krueger, 2014; Radtke et al., 2019; Raymakers et al., 2019; Shenderova and Ciftan Hens, 2013; Shenderova and McGuire, 2015; Shenderova et al., 2019; Sumant et al., 2014; Szunerits et al., 2014; Williams, 2011, 2014).

The following examples illustrate studies using high-resolution (S)TEM and EELS, combined with density functional theory (DFT) modeling, of the characterization of nanodiamond, of its structure and properties, of defects, of dopants, of nanodiamond-Au hybrid nanoparticles, and of surfaces of nanodiamond.

Size-Dependent Reactivity of DND particles. A method is presented that enables to purify and disperse nanodiamond powder at its core particle size. The hydrogenation process results in a breakdown of larger (>100 nm) aggregates of particles bonded in an sp^2 matrix into core particles with defined surface functional groups. This effect is based on the reaction of the sp^2 shells and bonding matrix between particles with hydrogen gas.

It is therefore restricted to particles with a high surface-to-volume fraction of sp^2 bonding of reconstructed surfaces. Dispersion of these particles into water via high power ultrasound and high speed centrifugation results in mono-disperse nanodiamond colloids with a narrow distribution of particles 3–4 nm in size (Fig. 11), with exceptional long time stability in a wide range of pH, and with high positive zeta potential (>60 mV). The method prevents, for instance, aggregate formation in buffer solutions which is a perfect starting condition for chemical surface modifications. Photonically active diamond nanoparticles and the surface functionalization of the particle surfaces are attractive for applications in drug delivery and in monitoring experiments or bio-labeling (Williams et al., 2010).

Atomic-scale imaging of H-treated DND nanodiamond particles and surfaces. Ultrahigh-resolution imaging with an aberration-corrected TEM is used to analyze on the atomic scale the microscopic structure of colloids with

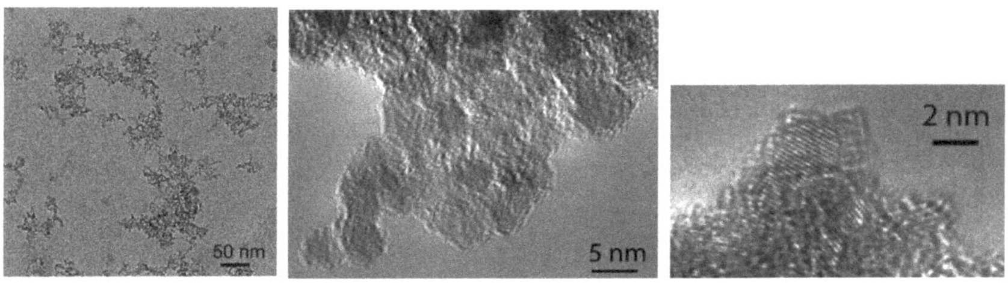

Fig. 11 H-treated DND powder particles. (Left) Bright-field TEM micrograph of particle aggregates. (Center, right) Atomic-scale images of H-treated DND particles and particle surfaces taken under so called Negative-Cs-Imaging (NCSI) conditions in an aberration-corrected HRTEM instrument. Individual C atoms and carbon in crystalline columns and in lattice planes are imaged in projection with bright contrast. The crystalline nature of the individual particle cores and, in many cases, distinct shell regions around crystalline core regions can be observed. After hydrogenation the particle aggregates are broken up into single particles typically 3–4 nm in size, whereas the untreated particles were aggregated into sizes of 100 nm and above. Shells are well-known to occur at the surfaces of detonation nanodiamond particles. They are related to the sp^2 bonding of carbon at the particle surface and are still present after H treatment. Note: Treated small particles aggregate on drying for TEM specimen preparation due to the surface tension of water (Fig. 11, left). *(Left): Copy of fig. 5 in: Williams, O.A., Hees, J., Dieker, C., Jäger, W., Kirste, L., Nebel, C.E., 2010. Size-dependent reactivity of diamond nanoparticles. ACS NANO 4 (8), 4824. (Center, right): Copies of figures in: Dieker, Ch., Tillmann, K., Williams, O.A., Jäger, W., Nebel, Ch.E., 2011. Microscopic characterisations of diamond nanoparticles by aberration-corrected HRTEM. In: Proc. Microscopy Conference MC 2011, Kiel, Germany, 28.08.-02.09.2011, ISBN 978-3 00-033910-3. Copyright 2011 DGE—German Society for Electron Microscopy e.V. Germany.*

H-treated nanodiamond particles 3–4 nm in size. The imaging is performed at 80 kV in order to reduce effects of beam damage to a minimum. Fig. 11 (right) shows examples of the structure and surfaces of H-treated DND particles. Distinct shell regions (depicted by different contrast) around crystalline core regions can be clearly identified in many cases. It is concluded that the assessment of light-element functional surface coatings and the monitoring of their modification, for instance, by annealing treatment in hydrogen are accessible with such methods. Spatially resolved high-resolution electron-energy loss spectroscopy in STEM may enable to analyze the bonding states and elements of light-element functional surface coatings (Dieker et al., 2011).

Positive zeta potential of ND. Electron microscopy and Raman spectroscopy methods have been used to show that origin of the *positive zeta potential* and the associated basicity exhibited by nanodiamond particles can be explained by the presence of sp^2 carbon at the surface, specifically graphene-like shells which promote the suppression of acidic functional groups. These shells can become protonated, generating a net positive charge and zeta potential. At the same time the extraction of protons from the water results in pronounced basicity (Ginés et al., 2017).

Structure, defects and N impurities of DND particles. The structure, morphology and the presence of nitrogen in detonation nanodiamond (DND) is investigated by applying aberration-corrected transmission electron microscopy at low voltages, spatially-resolved imaging with core-loss electrons in electron energy-loss spectroscopy, and density functional theory (DFT) calculations. Clean stable nanodiamond facets were imaged with atomic resolution. The nanodiamonds are stable under electron illumination chosen for these experiments (Fig. 12A). The core–shell model for nanodiamond is found to be not necessarily representative for materials that have been exposed to harsh surface cleaning treatments. Fig. 12B illustrates the results of high-resolution analyses of nanodiamond by spatially resolved EELS. Nitrogen is present within the core of DND particles and N enrichments at defective regions as well as at twin boundaries are observed, with a maximal nitrogen content around 4 at.%. The electron loss near-edge fine structure (ELNES) of the carbon K-edge in high resolution EELS spectra consists of three sub-peaks that are attributed to fullerene-like reconstructions at the nanoparticle surfaces and deviations from perfect sp^3 coordination at defects. Analyses of the N—K edge ELNES signatures and the comparison of experimental results with DFT calculations for the N_s, A, NV^0, and NV^- center allow the conclusion that the embedded nitrogen is likely to be a mix of small amounts of single substitutional and/or A-center nitrogen, combined with larger nitrogen clusters (Turner et al., 2013; Vlasov et al., 2012).

A

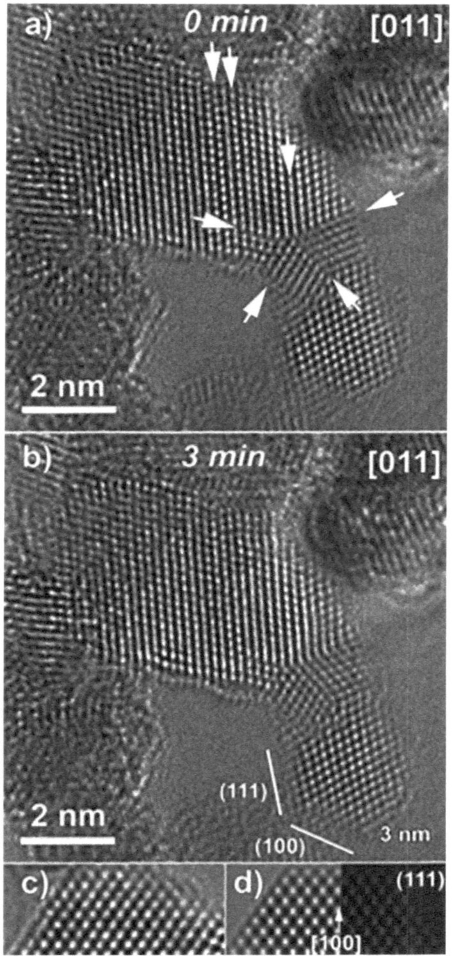

Fig. 12 Structure, defects, bonding and element distribution of DND nanoparticles. (A) Stability under electron beam illumination (a) HRTEM image of DND particles exhibiting coherent S3 {111} twin boundaries (indicated by white arrows). (b) The same particles after 3 min of electron beam illumination. (c) HRTEM image of a truncated nanodiamond surface, showing a minimal graphitic presence. (d) Image simulation performed for $C_s = 15\,\mu m$, $\Delta f = -4\,nm$. The right side has the projected potential overlaid; the white image contrast peaks at the atomic positions. (B) Spatially resolved EELS analysis of a multiply-twinned nanodiamond (a) HRTEM image of a typical DND particle showing the presence of a single coherent twin indicated by A and multiple parallel twin regions indicated by B. (b) ADF-STEM image of the same particle; the white rectangle depicts the STEM EELS scanned region (53 × 67 pixel spectrum image). (c) sp^3 C map, (d) non-sp^3 C map, (e) N map, (f) N K-edge spectra from region 1 and 2 indicated in (b). The N content in region 1 is 2.7 ± 0.2 at.%; defected region 2 shows a clear enrichment of nitrogen (3.8 ± 0.2 at.%). (g) Intensity profiles for non-sp^3 C (dashed line), sp^3 C (black line), and

(Continued)

B

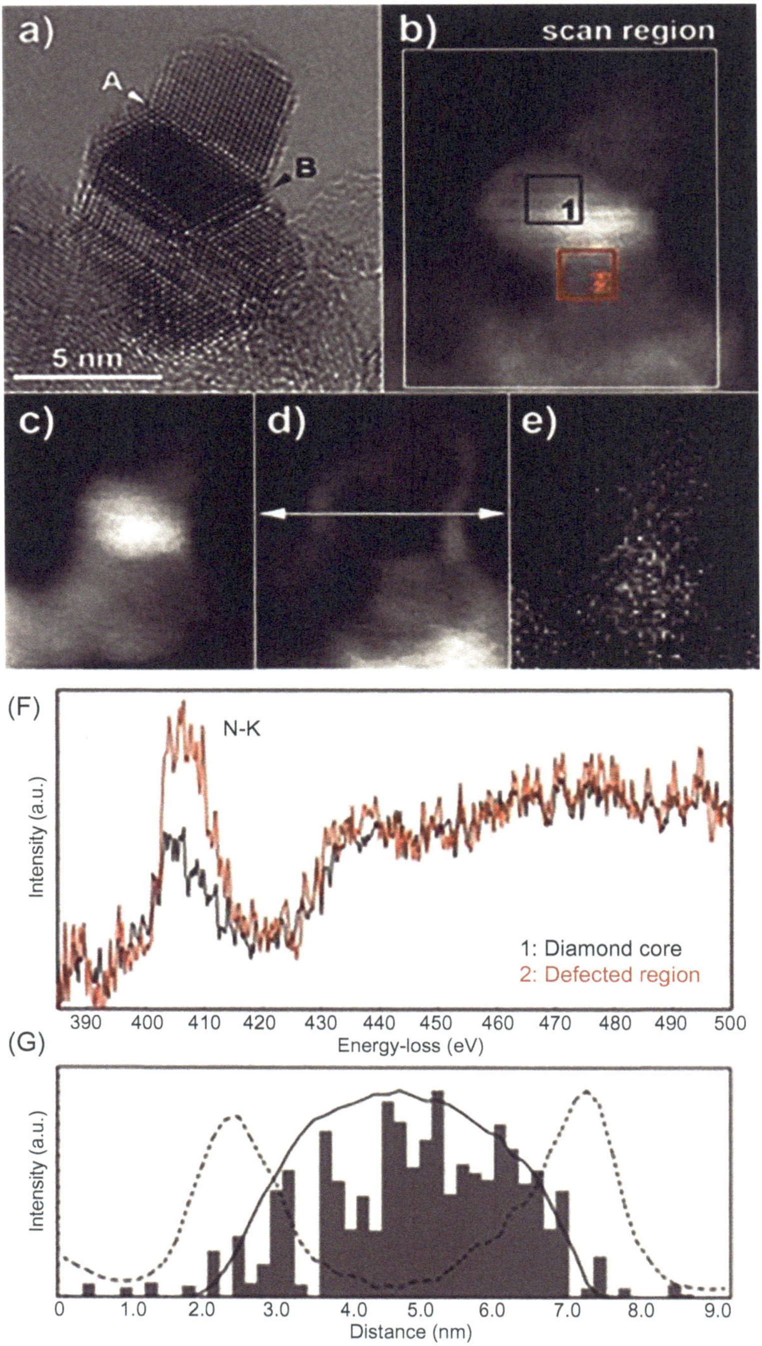

Fig. 12—Cont'd N (bars) taken over the region indicated by the arrow (d) (5 px integration width). *Copies of figs. 2, 4 in: Turner, S., Shenderova, O., Pieve, F. Da, Lu, Y.-G., Yücelen, E., Verbeeck, J., et al., 2013. Aberration-corrected microscopy and spectroscopy analysis of pristine, nitrogen containing detonation nanodiamond. Phys. Status Solidi A 210 (10), 1976–1984.*

Nitrogen-vacancy centers in DND. Motivated by their importance for applications in quantum information technology and biotechnology, the question of formation and concentration of nitrogen-vacancy (NV) centers in detonation nanodiamond particles has been addressed by simulation-aided EELS in a TEM. It is reported that vacancies in the small DND particles exist at concentrations of about 1 at.%. The ab initio calculations predict that about 20% of the vacancies form NV centers (Chang et al., 2016).

Stability of Boron in DND. Diamond particles ~4 nm in size obtained from a detonation process with added boron have been studied with a wide range of experimental techniques, such as CL, PL, Raman and IR spectroscopies, AFM, HRTEM and electrical measurements with impedance spectroscopy. Particle doping levels are determined to range from $\sim 10^{17}$ B cm^{-3} to $\sim 10^{20}$ B cm^{-3}, and substitutional doping with boron is considered to be possible (Afandi et al., 2018).

Fluorescent ND-Au hybrid nanoparticles for multimodal cellular imaging. Fluorescent nanodiamond (FND)–gold hybrid nanoparticles provide unique features for multimodal cellular imaging including confocal and optical coherence microscopy, photothermal imaging techniques, and (S)TEM analyses. Detailed studies of cells are enabled, ranging from statistical distributions of an entire cell at the μm scale to the tracking of individual particles in subcellular organelles on the nanometer scale. The analyses of such diamond-gold hybrid nanoparticles benefit from the enhanced (atomic number dependent) contrast in (S)TEM imaging as well as an extraordinary flickering behavior originating from the nanodiamonds. The processes of endosomal membrane uptake and release of FND could be investigated by imaging with single-particle resolution (Liu et al., 2016).

4. Defects in diamond

Crystal defects are often categorized according to their dimensional extension ("zero D," 1D, 2D, or 3D): point defects and impurity atoms belong to the category of zero-dimensional defects, dislocations are 1D defects, 2D defects are twins, grain boundaries and interfaces, and 3D defects are clusters or agglomerates of point defects, inclusions, or secondary phases. Defects are connected with changes in bonding (dangling bonds) and are known to determine the mechanical, thermal, electrical, electronic and optical properties of diamond. Dislocations, generated either during synthesis or by mechanical stresses, are detrimental for electronic device applications as they can act as trapping centers for recombination of electronic charge carriers. Inclusions and impurities are known to affect the optical appearance and coloration of diamond crystals.

The electronic properties of diamond, for instance, are determined by band structure, phonons, intrinsic and extrinsic defects, and dopant atoms (Nebel, 2017 for review). *Boron* in natural IIB semiconducting diamond acts as an acceptor, generating hole conductivity. The B dopant concentrations can reach values up to 1000 ppm for synthetic HPHT diamond and 10^{16}–10^{22} cm^{-3} for CVD diamond. *Nitrogen* as a dominant extrinsic impurity in diamond acts electronically as a deep donor when present as isolated substitutional nitrogen (P1 center). CVD diamond doped with phosphorous contains typical P concentrations of 10^{16}–10^{20} cm^{-3} (Nebel, 2017). *Impurity-vacancy defect complexes* called *color centers*, offer fascinating perspectives for applications in nanooptics, in quantum computing, or as sensors for electric, magnetic and optical near fields, for local temperature and strain. As an example, the *NV center in diamond* (Gruber et al., 1997; Wrachtrup et al., 2013) is highly luminescent, photo-stable and biocompatible at room temperature and has highly-coherent spin states. Applications of these centers as quantum emitters in single crystal diamond and in nanodiamonds are of interest for nanoscale sensing and imaging with ultimate resolution (Batzer et al., 2019; Radtke et al., 2019). Much research on the defect complexes NV, SiV, GeV, SnV, PbV or XeV center is dedicated to their generation, e.g., during CVD synthesis of diamond or by low-energy electron irradiation in the TEM, by ion implantation, or by laser writing (Edmonds et al., 2020; Lesik et al., 2016; Michl et al., 2014; Miyazaki et al., 2014; Nelz et al., 2018; Radtke et al., 2019; Smith et al., 2019), and to understanding their properties and applications (see, for instance, Aharonovich et al., 2011; Bradac et al., 2019; Görlitz et al., 2020; Hanlon et al., 2019; Häußler et al., 2017; Sandstrom et al., 2017; Wrachtrup et al., 2013).

The following examples illustrate studies of various types of extended defects in different diamond materials. The challenges for structural characterization of point defects and defect complexes are briefly addressed. Such investigations provide information on lattice strain, on the nature of dopants and their spatial distribution, and on the atomic structure of defects and contribute to understanding the origin of the defects and their impact on physical properties.

4.1 Inclusions and voidites

Voidites in polycrystalline natural diamond. Voidites as the principal inclusions in type Ia natural diamonds have been studied extensively. Polycrystalline natural diamond (carbonado, black diamond) reveals the existence of voidites that are different from the voidites in type Ia natural diamonds with respect to shape, content and origin. It is reported that voidites with the usual shape form along dislocation lines while voidites of unusual planar-like shape are formed at grain boundaries (Chen et al., 1998). Further TEM studies reveal the presence of a large number of defects, such as dislocations, dislocation networks, grain boundaries, and voidites. Structural details of different grain boundaries can be related to the direct bonding between diamond grains. It is concluded that polycrystalline natural diamonds have a combined growth-sintering genetic history in a high pressure and high temperature environment, rather than a simple growth or sintering process (Chen and Van Tendeloo, 1999).

Nitrogen inclusions in natural diamond. Inclusions in natural diamond have been characterized by combining TEM imaging methods with elemental mappings using EELS and XEDS and with Fourier transform infrared spectroscopy. Typical observations are {111}-faceted octahedral defect nanostructures and dislocation loops around larger inclusions, high nitrogen contents of inclusions, and a strong correlation between structurally bound nitrogen, hydrogen and the abundance of nanoinclusions. It is suggested that an opalescent 'milky' appearance of such diamonds is originating from nanoinclusions filled with NH-species (Rudloff-Grund et al., 2016).

Boron distributions and inclusions in HPHT polycrystalline superconducting diamond. Polycrystalline boron-doped superconducting diamond, synthesized at high pressure and high temperature (HPHT) via a reaction of a single piece of crystalline boron with monolithic graphite, has been investigated by analytical transmission electron microscopy including high resolution (S)TEM imaging and EELS elemental mapping. Largely crystalline diamond–diamond grain boundaries and the presence of substitutional boron dopants within individual diamond grains are observed. In addition, isolated boron-rich inclusions and triple points are observed, both amorphous and crystalline in nature, with a chemical composition close to boron carbide B_4C. It is concluded that the boron-rich inclusions are formed during the growth of polycrystalline diamond. Fig. 13 illustrates the STEM/EELS characterizations of grain junctions and boundaries in polycrystalline diamond (Lu et al., 2015).

Diamond and C phases in compressed graphite. The spatial distribution of different carbon allotropes (hexagonal diamond, cubic diamond, and graphite phases) in compressed graphite is analyzed, in support of assessing the synthesis method. Chemical mappings using EELS in STEM show the existence

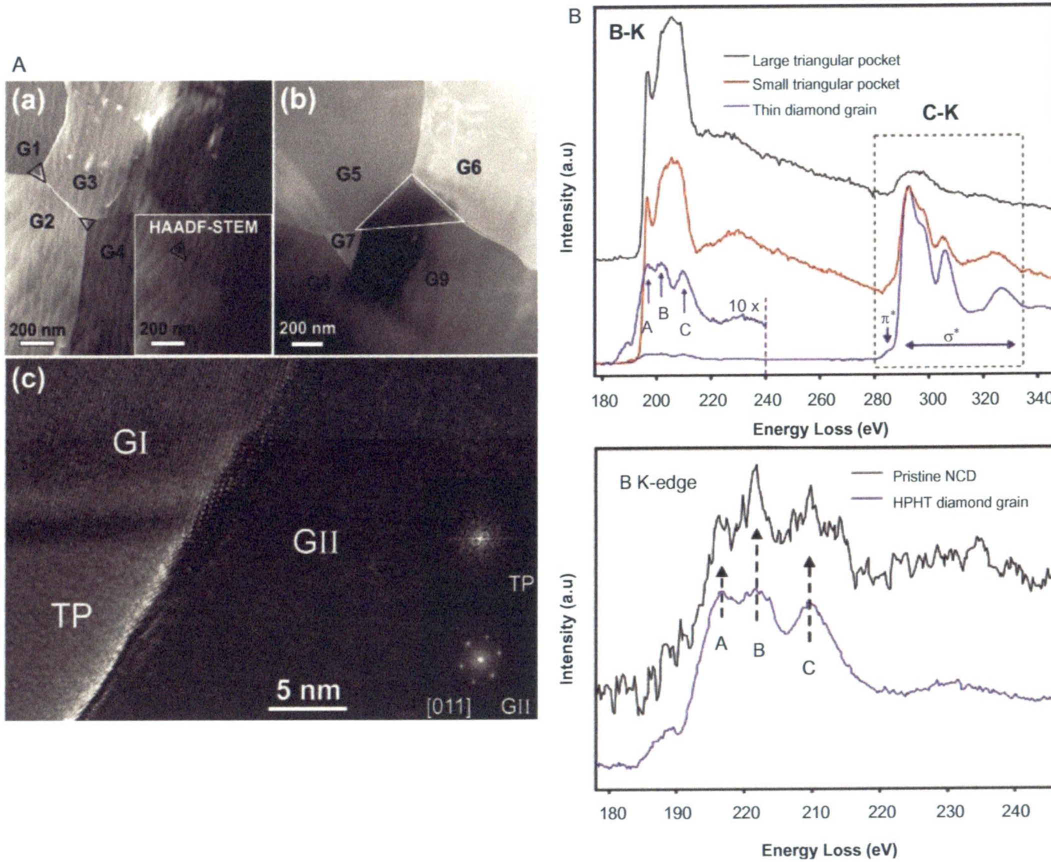

Fig. 13 See figure legend on opposite page.

of the individual carbon allotropes via the spatial distribution of the local bonding state for hexagonal and cubic diamond phases while the dark field imaging depicts the part of crystalline segments with long-range order. It is concluded that the method can be applied to synthesize diamond phases ~100 nm in size (Sato et al., 2016).

Defects in ion-implanted diamond. The nature of radiation damage produced during implantation of C ions into synthetic type Ib and natural diamonds and the effect of subsequent annealing treatments were investigated by cross-section TEM studies using electron diffraction, imaging and spectroscopy methods. It is reported that the implanted layer contains nano-regions with bent (002) graphitic planes, regions of amorphous carbon, and diamond nanocrystallites whereas no evidence is found for epitaxial regrowth (Nshingabigwi et al., 2014).

4.2 Grain boundaries, stacking faults, platelets

Grain boundaries formed by CVD synthesis of diamond films. Various types of grain boundaries have been identified in CVD synthesized diamond films: small-angle grain boundaries, large-angle grain boundaries and twin boundaries. Near-surface regions of highly-oriented diamond films and films grown by boron-assisted film deposition on Si(001) contain high fractions of small-angle grain boundaries extending predominantly along ⟨110⟩ crystal directions as well as defects formed preferentially near these grain boundaries (Jiang et al., 1998a, 1998b; Wittorf, 1999; Wittorf et al., 1997b, see also Section 2.1). The grain boundaries are largely free

Fig. 13 Grain junctions and boundaries in polycrystalline diamond. (A) STEM/TEM images of diamond grain junctions and boundaries. (a), (b) annular dark field STEM and HAADF-STEM (inset) images showing the presence of triangle-shaped pockets at grain junctions (indicated by the black and white triangles). (c) High resolution TEM image of the interface between two diamond grains (GI and GII) and a triangle-shaped pocket (TP). The pocket material is amorphous, as evidenced by the lack of reflections in the inset Fourier transform. No amorphous region is detected at the GI–GII diamond–diamond grain boundary. (B) (above) Boron and carbon EELS fine structure investigation. Black: background-subtracted EELS spectrum containing the boron and carbon K-edges from the large triangular pocket area in Fig. 13A(b). Red: background-subtracted EELS spectrum from a small triangular pocket area in Fig. 13A (a), being a mixed signal of boron carbide and boron: diamond. Blue: background-subtracted EELS spectrum from a typical thin diamond grain area. The amount of boron in the diamond grain area is around 3.5 at.%. (below) Comparison of the aligned boron ELNES acquired from the B: HPHT diamond grain (blue) with the boron ELNES from pristine CVD grown B: NCD (black). *Copies of figs. 3, 4, 5 in: Lu, Y.-G., Turner, S., Ekimov, E.A., Verbeeck, J., Tendeloo, G. Van., 2015. Boron-rich inclusions and boron distribution in HPHT polycrystalline superconducting diamond. Carbon 86, 156–162.*

from any amorphous carbon inclusions. Phenomena of the presence of amorphous carbon at grain boundaries were reported for CVD diamond films by Fallon and Brown (1993).

Defects near grain boundaries in thick (>100 mm) CVD diamond films. TEM contrast analyses reveal that the regions adjacent to columnar grain boundaries in high quality, thick (>100 mm) CVD diamond films are associated with considerable densities of defects, such as twins and dislocations. An attempt is made to correlate these observations with the anisotropy and temperature dependence of the thermal conductivity and also with their field-emission properties (Steeds et al., 1999).

Grain boundaries, junctions and dislocations in B-doped polycrystalline CVD diamond. Grain boundaries, relative grain misorientations, different kind of grain boundary junctions and variations in boron concentrations were investigated for high-quality polycrystalline B-doped ($\cong 10^{18}\,cm^{-3}$) CVD diamond films by methods of transmission electron microscopy. Coincident site lattice (CSL) Σ3, Σ9, and Σ27 grain boundaries and different types of symmetrical and asymmetrical junctions were found. Possible mechanisms for the formation of junctions and the role of the variation in B concentrations are discussed (Mora et al., 2002). Diffraction contrast analyses and electron diffraction in transmission electron microscopy revealed the dislocation structure of an individual {111}, Σ3 twin boundary in polycrystalline CVD diamond. It is concluded that the dislocations terminate monolayer steps on the grain boundary (Steeds et al., 2003).

Lonsdaleite in natural diamond. Lonsdaleite, also called hexagonal diamond, has been widely used as a marker of asteroidal impacts and is thought to play a central role during graphite-to-diamond transformation. Despite extensive efforts, it has never been produced as a separate, pure material. Electron diffraction and ultrahigh-resolution electron microscope analyses show that an explanation for the characteristic d-spacings and reflections reported for lonsdaleite can be attributed to the presence of defects in cubic diamond. Investigations by scanning TEM demonstrate that features attributed to lonsdaleite consist of cubic diamond, being dominated by extensive {113} twins and {111} stacking faults, and thus giving rise to the nanometer-scale structural complexity (Németh et al., 2014).

Defects and sp^2 hybridization in polycrystalline CVD diamond. For polycrystalline CVD diamond films, the microstructure defects are analyzed using transmission electron microscopy and energy loss near-edge spectroscopy. For fivefold twinned particles, the "incomplete filling of space" is found to be accommodated by stacking faults and a low-angle grain boundary.

ELNES analyses show that strain in low-angle as well as in non-coherent grain boundaries leads to sp^2 hybridization which is absent for coherent twin boundaries (Monteiro, 2019).

Platelet structure in natural type Ia diamond. Planar defects in {001} planes found in natural type-IaA/B diamonds, named "natural" platelets, and platelets formed in the laboratory by annealing type-IaA diamonds at temperatures above 2500 °C, called "man-made" platelets, show differences in their infrared emission. It has been suggested that high-temperature annealing changes the structure of the "natural" platelets. Electron-energy-loss spectroscopy shows that the platelets contain nitrogen as an impurity. High-temperature annealing does not change the concentration of the nitrogen in the platelets (Kiflawi et al., 1998).

Atomic-scale structure of platelets in natural type Ia diamond. Using aberration-corrected (scanning) transmission electron microscopy, image simulations and electron energy-loss spectroscopy, the nature of platelets could be clarified as an anisotropic atomic structure with a zigzag ordering of defect pairs within the platelet when viewed along ⟨110⟩ directions (Fig. 14). The electron energy-loss near-edge fine structure of both carbon K- and nitrogen K-edges obtained from the platelet core region is found to be consistent with a trigonal bonding arrangement at interstitial sites. The amount of N in the platelet core is determined to be between 5 and 8 at.%. Imaging the platelet

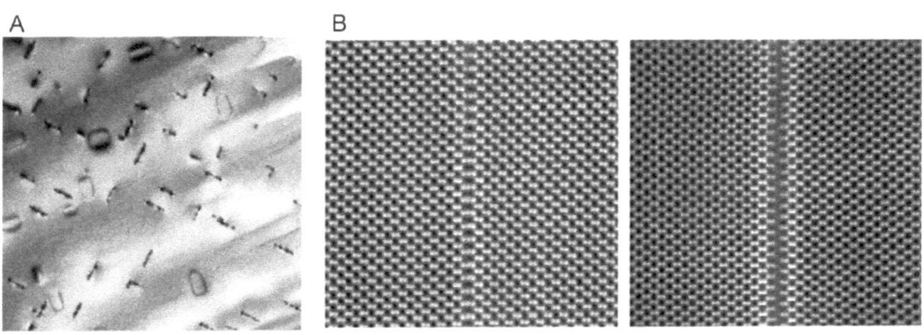

Fig. 14 Atomic-scale structure of platelets in natural diamond. (A) Bright-field TEM image of platelets in diamond. The platelets on (100)p and (010)p planes are viewed edge-on. Dislocation loops on inclined {111} planes are also visible. (B) High-resolution high-angle annular dark-field (HAADF) STEM images of two platelets viewed edge-on along ⟨110⟩ directions showing the asymmetric nature of the platelet structure. (Left) Image of platelet A. (Right) Image of platelet B. *Copies of figs. 1a, 3 in: Olivier, E.J., Neethling, J.H., Kroon, R.E., Naidoo, S.R., Allen, C.S., Sawada, H., et al., 2018. Imaging the atomic structure and local chemistry of platelets in natural type Ia diamond. Nat. Mater. 17, 243–248.*

at reduced accelerating voltage did show evidence for a periodic arrangement of nitrogen atoms, indicating ordering. These observations support an interstitial aggregate mode of formation for platelet defects in natural diamond (Olivier et al., 2018).

4.3 Dislocations

Dislocations in diamond. The dissociation of dislocations in high-temperature, high-pressure annealed natural brown diamond into partials and their glide motion has been investigated combining high-resolution electron microscopy and modeling using isotropic elasticity theory with ab initio–based tight-binding total-energy calculations. The majority of dislocations are found to be dissociated into 30° and 90° Shockley partials that are separated by a stacking fault. However, in some cases, undissociated 60° dislocations were also observed. Modeling shows the dissociation of 60° and screw dislocations in diamond to be associated with a substantial lowering of their line energies. For the 60° dislocation, an energy barrier to dissociation is found. The three types of core structures of a screw dislocation (glide, shuffle, and mixed) can be transformed into each other by the breaking and reforming of bonds. The glide motion of 90° and 30° partials is discussed involving the thermally activated formation and subsequent migration of kinks along the dislocation line. The calculated activation barriers to dislocation motion show that the 30° partial is less mobile than the 90° partial (Blumenau et al., 2003).

Boron segregation at dislocations in heteroepitaxial CVD diamond. For heavily B-doped diamond films on Ir/YSZ/Si(001), the concentration, the distribution and the local environment of boron in the dislocation-rich heteroepitaxial CVD diamond layer have been investigated by imaging and spectroscopic methods of STEM (Fig. 15). The density and nature of the dislocations were investigated by conventional and weak-beam dark-field transmission electron microscopy techniques, revealing the presence of two types of dislocations, namely edge and mixed-type 45° dislocations. From studies of the distribution of B using annular dark-field STEM and spatially resolved EELS in an aberration-corrected TEM instrument, it could be concluded that segregation of B occurs along the dislocations and toward the tensile strain field surrounding the core of edge-type dislocations. Analyses of the B—K edge fine structure in the EELS spectrum show that the boron is substitutionally incorporated into the diamond lattice. At the dislocation core, boron is present in a lower coordination, showing sp^2-like hybridization (Turner et al., 2016).

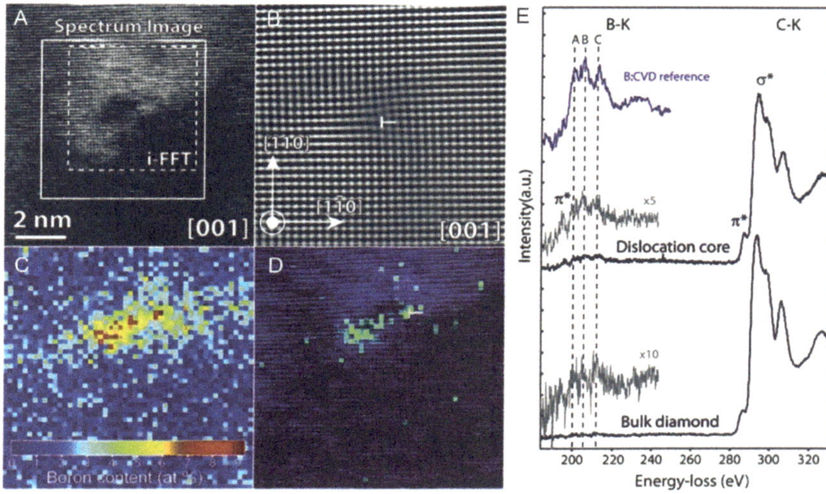

Fig. 15 Dislocation in CVD diamond—spectroscopic investigation of a single edge-type dislocation. (A) ADF-STEM image of a single edge dislocation. (B) Fourier filtered image of the dislocation core, showing the insertion of an extra half-plane. The white T marks the exact position of the edge dislocation. (C) Boron content map, showing the enrichment of B close to the dislocation core. (D) False-color image of the plotted B signal on top of the simultaneously acquired HAADF-STEM image. The white T marks the exact position of the edge dislocation. The boron segregates to area of tensile stress surrounding the dislocation. (E) EELS spectra containing the B-K and —C-—K edges. The boron signal in the dislocation core is clearly higher than in the bulk. A fine structure reference for tetrahedrally embedded boron in diamond is plotted in blue. *Copy of fig. 5 in: Turner, S., Idrissi, H., Sartori, A.F., Korneychuck, S., Lu, Y.-G., Verbeeck, J., et al., 2016. Direct imaging of boron segregation at dislocations in B: diamond heteroepitaxial films. Nanoscale 8, 2212.*

4.4 Point defects and defect complexes

Point defects in diamond. Information on type and distribution of point defects in diamond has not (yet) been accessible for direct imaging by (S)TEM methods. For the case of single-layer graphene, atomic-scale investigations employing 80 kV aberration-corrected high-resolution TEM were successful in imaging vacancy defects produced by ion irradiation (Lehtinen et al., 2014) and in observing different sp^2-coordinated self-interstitial configurations and interstitial aggregates (Lehtinen et al., 2015). Indirect information on the existence of point defects in diamond and their behavior during annealing has been obtained from experiments combining TEM with measurements of optical properties by photoluminescence spectroscopy (PL) studies. Different properties of point defects, like the migration behavior and the interaction with dislocations and grain boundaries during irradiation and during annealing were studied for a wide range of Ib and IIa diamond samples (Steeds and Kohn, 2014; Steeds et al., 2009). TEM instruments are also applied for generating defects and centers by electron irradiation (Charles et al., 2003) and to enhance the concentration of NV centers for enabling studies of, for instance, their optical properties (Farfurnik et al., 2017). Some studies of optical properties of defect centers in diamond with methods of (S)TEM, PL spectroscopy and CL spectroscopy are illustrated in Section 5.

Defect complexes in diamond. The information about spatial distribution, the atomic-scale structure and the spectroscopic properties of color centers are of interest for an assessment of their potential in applications, for instance, as sensors for nanoscale metrology (Radtke et al., 2019) or in research on the creation of color centers near the diamond surface, such as the nitrogen-vacancy center (Doherty et al., 2013; Wrachtrup et al., 2013). To the knowledge of the author, the experimental characterization of color centers in diamond by atomic scale imaging in (S)TEM has not yet been performed. However, by using computational simulations, the potential for high-resolution imaging of color centers in diamond has been explored for so called negative Cs imaging (NCSI) contrast conditions in aberration-corrected HRTEM. It is shown that SiV and GeV complexes should be clearly detectable in sufficiently thin TEM specimens and that, due to the higher atomic number of Ge, GeV centers should provide a slightly higher contrast than SiV centers (Leiter et al., 2016). Applying this argument to NV centers in diamond, an individual NV center is expected to have smaller contrast due to the smaller atomic number difference. Based on the same argument, the imaging of individual defect complexes involving heavier (higher atomic number) impurity atoms, like SnV and XeV centers, should be possible. To detect discernible contrast and to analyze the structure of NV centers will remain a challenge but should not be impossible when performing experiments with the methods of high-resolution lattice imaging in HRTEM and high-resolution STEM imaging and spectroscopy in aberration-corrected instruments under optimized conditions and with optimized TEM samples. As example, for the case of single-layer graphene containing Si atoms, the three-dimensional structure of Si atoms as well as the dynamics of the switching between up- and down-configurations of the Si atom with respect to the graphene plane could be visualized by using aberration-corrected STEM (Hofer et al., 2019).

5. TEM for characterization of physical properties

The following examples illustrate studies of the mechanical properties of diamond nanopillars, of electronic properties, of optical emission from defects and color centers, and of thermal conductivity properties of diamond films. Among the methods applied are the imaging and spectroscopy methods of aberration-corrected (S)TEM and of in situ (S)TEM, in combination with SEM methods, conventional TEM and electron diffraction methods, photoluminescence (PL) spectroscopy, and cathodoluminescence (CL) spectroscopy in STEM from nanoscale regions.

5.1 In situ TEM deformation of diamond nanopillars

The mechanical properties of various chemical vapor deposition diamond structures compared to the ideal single crystal have been reviewed by Hess (2012). The synthesis, structure properties and applications of diamond nanowires are described in a recent review by Yu and Zhi (2019).

Elastic deformation of diamond nanoneedles. "Ultralarge" elastic deformation of nanoscale single-crystalline and polycrystalline diamond needles (~300 nm) has been investigated by in situ scanning electron microscopy elastic bending experiments and TEM. Maximum tensile strains of up to 9% that approach the theoretical elastic limit are reported for single-crystalline diamond. The high strength and large elastic strain have been ascribed to the paucity of defects in the diamond nanoneedles and to their relatively smooth surfaces. While SEM assisted in monitoring the relative orientation of the diamond nanoneedles with respect to the indenter tip in the experimental configuration, TEM depicted the atomic-scale structure of a pristine diamond nanoneedle sample and the surface (Banerjee et al., 2018). The *dependency on size and crystal orientation* of the elastic deformation properties of nanoscale diamond pillars were investigated by in situ (S)TEM experiments with diamond nanoneedles of various diameters, oriented in $\langle 100 \rangle$, $\langle 110 \rangle$ and $\langle 111 \rangle$ crystal directions (Fig. 16). It is found that the fracture tensile elastic strain and strength of diamond nanoneedles depend strongly on diameter, orientation and surface conditions of the nanoneedles. $<100>$ – oriented nanoneedles with a diameter of 60 nm exhibit highest elastic tensile strain (13.4%) and tensile strength (125 GPa) (Nie et al., 2019).

Plastic deformation of nanoscale single-crystal diamond pillars. Diamond is known to generally fail by brittle fracture and does not deform plastically at room-temperature. However, applying in situ experiments in a scanning

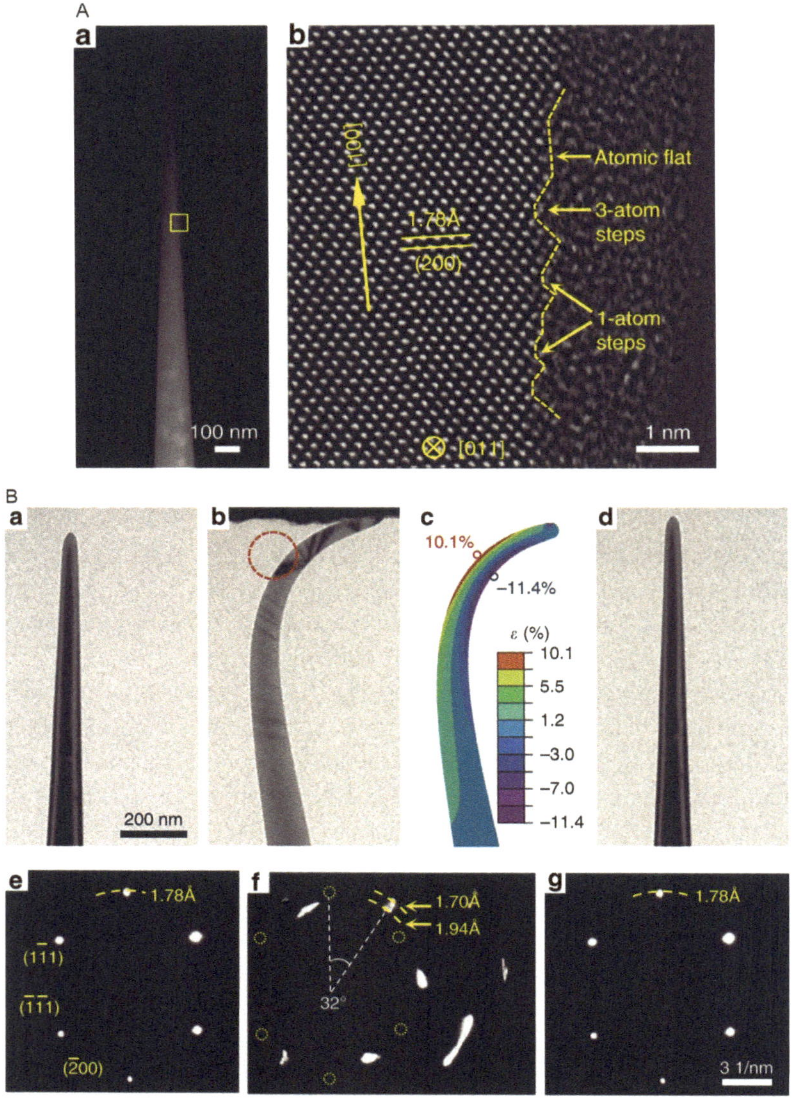

Fig. 16 In situ TEM deformation testing of a diamond nanoneedle. (A) Surface structure of a diamond ⟨100⟩ nanoneedle. (a) A low-magnitude high-angle annular dark-field scanning TEM (HAADF-STEM) image of a diamond nanoneedle after fabrication. The needle axis is parallel to [100]. (b) Atomically resolved annular bright-field scanning TEM (ABF-STEM) image of the free surface of the region marked by the yellow box in a. (B) In situ TEM elastic bending of a diamond nanoneedle—example of a reversible bending test and associated lattice expansion determined by electron diffraction. (a) TEM image of a ⟨100⟩ diamond nanoneedle with a tip diameter of ~50 nm. (b) The same nanoneedle during compression. (c) Finite element methods (FEM) simulation reproducing the shape in b revealing the maximum tensile strain (10.1%) located at the red circle. (d) The same nanoneedle, after unloading. The nanoneedle returns to its original shape. (e–g) Selected area electron diffraction (SAED) patterns taken during the bending test at strain states of a, b, d, respectively. (f) was taken from the highly curved region, circled in red, in (b). *Copies of figs. 1, 2 in: Nie, A., Bu, Y., Li, P., Zhang, Y., Jin, T., Liu, J., et al., 2019. Approaching diamond's theoretical elasticity and strength limits. Nat. Commun. 10, 5533.*

electron microscope and TEM investigations, room-temperature plasticity was reported for ⟨001⟩-oriented single-crystal diamond nanopillars with diameters smaller than ≈25 nm. It was suggested that the *plastic deformation is mediated by a phase transition* from sp^3 carbon to an ordered allotrope of carbon, termed O8-carbon, taking place spontaneously under strain in localized regions (Regan et al., 2020). In situ *TEM mechanical testing* studies with sub-μm sized diamond pillars in a transmission electron microscope and atom-resolved observations with tomographic reconstructions revealed that room-temperature *plasticity can be attributed to dislocation slip*, both for ⟨100⟩ and ⟨111⟩ crystal axis orientations. It could be observed that mixed-type dislocations with Burgers vectors of 1/2 ⟨110⟩ type are generated in the non-close-packed {001} planes of diamond under uniaxial compression of ⟨111⟩ and ⟨110⟩ directions, respectively. Under <100> directional loading, dislocations are activated in {111} planes, indicating an orientation-dependent dislocation behavior in diamond. The brittle-to-ductile transition is found to be highly dependent on the stress state (Bu et al., 2020; Nie et al., 2020).

Nanoscale wear mechanisms in tribology of ultrananocrystalline (UNCD) diamond. In-situ transmission electron microscopy has been used for tribology experiments with ultrananocrystalline diamond (UNCD) to visualize and quantify wear mechanisms on nanoscale asperities in sliding contacts. By comparing high-resolution TEM images of morphological and crystal structure changes as wear progresses and by measuring the normal force, the evolution of the specific wear rate is measured quantitatively. The experiments suggest that amorphization, gradual atomic-scale wear of crystalline grains, and fracture all play a role in the nanoscale wear process of UNCD nanocontacts and are in agreement with previous results of both atomistic simulations and macroscale experiments (Bernal and Carpick, 2019).

5.2 TEM in measurements of electronic properties

The electronic and transport properties of insulating and doped diamond are described in detail in the reviews by Nebel (2003, 2017). TEM studies have applied spectroscopic methods of EELS to investigate the local electronic properties of diamond, focusing on measuring the dielectric function, local variations in bond lengths of carbon, and the band gap of diamond. Characteristic signatures of different C bonding states in the ELNES of the C K edge are frequently used to assess the local presence of different C phases (see, e.g., Wittorf et al., 2000, Section 2.2).

Dielectric function of diamond. A methodological attempt that aims at minimizing the effect of retardation on the Kramers–Kronig analysis of valence electron energy-loss spectra has been taken to retrieve the *dielectric function of diamond* by valence EELS (VEELS) in a transmission electron microscope. Due to its high dielectric constant, diamond is a material that shows strong retardation effects and thus is regarded as a challenging material to be studied by VEELS. The resulting dielectric function shows deviations from the optical result which can be most likely attributed to the residual retardation contributions and nonzero momentum transfers (Zhang et al., 2008).

Local bond length variations in boron-doped NCD diamond. Spatially resolved variations in local bond length and coordination have been studied through changes in the ELNES of B—K and C K-edges for boron-doped nanocrystalline diamond (NCD) films in an aberration-corrected scanning transmission electron microscope. For defective regions of B-doped grains, an energy shift of the C π^* edge which is indicative of bond elongations at sites with local high B content, and a significant difference in the B ELNES have been experimentally observed. It is concluded that the position of the C σ^* edge can be used as a probe of bond length distribution in doped diamond materials (Lu et al., 2013).

Electronic band gap of diamond. Electron energy loss spectroscopy (EELS) has proven to be able to map direct band gaps with nanometer resolution. By preparing targeted specimens, EELS allows to analyze the band structure not only at the surface of a layer but also at a chosen place in the bulk material, or from an individual nanoparticle. For diamond, measurements of the local indirect band gap were performed by EELS in scanning transmission electron microscopy. The momentum space resolution achievable with EELS and the emerging possibility to partly reconstruct the band structure with EELS are discussed. The method can be extended to indirect band gaps and can help to link the structure of, for instance, indirect semiconductors with defects, strain or impurities to the properties of their band structure (Korneychuk et al., 2018a,b). An EELS-based approach was used to measure the band gaps of reference samples of diamond, GaN and AlN with good signal to noise ratio and interpretable band gap signals while offering spatial resolution in the nm range (Korneychuk et al., 2018a,b, check).

5.3 (S)TEM, PL and CL spectroscopy in characterizations of optical emission properties

Methods of TEM in combination with photoluminescence (PL) or cathodoluminescence (CL) spectroscopy were used to investigate the correlation between optical properties of diamond and defects, such as grain boundaries, dislocations, twins, and intrinsic point defects, optical centers and dopants.

Point defects in diamond. The PL properties of intrinsic dopant atoms and *point defects* generated by electron irradiation in a transmission electron microscope were studied using a cryogenic stage in a transmission electron microscope (Steeds et al., 2000).

Optical properties of dislocations and grain boundaries. For polycrystalline CVD diamond the role of the microstructure for the optical properties of diamond was established by experiments combining micro-PL spectroscopy and TEM of B-doped (10^{-18} cm^3) and unintentionally N-doped (nominally undoped) CVD diamond films, revealing the correlation between broad-band luminescence and grain boundaries (Mora et al., 2003). The correlation of TEM imaging and micro-photoluminescence studies in electron irradiated areas of diamond has been used to study optical centers and defects. Optical centers at wavelengths of 518.6 and 518.8 nm are found to be directly correlated with straight and isolated dislocations with $a/2\langle 110\rangle$ Burgers vectors, and an optical center at 519.1 nm is found to be correlated with a high density of nano-twins emanating from a grain boundary (Mora et al., 2005).

Emission from N—V defects in DND. Prospects for production of nanodiamond photoemitters with intense and stable emission from N—V centers are reported for diamond crystallites (>30 nm) produced by detonation synthesis. High nitrogen concentrations (up to 3.5 at.%) in detonation nanodiamond (DND) with primary particle sizes smaller than 6 nm are confirmed by a detailed EELS analysis. For larger DND particles, a non-uniform distribution of nitrogen is observed, indicating a preference for location within the defective central part or at twin boundaries. Combined Raman and PL spectroscopy of DND revealed emission from $(N—V)^0$ and $(N—V)^-$ centers in the larger crystallites (>30 nm) after electron irradiation and annealing of the DND material (Vlasov et al., 2010).

Emission from Si—V defects in UNCD. Strong photoluminescence intensity from Si—V defects in CVD diamond nanocrystals with a typical size of 2–5 nm in a UNCD film is observed, being comparable to that in CVD diamond microcrystals (Vlasov et al., 2009). These results are supported theoretically by DFTB calculations.

Stability and fluorescence properties of Si—V complexes in ND. The stability and fluorescence emission from photoluminescent SiV color centers in nanodiamonds from meteorites below 2 nm in size are investigated by combining various characterization methods. The SiV fluorescence emission is found to be size-dependent and highly stable over time, and few or only single color centers are found per nanocrystal. Calculations suggest that the results can be attributed to a quantum confinement effect mediated by the valence band electrons of the nanodiamond, and that the SiV excited-state wavefunction is strongly localized on the defect site, resulting in high stability of the photoelectron on the defect site. Such small molecular-sized fluorescent nanodiamonds are considered to be promising as, for example, fluorescent markers in cell biology and in sensing (Vlasov et al., 2014).

SiV centers in diamond nanopillars. Diamond nanopillars with diameters around 100 nm were structured from nanocrystalline diamond films by applying electron beam lithography and inductively coupled plasma reactive ion etching. SiV color centers were incorporated in the nanopillars during an overgrowth step. The optical investigations indicated the creation of ensembles of color centers. (Felgen et al., 2016).

SiV color centers in diamond crystallites. The optical emission properties of SiV centers have been investigated using the complementary methods of (S) TEM together with PL and electroluminescence spectroscopy. It is concluded that Si impurities lead to formation of twins on {100} facets of octahedral grains and, in turn, that twins act through their atomic order as host favoring the positioning of the optically active SiV complexes which results in steady optical emission properties (Lebedev et al., 2020).

CL from NV0 centers in nanodiamond and spatially resolved quantum nano-optics of single photons. Spectrally and spatially resolved cathodoluminescence of diamond nanoparticles using focused fast electron beams at 80 kV in a scanning transmission electron microscope enabled to detect various individual color centers in diamond nanoparticles (Fig. 17). The neutral nitrogen-vacancy (NV0) intensity maxima are attributed to individual emitters in the nanoparticles. Spectra of two individual NV0 emitters separated by 80 nm inside a nanoparticle could be spatially discerned, the spatial resolution being solely limited by charge carrier diffusion. A shift of the zero phonon line between the two emitters within the same nanoparticle is attributed to internal stress. Details of emission spectra from different particles could be measured for, in total, about 50 individual NV0 centers (Tizei and Kociak, 2012). Using low intensity electrons (with energy between 60 and 100 keV) focused to an electron beam probe of about 1 nm diameter, the single-photon state generation and characterization in an electron microscope is demonstrated experimentally. The electron beam probe was applied to excite diamond nanoparticles and to trigger individual neutral nitrogen-vacancy centers to emit photons. The photons are collected in an intensity interferometer. It is concluded that CL spectroscopy in STEM can be viewed as the nanometer scale analog of PL spectroscopy (Tizei and Kociak, 2013).

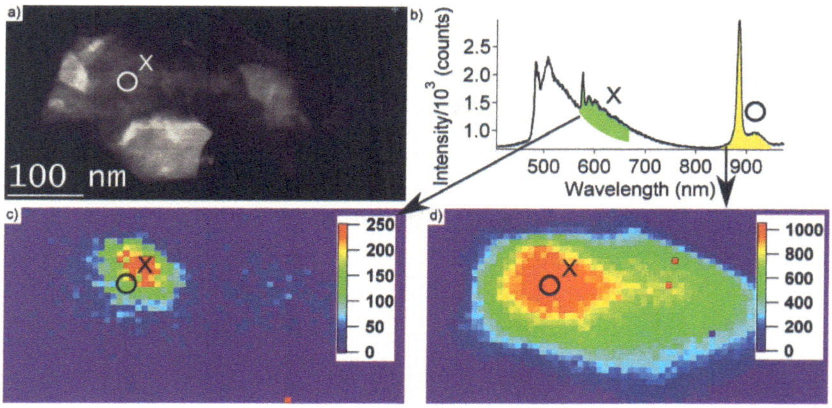

Fig. 17 CL spectroscopy analysis of a single diamond particle containing a NV0 center. (A) HAADF STEM image of a single diamond particle. (B) Typical spectrum of a region containing the NV0 center. The marked regions show where the intensity was integrated (with background subtraction by interpolation) to construct the maps of the NV0 (green, (C)) and the 1.40 eV (yellow, (D)) emission intensities. (C)–(D) Intensity maps of the NV0 and the 1.40 eV color centers. Note that the maxima of each emission do not occur at the same position within the same particle (crosses in the maps indicate the maximum of the NV0 emission). This indicates that they do not have the same physical origin. *Copy of fig. 2 in: Tizei, L.H.G., Kociak, M., 2012. Spectrally and spatially resolved cathodoluminescence of nanodiamonds: local variations of the NV0 emission properties. Nanotechnology 23, 175702.*

Photon bunching from CL of NV centers. Using a special interferometer adapted for a CL detection system in a scanning electron microscope, CL studies of the emission statistics of photons created by the electron irradiation of nanocrystals of diamond and hexagonal boron nitride show evidence for photon bunching which can be attributed to the synchronized emission from several defect centers excited by the same electron through deexcitation of a bulk plasmon into few electron-hole pairs (Meuret et al., 2015).

Efficiency of CL emission from NV centers in nanodiamonds. Correlative experiments using scanning electron microscopy, PL spectroscopy, and atomic force microscope measurements, and numerical simulations enabled to investigate the relative importance of the various factors determining the CL emission intensity from nanodiamonds containing nitrogen-vacancy (NV) color centers. Factors include particle size, density of color centers, efficiency of energy deposition by electrons passing through the particle, and conversion efficiency from deposited energy to CL emission. The

CL emission decreases significantly for smaller NDs. It is found that CL can be detected from NV-doped nanodiamonds with dimensions as small as ≈40 nm (Zhang et al., 2017).

Probing plasmon coupling with NV0 centers from CL. The coupling at the nanometer scale between nitrogen-vacancy centers in nanodiamonds and surface plasmons for silver nanocubes is probed by combining electron-based imaging, spectroscopies, and photon-based correlation spectroscopy. Measurements of the nitrogen-vacancy excited state lifetimes in a large number of either isolated nanodiamonds or nanodiamond-nanocube dimers demonstrated a significant lifetime reduction for dimers (Lourenço-Martins et al., 2018).

Electron-induced state conversion in diamond NV centers from pump-probed CL in ultrafast TEM (UTEM). Ultrafast pump – probe cathodoluminescence spectroscopy in TEM is used to prepare and read out the NV states in diamond NV centers by electron-induced state conversion and to contribute to an understanding of the dynamics between the different states. Two charge states are known for NV centers, NV0 and NV −, with the latter being mostly studied due to its long electron spin coherence time. The conversion from NV − to NV0 due to electron-induced carrier generation is achieved with electron pulses as pump and laser pulses as probe. It is shown that NV − centers are effectively converted to NV0 centers by electron irradiation and why NV − emission is not observed in CL measurements. A model describes the experimental data, providing insights into the NV − → NV0 conversion dynamics and into the use of pump-probe cathodoluminescence as a nanoscale NV characterization tool (Sola-Garcia et al., 2020).

5.4 Thermal conductivity of thin diamond films

For measurements of thermal properties, methods of TEM were combined with photothermal techniques, Raman thermography, and picosecond time-domain thermoreflectance measurements.

Thermal properties of thin polycrystalline diamond films. Films with amorphous interlayers or with a high fraction of amorphous secondary phases show remarkably increased thermal interface resistance. The influence of the microstructure on the thermal properties was investigated by comparing highly oriented with columnar grained diamond films only few microns thick. The films were deposited by microwave-assisted CVD on Si(001) at different substrate temperatures (500–800 °C) and were investigated by photothermal techniques and by HRTEM. For thermal conduction perpendicular to the diamond-silicon interface, small boundary resistances

$<4 \times 10^{-9}\,m^2K/W$ are found. Thermal conductivities normal to the interface are found to be larger than conductivities parallel to the interface by about an order of magnitude. Highly oriented films show a smaller increase in total thermal resistance with increasing layer thickness than the columnar-grained films deposited at low temperatures. In comparison to a nearly isotropic conductivity of a nanocrystalline layer, the vertical conductivity is an order of magnitude larger than the lateral conductivity, reflecting the anisotropic behavior of such films. This thermal anisotropy can be explained by the presence of grain boundaries which induce significant thermal barriers for the lateral heat transport. The knowledge of the near-interface microstructure as observed by HRTEM enables to understand the thermal film properties and assists in assessing the benefits of thin diamond layers in thermal management applications (Verhoeven et al., 1997). Fig. 18 shows examples of thermal resistance measurements and of thermal anisotropy measurements for thin CVD diamond layers.

Thermal conductivity of NCD films. The thermal conductivities of homogeneous nanocrystalline diamond films fabricated by hot filament CVD are reported to be smaller as compared with single crystal diamond. This behavior is attributed to the influence of the grain boundaries. A theoretical approach is used to describe thermal grain boundary conductivity (Mohr et al., 2016).

Thermal conductivity of polycrystalline diamond films. Using transmission electron microscopy (TEM) in combination with Raman thermography and picosecond time-domain thermoreflectance, an anisotropy of thermal conductivities of polycrystalline diamond films is observed and attributed to its grain structure and the thermal resistance at grain boundaries. The thermal resistance between grains and the lattice thermal conductivity is assessed by an approach using finite element thermal modeling (Anaya et al., 2017). A remarkably large reduction in the measured thermal conductivity by dilute levels of defects induced through ion beam modification is reported for polycrystalline diamond implanted with different ions (C, N, and O) at an energy of 16.5 MeV (Scott et al., 2020).

6. TEM for process and device technology

The applications of TEM methods in support of process and device technology are numerous. The following selection illustrates various areas of topical interest: electronics, sensors and microelectromechanical systems, power devices, capacitors, field emitters, heat dissipation systems, electrochemistry, laser machining, coatings for medical devices, and the synthesis

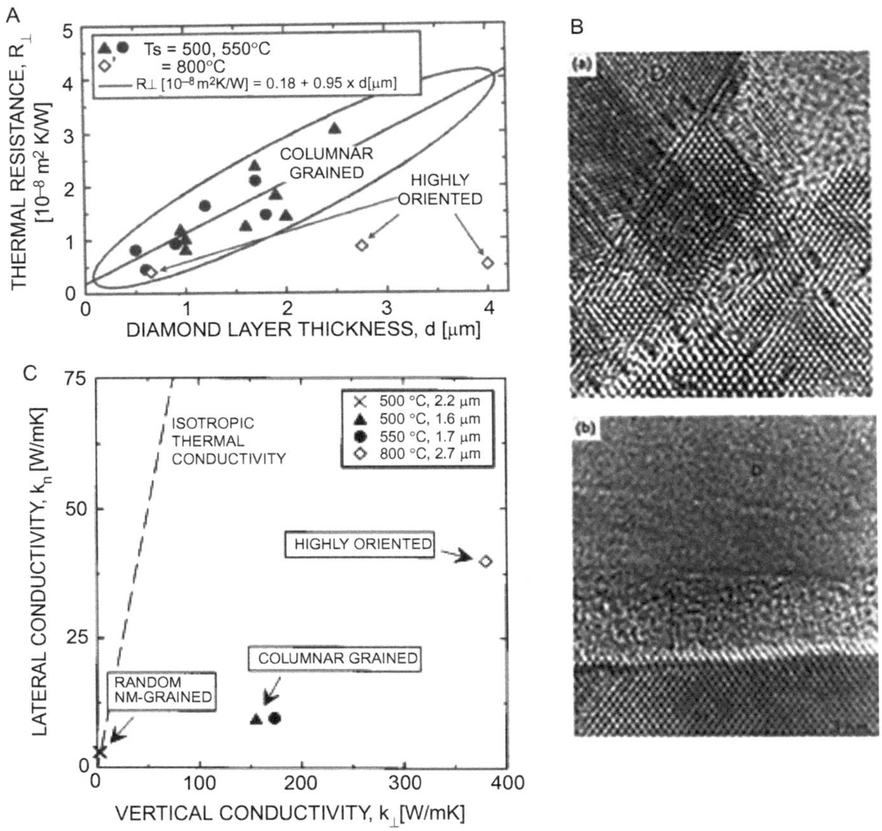

Fig. 18 CVD diamond layers on Si(001)—thermal resistance measurements. (A) Thermal resistance for conduction normal to distinct grained CVD diamond layers, which were deposited with different nucleation pretreatments and substrate temperatures on silicon, as a function of diamond layer thickness. The solid line is a linear regression of the thermal data of the layer series fabricated under identical growth conditions at 550 °C. (B) Interface structures of CVD diamond (D) deposited on (001)Si. ⟨110⟩-zone axis HRTEM lattice images of the 0.6-mm-thick, highly-oriented CVD diamond film deposited at 800 °C (a) and the 1.8-mm-thick, columnar grained diamond film deposited at 550 °C (b). The arrows in (a) show positions of terminating (111)-lattice planes of diamond, silicon, and interfacial β-SiC. The arrow in (b) marks a 3-nm-thick amorphous interfacial layer between a diamond grain and Si. Parallel lines mark {111}-lattice plane spacings in diamond and silicon. (C) Thermal anisotropy of thin CVD diamond layers with distinct grain structures. The broken line indicates the case of isotropic thermal conductivity. The substrate temperature during diamond deposition, as well as the diamond layer thickness are given in the legend. *Copies of figs. 1, 2, 3 in: Verhoeven, H., Flöter, A., Reiß, H., Zachai, R., Wittorf, D., Jäger, W., 1997, Influence of the microstructure on the thermal properties of polycrystalline thin diamond films. Appl. Phys. Lett. 71, 1324–1331.*

of porous diamond for supercapacitors and water treatment applications. The characterizations contribute to understanding of processing, functionalities and degradation as well as to the development, control and optimization of devices.

CVD diamond films for electronics, sensors and microelectromechanical systems. CVD diamond films about 10 μm thick grown on Si substrates under optimized conditions possess improved physical properties, such as, e.g., high thermal transport properties (Verhoeven et al., 1997) and excellent elastic film properties, with values of 830 GPa for the elasticity (Young's) modulus, reaching almost the value for natural diamond. The favorable elastic properties are correlated with the improved microstructure of these CVD diamond films, consisting of grains with predominantly small angle grain boundaries and containing a negligible amount of secondary phases. The fabrication and properties of devices in electronics, sensor and microelectromechanical systems technology (Gluche et al., 1998a,b) were assessed, using characterization protocols that employed various methods of material analyses, including the methods of TEM. Examples are transistors, UV and particle detectors, diamond surface p-channel MESFETs (Kawarada, 1998; Kawarada et al., 1994), and thermal microactuators (see, e.g., early reviews Güttler et al., 1999, Werner et al., 1999, Sussmann, 2009). The superior mechanical properties of the optimized diamond microstructure, such as reduced layer strain and improved film adhesion, have been applied in microsystems technology for high-temperature sensors based on diamond technology. The development of diamond substrates with low defect density and of bipolar doping has been accompanied by an increase number of diamond device studies (Tsao et al., 2018 for review). The progress toward diamond power field-effect transistors and substrate development is reviewed by Geis et al. (2018). Diamond field-effect transistors are expected to have superior high-power high-frequency performance favored by high thermal conductivity, high hole and electron mobilities, and high electric breakdown field of diamond. Limitations are the lack of high quality substrates and the high ionization energy of the primary dopant boron.

Schottky power devices. Aiming at the development of Schottky power devices, diamond metal–oxide–semiconductor (MOS) structures based on an Al/Al_2O_3/diamond MOS structure and a Zr/ZrO_2/diamond Schottky diode were characterized. The variations in the oxygen distribution, the oxide/metal mixtures and the oxide thickness and their correlation with the electrical behavior of the associated electric device could be assessed with atomic resolution by several TEM methods (Piñero et al., 2014). Similarly, the oxide-related effects on the electrical and microstructure properties of diamond/metal interface were analyzed for improving Zr- and WC-based Schottky power diodes by TEM and STEM-EELS (Pinero et al., 2017). The fabrication of lateral Schottky diodes based on a diamond growth process on Ir/STO/Si(001) was monitored using HRTEM and STEM-EELS, combined with UV Raman measurements for assessment of film strains. It is concluded that the device characteristics are highly uniform and support the use of heteroepitaxial diamond for large area electronic devices (Arnault et al., 2020).

Diamond metal–oxide–semiconductor capacitors (MOSCAPs). By combining electrical characterizations and local probe measurements coupled to SEM and TEM instruments, the origin of the leakage current in the depletion regime of homoepitaxial B-doped oxygen-terminated (001) diamond metal–oxide–semiconductor capacitors (MOSCAPs) was identified. Transmission electron microscopy showed that the regions of leakage are correlated with dislocations, described as acting as so called "killer defects" (Pham et al., 2018).

Diamond for field emission and displays. Cross-sectional TEM provided insight into the tip structure of ultra-sharp diamond pyramidal tips with ~5 nm tip radius and growth of CVD diamond cold-cathode devices for electron field emitters. The voids and nanocrystalline diamond observed at the tip surface are considered to be relevant for the assessment of electron conduction mechanisms and for strategies of improving field emission (Wade et al., 2012). The superior field electron emission behavior of *diamond-graphite nanorod arrays* with a low turn-on field, long lifetime stability, and large field enhancement factor is attributed to the nanocomposite nature of the nanorods 15–20 nm in diameter, which contain sp^2-graphitic phases in the boundaries of nano-sized diamond grains (Ramaneti et al., 2017). Vertically aligned nitrogen-doped nanocrystalline diamond nanorods fabricated from N-doped nanocrystalline diamond films show enhanced thermionic electron emission characteristics which is ascribed to abundant sp^2-graphitic phases at the grain boundaries and the field penetration effect through the local field enhancement from nanorods (Sankaran et al., 2018). The field electron emission applications of nitrogen-incorporated (ultra) nanocrystalline diamond films are discussed in an article by Sankaran and Haenen (2019). For use as high-performance electron field emitters based on hybrid sp^2/sp^3-bonded carbon, a novel hierarchical diamond nanostructure material is developed from components consisting of diamond nanoplatelets sandwiched into carbon nanowalls. Hydrogen plasma treatment plays an effective role in the transformation of sp^2-bonded carbon of carbon nanowalls to sp^3-bonded diamond (Zhai et al., 2020). The formation of diamond tips on Si and the effects of ion-beam-treatment were investigated by methods of TEM. The field electron emission characteristics were improved due to ion treatment that results in formation of an amorphous cover layer containing nanometric size fragments of diamond (Kiselev et al., 2005).

Heat dissipation systems for power devices. In order to enhance the performance of devices for high power applications, e.g., of GaN-based HEMTs, the integration of materials with high thermal conductivity provides a solution by improving the heat extraction and lowering the operating temperature. The *growth* and the *interfaces* of single crystalline (0001)-oriented GaN thin films on (100), (110) and (111) diamond single crystals were studied on the atomic scale using aberration-corrected scanning TEM combined with energy dispersive X-ray and electron energy-loss spectroscopy (Pécz et al., 2015). For developing a heat dissipation system for power devices, the room-temperature *bonding* of a single-crystal diamond chip to a Si substrate using Ti/Au intermediate layers was investigated. The TEM observations supported conclusions that the atomic diffusion and grain growth across the bonding interface and the chemical reaction at the diamond contribute to the high bonding strength (Matsuma et al., 2018). For use in high-power and high-frequency electronics applications of GaN, room-temperature surface activated *bonding interfaces* of GaN with single crystal diamond and the interfacial thermal conductance have also been investigated (Cheng et al., 2020).

Electrochemistry on the atomic scale. The dynamics of Au metal atom transport processes, the clustering of the atoms, the formation of nanoparticles, and the interaction mechanisms between particles during an electrochemical process under applied pulsed potential conditions are investigated by using identical-location STEM. The method uses boron-doped conductive diamond (BDD) that serves both as an electron-transparent TEM substrate and as electrode, holding promise also for wide-ranging further applications, for instance, in electrocatalysis or in magnetism. It enables to obtain time-resolved (snap-shot like) information on the atomic scale of the dynamics of electrodeposition on the same location of the electrode from multiple deposition and imaging cycles. Fig. 19 depicts the processes and morphological changes during the early stages of electrochemically driven nucleation and growth of Au (5–30 ms), providing direct evidence for potential-induced atom movement, atom clustering, and cluster

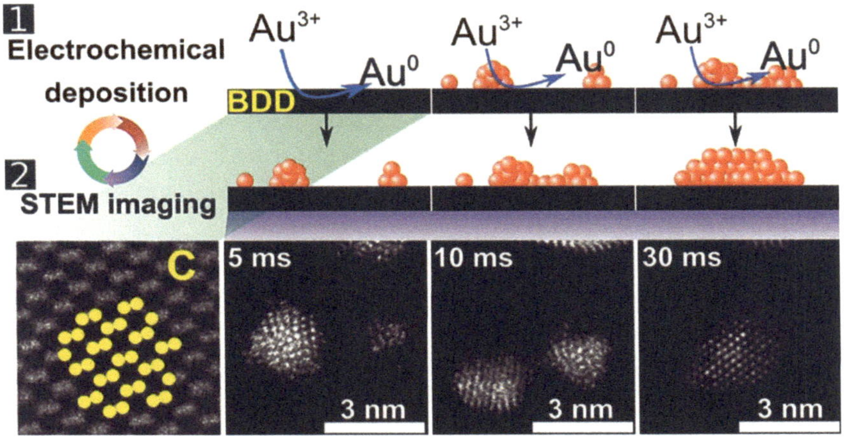

Fig. 19 Dynamics of metal electrodeposition by identical-location STEM. (1) Processes observed on the surface of the boron-doped diamond (BDD) electrode (schematic). The BDD serves as both an electron-transparent TEM substrate and electrode. (2) Atomic-resolution image of the BDD surface and of an identical electrode area imaged by ADF STEM after 5, 10 and 30 ms of pulsed potential electrodeposition. The snapshot-like image sequence depicts the dynamics of Au atoms and the nucleation and growth of clusters and of a crystalline nanoparticle. *Copy of Abstract figure in: Hussein, H.E.M., Maurer, R.J., Amari, H., Peters, J.J.P., Meng, L., Beanland, R., et al., 2018. Tracking metal electrodeposition dynamics from nucleation and growth of a single atom to a crystalline nanoparticle. ACS Nano 12 (7), 7388–7396.*

transformation to crystalline nanoparticles (Hussein et al., 2018). *Conductive diamond* is widely used as an enabling electrode in chemical and biochemical sensing, environmental degradation, electrosynthesis, electrocatalysis, and energy storage and conversion (Yang et al., 2019 for review).

Laser processing of diamond. Laser writing with ultrashort pulses as a potential route for the manufacturing of three-dimensional wires, waveguides and laser-induced defects within diamond are monitored by TEM and EELS. Electrically-conductive paths are attributed to formation of nano-scale sp^2-bonded graphitic wires and a network of strain-relieving micro-cracks (Salter et al., 2017). Nanosecond pulsed laser-processing of CVD diamond was monitored combining TEM and Raman analyses in order to assess the effect of laser fluence on the diamond microstructure. The subsurface microstructure is a mixture of both graphite and amorphous carbon. The quantitative results contribute to developing and testing a simulation model for predicting the residual graphitization thickness formed in CVD diamond by a laser beam (Mouhamadali et al., 2020).

Coatings for medical devices. Cutting blades with CVD-coated ultra-sharp diamond cutting edges are durable and remain sharp even after very long cutting times. Such blades are mainly used as medical scalpels in eye and neurosurgery. TEM methods enabled to characterize the edge microstructure and to measure the radii of curvature of typically few nanometers (Dieker et al., 2001). Biocompatible UNCD coatings are discussed for applications in various implantable medical devices, such as microchips for use as artificial retina or intraocular devices for treatment of glaucoma in the eye, and as UNCD-coated metal dental implants. The TEM methods are monitoring the processing and functionalities of the coatings in UNCD film technology (Auciello et al., 2014).

Porous diamond for supercapacitors. Methods of electron microscopy were also used to monitor and characterize the fabrication of porous diamond which has various promising applications, such as water treatment, energy conversion and storage, as well as electrochemical and biochemical sensing. Boron-doped diamond foam for applications as a thin-film microsupercapacitor electrode material has been synthesized on wafer scale by CVD (Gao et al., 2014). The fabrication of porous diamond networks with tunable porosity in the range of 15%–68% and excellent electrochemical properties are described by Zhuang et al. (2015). The material development for surface-enlarged supercapacitor electrodes, based on boron-doped polycrystalline diamond, and their typical properties are summarized by Gao and Nebel (2016). This included methods of surface enlargement by top-down plasma etching and by templated growth and the performance evaluation of a prototype cell based on free-standing "diamond paper." The synthesis of boron-doped diamond electrodes for *supercapacitors* of high specific capacitance, consisting of porous nanostructure layers or multilayers is described by Wang et al. (2020).

7. Conclusions and outlook

The imaging, diffraction and spectroscopy methods of advanced and in situ transmission electron microscopy (TEM) are pivotal for the materials science of diamond and for the development of applications in device technology. Their contribution to the progress in the topical areas of crystal growth and the defect and surface phenomena in diamond epitaxy, in research on nanodiamond, in defect characterizations, in correlating structure with physical properties, and for the monitoring, optimization, and control of processing and failure in device technology has been reviewed and illustrated in this article.

The future research for diamond will benefit from the potential for nanocharacterization provided by the new generation of aberration-corrected advanced and in situ TEM instrumentation that also enables experiments at low electron energy and with minimal electron beam exposures of the samples. Experimental approaches can be taken now that utilize novel and fast electron detectors, fast data acquisition, data processing and data evaluation, energy monochromators for high-resolution spectroscopy, large area detectors for improved x-ray spectroscopy, or the methodology for determining magnetic and electric fields on the nanoscale. The application areas for the materials science of diamond are manifold, ranging from research on interfaces, defects and surfaces or exploring photonic emission processes to in situ studies of physical properties and of materials reactions.

Acknowledgments

It is my great pleasure to acknowledge the numerous discussions during the collaborations with my former and present colleagues and with all colleagues who have contributed as coauthors to some of the research presented in this article. The author would like to thank Professor Christoph Nebel and the colleagues from the editorial board for their kind invitation to contribute with this review to this topical and fascinating research area.

References

Achard, J., Tallaire, A., 2018. Growth of thick CVD diamond films on different crystalline orientations: defects and doping. In: Koizumi, S., Umezawa, H., Pernot, J., Suzuki, M. (Eds.), Power Electronics Device Applications of Diamond Semiconductors. Elsevier Ltd. Ch. 1.1.

Achard, J., Jacques, V., Tallaire, A., 2020. Chemical vapour deposition diamond single crystals with nitrogen-vacancy centres: a review of material synthesis and technology for quantum sensing applications. J. Phys. D Appl. Phys. 53, 313001.

Adamschik, M., Hinz, M., Maier, C., Schmid, P., Seliger, H., Hofer, E.P., Kohn, E., 2001. Diamond micro system for bio-chemistry. Diamond Relat. Mater. 10, 722.

Afandi, A., Howkins, A., Boyd, I.W., Jackman, R.B., 2018. Nanodiamonds for device applications: an investigation of the properties of boron-doped detonation nanodiamonds. Sci. Rep. 8, 3270.

Aharonovich, I., Greentree, A.D., Prawer, S., 2011. Diamond photonics. Nat. Photonics 5, 397.

Alegre, M.P., Araújo, D., Fiori, A., Pinero, J.C., Lloret, F., Villar, M.P., Achatz, P., Chicot, G., Bustarret, E., Jomard, F., 2014. Critical boron-doping levels for generation of dislocations in synthetic diamond. Appl. Phys. Lett. 105, 173103.

Anaya, J., Bai, T., Wang, Y., Li, C., Goorsky, M., Bougher, T.L., Yates, L., Cheng, Z., Graham, S., Hobart, K.D., Feygelson, T.I., Tadjer, M.J., Anderson, T.J., Pate, B.B., Kuballet, M., 2017. Simultaneous determination of the lattice thermal conductivity and grain/grain thermal resistance in polycrystalline diamond. Acta Mater. 139, 215–225.

Ando, Y., Kamano, T., Suzuki, K., Sawabe, A., 2012. Epitaxial lateral overgrowth of diamonds on iridium by patterned nucleation and growth method. Jpn. J. Appl. Phys. 51, 090101.

Araújo, D., Paz Alegre, M.A., García, A.J., Pilar Villar, M., Bustarret, E., Achatz, P., Volpe, P.N., Omnès, F., 2011. Cross sectional evaluation of boron doping and defect distribution in homoepitaxial diamond layers. Phys. Status Solidi C 8 (4), 1366–1370.

Arenal, R., 2012. n-type nanocrystalline diamond films: synthesis, structure, composition, properties, and applications. In: Ultrananocrystalline Diamond. Elsevier Inc, p. 103. Ch. 4.

Arenal, R., Stephan, O., Bruno, P., Gruen, D.M., 2009. Spatially resolved electron energy loss spectroscopy on n-type ultrananocrystalline diamond films. Appl. Phys. Lett. 94, 111905.

Arnault, J.C. (Ed.), 2017. Nanodiamonds: Advanced Material Analysis, Properties and Applications. Elsevier.

Arnault, J.C., 2019. Surface modifications of nanodiamonds and current issues for their biomedical applications. In: Yang, N. (Ed.), Novel Aspects of Diamond—From Growth to Applications. Topics in Applied Physics, vol. 121. Springer, pp. 85–122.

Arnault, J.C., Girard, H.A., 2014. Ch. 10, Diamond nucleation and seeding techniques: two complementary strategies for the growth of ultra-thin diamond films. In: Williams, O. (Ed.), RSC Nanoscience & Nanotechnology No. 31, Nanodiamond. The Royal Society of Chemistry, pp. 221–252.

Arnault, J.C., Lee, K.H., Delchevalrie, J., Penuelas, J., Mehmel, L., Brinza, O., Temgoua, S., Stenger, I., Letellier, J., Saint-Girons, G., Bachelet, R., Issaoui, R., Tallaire, A., Achard, J., Barjon, J., Eon, D., Ricolleau, C., Saada, S., 2020. Epitaxial diamond on Ir/SrTiO3/Si (001): from sequential material characterizations to fabrication of lateral Schottky diodes. Diamond Relat. Mater. 105, 107768.

Ascarelli, P., Capelli, E., Pinzari, F., 1997. Nearest-neighbor distance distribution of diamond nuclei on substrate surfaces. Appl. Phys. Lett. 70, 1697.

Asmussen, J., Grotjohn, T.A., Schuelke, T., Becker, M.F., Yaran, M.K., King, D.J., Wicklein, S., Reinhard, D.K., 2008. Multiple substrate microwave plasma-assisted chemical vapor deposition single crystal diamond synthesis. Appl. Phys. Lett. 93, 031502.

Auciello, O., Gurman, P., Guglielmotti, M.B., Olmedo, D.G., Berra, A., Saravia, M.J., 2014. Biocompatible ultrananocrystalline diamond coatings for implantable medical devices. MRS Bull. 39, 621.

Balluffi, R.W., Brokman, A., King, A.H., 1982. CSL/DSC lattice model for general crystal boundaries and their line defects. Acta Metall. 30, 1453.

Banerjee, A., Bernoulli, D., Zhang, H., Yuen, M.-F., Liu, J., Dong, J., Ding, F., Lu, J., Dao, M., Zhang, W., Lu, Y., Suresh, S., 2018. Ultralarge elastic deformation of nanoscale diamond. Science 360, 300–302.

Banhart, F., Ajayant, P.M., 1996. Carbon onions as nanoscopic pressure cells for diamond formation. Nature 382, 433.

Barbay, C., Saada, S., Mer-Calfati, C., Temgoua, S., Barjon, J., Arnault, J.C., 2019. Dislocations imaging in low boron doped diamond epilayers using Field Emission Scanning Electron Microscopy (FE-SEM). Appl. Surf. Sci. 495, 143564.

Batzer, M., Shields, B., Neu, E., Widmann, C., Giese, C., Nebel, C., Maletinsky, P., 2019. Single crystal diamond pyramids for applications in nanoscale quantum sensing. arXiv 1910.10737, Opt. Mater. Express 10, 492–500.

Bauer, T., Gsell, S., Schreck, M., Goldfuß, J., Lettieri, J., Schlom, D.G., Stritzker, B., 2005. Growth of epitaxial diamond on silicon via iridium/$SrTiO_3$ buffer layers. Diamond Relat. Mater. 14, 314.

Bell, D.C., Mankin, M., Day, R.W., Erdman, N., 2014. Successful application of low voltage electron microscopy to practical materials problems. Ultramicroscopy 145, 56–65.

Bernal, R.A., Carpick, R.W., 2019. Visualization of nanoscale wear mechanisms in ultrananocrystalline diamond by in-situ TEM tribometry. Carbon 154, 132–139.

Blumenau, A.T., Jones, R., Frauenheim, T., Willems, B., Lebedev, O.I., Van Tendeloo, G., Fisher, D., Martineau, P.M., 2003. Dislocations in diamond: dissociation into partials and their glide motion. Phys. Rev. B 68, 014115.
Bradac, C., Gao, W., Forneris, J., Trusheim, M.E., Aharonovich, I., 2019. Quantum nanophotonics with group IV defects in diamond. Nat. Commun. 10, 5625.
Brescia, R., Schreck, M., Gsell, S., Fischer, M., Stritzker, B., 2008. Transmission electron microscopy study of the very early stages of diamond growth on iridium. Diamond Relat. Mater. 17, 1045–1050.
Brown, H.G., Shibata, N., Sasaki, H., Petersen, T.C., Paganin, D.M., Morgan, M.J., Findlay, S.D., 2017. Measuring nanometre-scale electric fields in scanning transmission electron microscopy using segmented detectors. Ultramicroscopy 182, 169–178.
Bu, Y., Wang, P., Nie, A., Wang, H., 2020. Dislocation Slip or Phase Transformation Lead to Room-Temperature Plasticity in Diamond: Comment on Plastic Deformation of Single-Crystal Diamond Nanopillars. https://arxiv.org/abs/2002.01104v1. (Submitted on 4 Feb 2020).
Chang, S.L.Y., Barnard, A.S., Dwyer, C., Boothroyd, C.B., Hocking, R.K., Ōsawa, E., Nicholls, R.J., 2016. Counting vacancies and nitrogen-vacancy centers in detonation nanodiamond. Nanoscale 8, 10548–10552.
Charles, S.J., Steeds, J.W., Butler, J.E., Evans, D.J.F., 2003. Optical centers introduced in boron-doped synthetic diamond by near-threshold electron irradiation. J. Appl. Phys. 94, 3091.
Chen, J.H., Van Tendeloo, G., 1999. Microstructure of tough polycrystalline natural diamond. J. Electron Microsc. 48, 121–129.
Chen, J.H., Bernaerts, D., Seo, J.W., Van Tendeloo, G., Kagi, H., 1998. Voidites in polycrystalline natural diamond. Philos. Mag. Lett. 77 (3), 135–140.
Cheng, Z., Mu, F., Yates, L., Suga, T., Graham, S., 2020. Interfacial thermal conductance across room-temperature-bonded GaN/diamond interfaces for GaN-on-diamond devices. ACS Appl. Mater. Interfaces 12, 8376–8384.
Colliex, C., Gloter, A., March, K., Mory, C., Stephan, O., Suenaga, K., Tencé, M., 2012. Capturing the signature of single atoms with the tiny probe of a STEM. Ultramicroscopy 123, 80–89.
Degutis, G., Pobedinskas, P., Turner, S., Luc, Y.-G., Al Riyami, S., Ruttens, B., Yoshitake, T., D'Haen, J., Haenen, K., Verbeeck, J., Hardy, A., Van Bael, M.K., 2016. CVD diamond growth from nanodiamond seeds buried under a thin chromium layer. Diamond Relat. Mater. 64, 163–168.
Denisenko, A., Aleksov, A., Kohn, E., 2001. pH sensing by surface-doped diamond and effect of the diamond surface termination. Diamond Relat. Mater. 10, 667.
Dieker, C., Jäger, W., Ertl, S., Strobel, S., 2001. High-precision diamond surgical blades—TEM characterization of edge microstructure and radii of curvature. In: Proc. 'Dreiländertagung für Elektronenmikroskopie', A Conference on Modern Microscopical Methods, 9–14 Sept. 2001, Innsbruck, Österreich, p. 114.
Dieker, C., Tillmann, K., Williams, O.A., Jäger, W., Nebel, C.E., 2011. Microscopic characterisations of diamond nanoparticles by aberration-corrected HRTEM. In: Proceeding Microscopy Conference MC 2011, Kiel, Germany. German Society for Electron Microscopy e.V, Germany. 28.08.-02.09.2011, ISBN 978-3 00-033910-3 Copyright 2011 DGE.
Dischler, B., Wild, C. (Eds.), 1998. Low-Pressure Synthetic Diamond. Springer, Berlin.
Doherty, M.W., Manson, N.B., Delaney, P., Jelezko, F., Wrachtrup, J., Hollenberg, L.C.L., 2013. The nitrogen-vacancy colour centre in diamond. arXiv 1302.3288v1, Phys. Rep. 528, 1–45.
Drijkoningen, S., Pobedinskas, P., Korneychuk, S., Momot, A., Balasubramaniam, Y., Van Bael, M.K., Turner, S., Verbeeck, J., Nesládek, M., Haenen, K., 2017. On the origin of diamond plates deposited at low temperature. Cryst. Growth Des. 17, 4306–4314.

Dubochet, J., Frank, J., Henderson, R., 2017. Nobel Prize in Chemistry 2017 "for Developing Cryo-Electron Microscopy for the High-Resolution Structure Determination of Biomolecules in Solution." The Nobel Prize in Chemistry 2017. NobelPrize.org. Nobel Media AB 2020. Sat. 16 May 2020. https://www.nobelprize.org/prizes/chemistry/2017/summary/.

Dunin-Borkowski, R.E., Kovács, A., Kasama, T., McCartney, M.R., Smith, D.J., 2019. Electron holography. In: Hawkes, P.W., Spence, J.C.H. (Eds.), Handbook of Microscopy. Springer Nature, Switzerland AG, pp. 767–818. Ch. 16.

Edmonds, A.M., Hart, C.A., Turner, M.J., Colard, P.-O., Schloss, J.M., Olsson, K., Trubko, R., Markham, M.L., Rathmill, A., Horne-Smith, B., Lew, W., Manickam, A., Bruce, S., Kaup, P.G., Russo, J.C., DiMario, M.J., South, J.T., Hansen, J.T., Twitchen, D.J., Walsworth, R.L., 2020. Generation of nitrogen-vacancy ensembles in diamond for quantum sensors: optimization and scalability of CVD processes. arXiv. 2004.01746.

Einaga, Y., Foord, J.S., Swain, G.M., 2014. Diamond electrodes: diversity and maturity. MRS Bull. 39, 525.

Ekimov, E.A., Lyapin, S.G., Grigoriev, Y.V., Zibrov, I.P., Kondrina, K.M., 2019. Size-controllable synthesis of ultrasmall diamonds from halogenated adamantanes at high static pressure. Carbon 150, 436–438.

Ekimova, E.A., Kondrin, M.V., Lyapin, S.G., Grigoriev, Y.V., Razgulov, A.A., Krivobok, V.S., Gierlotka, S., Stelmakh, S., 2020. High-pressure synthesis and optical properties of nanodiamonds obtained from halogenated adamantanes. Diamond Relat. Mater. 103, 107718.

Fallon, P.J., Brown, L.M., 1993. Analysis of chemical-vapour-deposited diamond grain boundaries using transmission electron microscopy and parallel electron energy loss spectroscopy in a scanning transmission electron microscope. Diamond Relat. Mater. 2, 1004.

Farfurnik, D., Alfasi, N., Masis, S., Kaufmann, Y., Farchi, E., Romach, Y., Hovav, Y., Buks, E., Bar-Gill, N., 2017. Enhanced concentrations of nitrogen-vacancy centers in diamond through TEM irradiation. arXiv: 1702.05332v2 [quant-ph] 25 Apr 2017, Appl. Phys. Lett. 111, 123101.

Felgen, N., Naydenov, B., Turner, S., Jelezko, F., Reithmaier, J.P., Popov, C., 2016. Incorporation and study of SiV centers in diamond nanopillars. Diamond Relat. Mater. 64, 64–69.

Fischer, M., Brescia, R., Gsell, S., Schreck, M., Brugger, T., Greber, T., Osterwalder, J., Stritzker, B., 2008. Growth of twin-free heteroepitaxial diamond on Ir/YSZ/Si(111). J. Appl. Phys. 104, 123531.

Gallheber, B.-C., Klein, O., Fischer, M., Schreck, M., 2017. Propagation of threading dislocations in heteroepitaxial diamond films with (111) orientation and their role in the formation of intrinsic stress. J. Appl. Phys. 121, 225301.

Gallheber, B.-C., Fischer, M., Mayr, M., Straub, J., Schreck, M., 2018. Growth, stress, and defects of heteroepitaxial diamond on Ir/YSZ/Si(111). J. Appl. Phys. 123, 225302.

Gao, F., Nebel, C.E., 2016. Diamond-based supercapacitors: realization and properties. ACS Appl. Mater. Interfaces 8, 28244–28254.

Gao, F., Wolfer, M.T., Nebel, C.E., 2014. Highly porous diamond foam as a thin-film micro-supercapacitor material. Carbon 80, 833–840.

Geis, M.W., Wade, T.C., Wuorio, C.H., Fedynyshyn, T.H., Duncan, B., Plaut, M.E., Varghese, J.O., Warnock, S.M., Vitale, S.A., Hollis, M.A., 2018. Progress toward diamond power field-effect transistors. Phys. Status Solidi A 215, 1800681.

Ginés, L., Mandal, S., Ahmed, A.I., Cheng, C.-L., Sow, M., Williams, O.A., 2017. Positive zeta potential of nanodiamonds. Nanoscale 9, 12549–12555. https://doi.org/10.1039/c7nr03200e.

Gluche, P., Adamschik, M., Vescan, A., Ebert, W., Szücs, F., Fecht, H.J., Flöter, A., Zachai, R., Kohn, E., 1998a. Application of highly oriented, planar diamond (HOD) films of high mechanical strength in sensor technologies. Diamond Relat. Mater. 7, 779–782.

Gluche, P., Leuner, R., Vescan, A., Ebert, W., Kohn, E., Rembe, C., aus der Wiesche, S., Hofer, E.P., 1998b. Actuator—sensor technology on electronic grade diamond films. Microsyst. Technol. 5, 38–43.

Görlitz, J., Herrmann, D., Thiering, G., Fuchs, P., Gandil, M., Iwasaki, T., Taniguchi, T., Kieschnick, M., Meijer, J., Hatano, M., Gali, A., Becher, C., 2020. Spectroscopic investigations of negatively charged tin-vacancy centres in diamond. arXiv 1909.09435, New J. Phys. 22, 013048.

Gruber, A., Dräbenstedt, A., Tietz, C., Fleury, L., Wrachtrup, J., von Borczyskowski, C., 1997. Scanning confocal optical microscopy and magnetic resonance on single defect Centers. Science 276, 2012.

Güttler, H., Flöter, A., Kohn, E., Gluche, P., Jiang, X., Klages, C.P., Dollinger, G., Bergmaier, A., Wittorf, D., Jäger, W., 1999. Evaluation of semiconducting CVD diamond films for electronic and sensor applications. In: Kempter, K., Haußelt, J. (Eds.), Werkstoffwoche '98, Symposium. 1 Materials for Information Technology. Wiley-VCH, Weinheim, pp. 189–194 (ISBN 3 527 29938 6).

Haider, M., Rose, H., Uhlemann, S., Kabius, B., Urban, K., 1998a. Electron microscopy image enhanced. Nature 392, 768–769.

Haider, M., Rose, H., Uhlemann, S., Kabius, B., Urban, K., 1998b. Towards 0.1 nm resolution with the first spherically corrected transmission electron microscope. J. Electron Microsc. (Tokyo) 47, 395–405.

Haider, M., Rose, H., Uhlemann, S., Schwan, E., Kabius, B., Urban, K., 1998c. A spherical-aberration-corrected 200 kV transmission electron microscope. Ultramicroscopy 75, 53–60.

Hall, L.T., Simpson, D.A., Hollenberg, L.C.L., 2013. Nanoscale sensing and imaging in biology using the nitrogen-vacancy center in diamond. MRS Bull. 38, 162.

Hanlon, L., Gautam, V., Wood, J.D.A., Reddy, P., Barson, M.S.J., Niihori, M., Silalahi, A.R.J., Corry, B., Wrachtrup, J., Sellars, M.J., Daria, V.R., Maletinsky, P., Stuart, G.J., Doherty, M.W., 2019. Diamond nanopillar arrays for quantum microscopy of neuronal signals. arXiv 1901.08743 (2019), Neurophotonics 7.

Hatano, Y., Sekiguchi, T., Iwasaki, T., Hatano, M., Harada, Y., 2018. Magnetic field imaging of superparamagnetic particles using high-density, perfectly oriented NV centers in diamond CVD film. Phys. Status Solidi A 215, 1800254.

Häußler, S., Thiering, G., Dietrich, A., Waasem, N., Teraji, T., Isoya, J., Iwasaki, T., Hatano, M., Jelezko, F., Gali, A., Kubanek, A., 2017. Photoluminescence excitation spectroscopy of SiV− and GeV− color center in diamond. arXiv 1705.10486 (2017), New J. Phys. 19.

Hawkes, P.W., Spence, J.C.H. (Eds.), 2019. Handbook of Microscopy. Springer Nature, Switzerland AG. https://doi.org/10.1007/978-3-030-00069-1.

Hess, P., 2012. The mechanical properties of various chemical vapor deposition diamond structures compared to the ideal single crystal. J. Appl. Phys. 111, 051101.

Heyer, S., Janssen, W., Turner, S., Lu, Y.-G., Yeap, W.S., Verbeeck, J., Haenen, K., Krueger, A., 2014. Toward deep blue nano hope diamonds: heavily boron-doped diamond nanoparticles. ACS Nano 8 (6), 5757–5764.

Hofer, C., Skakalova, V., Monazam, M.R.A., Mangler, C., Kotakoski, J., Susi, T., Meyer, J.C., 2019. Direct visualization of the 3D structure of silicon impurities in graphene. Appl. Phys. Lett. 114. https://doi.org/10.1063/1.5063449.

Hörmann, F., Peng, H.Y., Bauer, T., Li, Q., Schreck, M., Lifshitz, Y., Lee, S.T., Stritzker, B., 2002. Flat epitaxial diamond/Ir (001) interface visualized by high resolution transmission electron microscopy. Surf. Sci. 513, 525–529.

Hussein, H.E.M., Maurer, R.J., Amari, H., Peters, J.J.P., Meng, L., Beanland, R., Newton, M.E., Macpherson, J.V., 2018. Tracking metal electrodeposition dynamics from nucleation and growth of a single atom to a crystalline nanoparticle. ACS Nano 12 (7), 7388–7396.

Jäger, W., Jiang, X., 2001. Diamond heteroepitaxy: nucleation, interface structure, film growth. Acta. Metall. Sin. 14 (6), 425–434.

Jäger, W., Wittorf, D., Jia, C.L., Urban, K., 1999. High-resolution electron microscopy of grain boundaries and interfaces in heteroepitaxial diamond films. Adv. Sci. Tech. 21, 137–144.

Janssen, W., Turner, S., Sakr, G., Jomard, F., Barjon, J., Degutis, G., Lu, Y.-G., D'Haen, J., Hardy, A., Van Bael, M., Verbeeck, J., Van Tendeloo, G., Haenen, K., 2014. Substitutional phosphorus incorporation in nanocrystalline CVD diamond thin films. Phys. Status Solidi (RRL) 8 (8), 705–709.

Jia, C.L., Urban, K., Jiang, X., 1995. Heteroepitaxial diamond films on silicon (001): interface structure and crystallographic relations between film and substrate. Phys. Rev. B 52, 5164.

Jiang, X., 2019. Ultra-Wide Bandgap Semiconductor Materials Ch. 2.3 Heteroepitaxial Growth of Diamond. Elsevier Inc.

Jiang, X., Jia, C.L., 1996. The coalescence of [001] diamond grains heteroepitaxially grown on (001) silicon. Appl. Phys. Lett. 69, 3902.

Jiang, X., Jia, C.L., 2000. Direct local epitaxy of diamond on Si(100) and surface-roughening-induced crystal misorientation. Phys. Rev. Lett. 84 (16), 3658–3661.

Jiang, X., Klages, C.-P., Zachai, R., Hartweg, M., Fusser, H.-J., 1993. Epitaxial diamond thin films on (001) silicon substrates. Appl. Phys. Lett. 62, 3438.

Jiang, X., Schiffmann, K., Klages, C.-P., 1994. Nucleation and initial growth phase of diamond thin films on (100) silicon. Phys. Rev. B 50, 8402.

Jiang, X., Zhang, W.J., Paul, M., Klages, C.-P., 1996. Diamond film orientation by ion bombardment during deposition. Appl. Phys. Lett. 68, 1927.

Jiang, X., Schiffmann, K., Klages, C.-P., Wittorf, D., Jia, C.L., Urban, K., Jäger, W., 1998a. Coalescence and overgrowth of diamond grains for improved heteroepitaxy on Si(001). J. Appl. Phys. 83, 2511.

Jiang, X., Zhang, W.J., Klages, C.-P., 1998b. Effects of ion bombardment on the nucleation and growth of diamond films. Phys. Rev. B 58, 7064.

Jiang, X., Fryda, M., Jia, C.L., 2000. High quality heteroepitaxial diamond films on silicon: recent progresses. Diamond Relat. Mater. 9, 1640.

Jin, P., Yu, W., Zhang, Y., Wu, J., Wang, Z., 2019. Homoepitaxial growth of single-crystal diamond. In: Ultra-Wide Bandgap Semiconductor Materials. Elsevier Inc. Ch. 2.4.

Kaiser, U., Stöger-Pollach, M. (Eds.), 2014. Low-voltage electron microscopy. In: Ultramicroscopy. vol. 145. (Special Issue).

Kato, Y., Teraji, T., 2018. Characterization methods of defects in single crystalline diamond for power device application. In: Koizumi, S., Umezawa, H., Pernot, J., Suzuki, M. (Eds.), Power Electronics Device Applications of Diamond Semiconductors. Elsevier. Ch. 4.1.

Kawarada, H., 1998. In: Dischler, B., Wild, C. (Eds.), Low-Pressure Synthetic Diamond. Springer, Berlin, pp. 140–162.

Kawarada, H., Aoki, M., Ito, M., 1994. Enhancement mode metal-semiconductor field effect transistors using homoepitaxial diamonds. Appl. Phys. Lett. 65, 1563.

Kawarada, H., Suesada, T., Nagasawa, H., 1995. Heteroepitaxial growth of smooth and continuous diamond thin films on silicon substrates via high quality silicon carbide buffer layers. Appl. Phys. Lett. 66, 583.

Kawarada, H., Wild, C., Herres, N., Locher, R., Koidl, P., Nagasawa, H., 1997. Heteroepitaxial growth of highly oriented diamond on cubic silicon carbide. J. Appl. Phys. 81, 3490.

Kiflawi, I., Bruley, J., Luyten, W., Van Tendeloo, G., 1998. 'Natural' and 'man-made' platelets in type-Ia diamonds. Philos. Mag. B 78 (3), 299–314.
Kiselev, N.A., Hutchison, J.L., Roddatis, V.V., Stepanova, A.N., Aksenova, L.L., Rakova, E.V., Mashkova, E.S., Molchanov, V.A., Givargizov, E.I., 2005. TEM and HREM of diamond crystals grown on Si tips: structure and results of ion-beam-treatment. Micron 36, 81–88.
Knoll, M., Ruska, E., 1932. Das Elektronenmikroskop. Z. Phys. 78, 318–339.
Kohl, R., Wild, C., Herres, N., Koidl, P., Stoner, B.R., Glass, J.T., 1993. Oriented nucleation and growth of diamond films on β-SiC and Si. Appl. Phys. Lett. 63, 1792.
Koizumi, S., Umezawa, H., Pernot, J., Suzuki, M. (Eds.), 2018. Power Electronics Device Applications of Diamond Semiconductors. Elsevier Ltd.
Korneychuk, S., Guzzinati, G., Verbeeck, J., 2018a. Measurement of the indirect band gap of diamond with EELS in STEM. Phys. Status Solidi A 215, 1800318.
Korneychuk, S., Partoens, B., Guzzinati, G., Ramaneti, R., Derluyn, J., Haenen, K., Verbeeck, J., 2018b. Exploring possibilities of band gap measurement with off-axis EELS in TEM. Ultramicroscopy 189, 76–84.
Krivanek, O., Delby, N., Lupini, A., 1999. Towards sub-Å electron beams. Ultramicroscopy 78, 1–11.
Krueger, A., 2014. The chemistry of nanodiamond. Ch. 3. In: Williams, O. (Ed.), RSC Nanoscience & Nanotechnology No. 31 Nanodiamond. The Royal Society of Chemistry, p. 49.
Lebedev, V., Yoshikawa, T., Giese, C., Kirste, L., Žukauskaitė, A., Graff, A., Meyer, F., Burmeister, F., Ambacher, O., 2019. Formation of icosahedron twins during initial stages of heteroepitaxial diamond nucleation and growth. J. Appl. Phys. 125, 075305.
Lebedev, V., Yoshikawa, T., Schreyvogel, C., Kirste, L., Weippert, J., Kunzer, M., Graff, A., Ambacher, O., 2020. Microstructural and optical emission properties of diamond multiply twinned particles. J. Appl. Phys. 127, 025303.
Lehtinen, O., Tsai, I.-L., Jalil, R., Nair, R.R., Keinonen, J., Kaiser, U., Grigorieva, I.V., 2014. Non-invasive transmission electron microscopy of vacancy defects in graphene produced by ion irradiation. Nanoscale 6, 6569–6576.
Lehtinen, O., Vats, N., Algara-Siller, G., Knyrim, P., Kaiser, U., 2015. Implantation and atomic-scale investigation of self-interstitials in graphene. Nano Lett. 15, 235–241.
Leiter, R., Qi, H., Biskupek, J., Naydenov, B., Jelezko, F., Kaiser, U., 2016. Towards imaging of defects in diamond by high-resolution TEM. In: Proc. 16th European Microscopy Congress 2016, Lyon, France. https://doi.org/10.1002/9783527808465.EMC2016.6553.
Lesik, M., Raatz, N., Tallaire, A., Spinicelli, P., John, R., Achard, J., Gicquel, A., Jacques, V., Roch, J.-F., Meijer, J., Pezzagna, S., 2016. Production of bulk NV Centre arrays by shallow implantation and diamond CVD overgrowth. Phys. Status Solidi A 213 (10), 2594–2600.
Lettingdon, A., Steeds, J.W. (Eds.), 1994. Thin Film Diamond. Chapman & Hall, London. ISBN 0-412-49630-5.
Lifshitz, Y., Köhler, T., Frauenheim, T., Guzmann, I., Hoffman, A., Zhang, R.Q., Zhou, X.T., Lee, S.T., 2002. The mechanism of diamond nucleation from energetic species. Science 297, 1531.
Lin, I.-N., Koizumi, S., Yater, J., Koeck, F., 2014. Diamond electron emission. MRS Bull. 39, 533.
Liu, W., Naydenov, B., Chakrabortty, S., Wuensch, B., Hübner, K., Ritz, S., Cölfen, H., Barth, H., Koynov, K., Qi, H., Leiter, R., Reuter, R., Wrachtrup, J., Boldt, F., Scheuer, J., Kaiser, U., Sison, M., Lasser, T., Tinnefeld, P., Jelezko, F., Walther, P.,

Wu, Y., Weil, T., 2016. Fluorescent nanodiamond–gold hybrid particles for multimodal optical and electron microscopy cellular imaging. Nano Lett. 16, 6236–6244.
Lloret, F., Araujo, D., Alegre, M.P., Gonzalez-Lea, J.M., Villar, M.P., Eon, D., Bustarret, E., 2015. TEM study of defects versus growth orientations in heavily boron-doped diamond. Phys. Status Solidi A 212 (11), 2468–2473.
Lloret, F., Eon, D., Bustarret, E., Fiori, A., Araujo, D., 2018. Boron-doping proximity effects on dislocation generation during non-planar MPCVD homoepitaxial diamond growth. Nanomaterials 8, 480.
Locher, R., Wild, C., Herres, N., Behr, D., Koidl, P., 1994. Nitrogen stabilized <100> texture in chemical vapor deposited diamond films. Appl. Phys. Lett. 65, 34.
Loncar, M., Faraon, A., 2013. Quantum photonic networks in diamond. MRS Bull. 38, 144.
Lourenço-Martins, H., Kociak, M., Meuret, S., Treussart, F., Lee, Y.H., Ling, X.Y., Chang, H.-C., Tizei, L.H.G., 2018. Probing Plasmon-NV0 coupling at the Nanometer scale with photons and fast electrons. ACS Photonics 5, 324–328.
Lu, Y.-G., Turner, S., Verbeeck, J., Janssens, S.D., Haenen, K., Van Tendeloo, G., 2013. Local bond length variations in boron-doped nanocrystalline diamond measured by spatially resolved electron energy-loss spectroscopy. Appl. Phys. Lett. 103, 032105.
Lu, Y.-G., Turner, S., Ekimov, E.A., Verbeeck, J., Van Tendeloo, G., 2015. Boron-rich inclusions and boron distribution in HPHT polycrystalline superconducting diamond. Carbon 86, 156–162.
Lyutovich, Y., Banhart, F., 1999. Low-pressure transformation of graphite to diamond under irradiation. Appl. Phys. Lett. 74, 659.
Matsuma, T., Kurashima, Y., Umezawa, H., Mokuno, Y., Takagi, H., 2018. Room-temperature bonding of single-crystal diamond and Si using Au/Au atomic diffusion bonding in atmospheric air. Microelectron. Eng. 195, 68–73.
May, P.W., Pobedinskas, P., Nicle, S.S. (Eds.), 2018. Recent advances in diamond science and technology. Phys. Status Solidi A 215 (22), 1800787, (Topical section).
Meuret, S., Tizei, L.H.G., Cazimajou, T., Bourrellier, R., Chang, H.C., Treussart, F., Kociak, M., 2015. Photon bunching in cathodoluminescence. Phys. Rev. Lett. 114, 197401.
Michl, J., Teraji, T., Zaiser, S., Jakobi, I., Waldherr, G., Dolde, F., Neumann, P., Doherty, M.W., Manson, N.B., Isoya, J., Wrachtrup, J., 2014. Perfect alignment and preferential orientation of nitrogen-vacancy centers during chemical vapor deposition diamond growth on (111) surfaces. Appl. Phys. Lett. 104, 102407. https://doi.org/10.1063/1.4868128.
Miyazaki, T., Miyamoto, Y., Makino, T., Kato, H., Yamasaki, S., Fukui, T., Doi, Y., Tokuda, N., Hatano, M., Mizuochi, N., 2014. Atomistic mechanism of perfect alignment of nitrogen-vacancy centers in diamond Appl. Phys. Lett. 105, 261601. https://doi.org/10.1063/1.4904988.
Mohr, M., Brühne, K., Fecht, H.-J., 2016. Thermal conductivity of nanocrystalline diamond films grown by hot filament chemical vapor deposition. Phys. Status Solidi A 213 (10), 2590–2593.
Mokuno, Y., Yamada, H., Chayahara, A., 2018. Single crystal diamond wafers. In: Koizumi, S., Umezawa, H., Pernot, J., Suzuki, M. (Eds.), Power Electronics Device Applications of Diamond Semiconductors. Elsevier Ltd, p. 1. Ch. 1.3.
Monteiro, O.R., 2019. Structural defects and sp^2 localization in CVD diamond. J. Mater. Sci. 54, 2300–2306.
Mora, A.E., Steeds, J.W., Butler, J.E., 2002. Grain boundaries in boron-doped CVD diamond films. Diamond Relat. Mater. 11, 697–702.
Mora, A.E., Steeds, J.W., Butler, J.E., 2003. Relationship between grain boundaries and broad luminescence peaks in CVD diamond films. Diamond Relat. Mater. 12, 310–317.

Mora, A.E., Steeds, J.W., Butler, J.E., Yan, C.-S., Mao, H.K., Hemley, R.J., Fisher, D., 2005. New direct evidence of point defects interacting with dislocations and grain boundaries in diamond. Phys. Status Solidi A 202 (15), 2943–2949.

Morishita, H., Kobayashi, S., Fujiwara, M., Kato, H., Makino, T., Yamasaki, S., Mizuochi, N., 2020. Room temperature electrically detected nuclear spin coherence of NV centres in diamond. Sci. Rep. 10, 792.

Morris, A.A., Pehrsson, P.E., 1991. Effects of surface pretreatments on nucleation and growth of diamond films on a variety of substrates. Appl. Phys. Lett. 59, 417.

Mouhamadali, F., Equis, S., Saeidi, F., Best, J.P., Cantoni, M., Hoffmann, P., Wasmer, K., 2020. Nanosecond pulsed laser-processing of CVD diamond. Opt. Lasers Eng. 126, 105917.

Müller, K., Krause, F.F., Béché, A., Schowalter, M., Galioit, V., Löffler, S., Verbeeck, J., Zweck, J., Schattschneider, P., Rosenauer, A., 2014. Atomic electric fields revealed by a quantum mechanical approach to electron picodiffraction. Nat. Commun. 5, 5653.

Nebel, C.E., 2003. Electronic properties of CVD diamond. Semicond. Sci. Technol. 18, S1–S11.

Nebel, C.E., 2017. Ch. 1 General properties of diamond. In: Nanodiamonds. Elsevier Inc.

Nebel, C.E., Munz, J., Stutzmann, M., Zachai, R., Güttler, H., 1997. Electronic properties of CVD and synthetic diamond. Phys. Rev. B 55 (15), 9786–9791.

Nelz, R., Görlitz, J., Herrmann, D., Slablab, A., Challier, M., Radtke, M., Fischer, M., Gsell, S., Schreck, M., Becher, C., Neu, E., 2018. Towards wafer-scale diamond nano- and quantum technologies. arXiv 1810.09350, APL Mater. 7, 011108.

Nemanich, R.J., Carlisle, J.A., Harata, A., Haenen, K., 2014. CVD diamond—research, applications, and challenges. MRS Bull. 39, 490.

Németh, P., Garvie, L.A.J., Aoki, T., Dubrovinskaia, N., Dubrovinsky, L., Buseck, P.R., 2014. Lonsdaleite is faulted and twinned cubic diamond and does not exist as a discrete material. Nat. Commun. 5, 5447.

Nie, A., Bu, Y., Li, P., Zhang, Y., Jin, T., Liu, J., Su, Z., Wang, Y., He, J., Liu, Z., Wang, H., Tian, Y., Yang, W., 2019. Approaching diamond's theoretical elasticity and strength limits. Nat. Commun. 10, 5533.

Nie, A., Bu, Y., Huang, J., Shao, Y., Zhang, Y., Hu, W., Liu, J., Wang, Y., Xu, B., Liu, Z., Wang, H., Yang, W., Tian, Y., 2020. The Room-Temperature Dislocation Plasticity in Diamond. https://arxiv.org/abs/2002.05913. (Submitted on 14 Feb 2020).

Nshingabigwi, E.K., Derry, T.E., Naidoo, S.R., Neethling, J.H., Olivier, E.J., O'Connell, J.H., Levitt, C.M., 2014. Electron microscopy profiling of ion implantation damage in diamond: dependence on fluence and annealing. Diamond Relat. Mater. 49, 1–8.

Ohmagari, S., Yamada, H., Tsubouchi, N., Umezawa, H., Chayahara, A., Mokuno, Y., Takeuchi, D., 2019. Toward high-performance diamond electronics: control and annihilation of dislocation propagation by metal-assisted termination. Phys. Status Solidi A 216, 1900498.

Ohtsuka, K., Suzuki, K., Sawabe, A., Inuzuka, T., 1996. Epitaxial growth of diamond on iridium. Jpn. J. Appl. Phys. 35, L1072.

Olivier, E.J., Neethling, J.H., Kroon, R.E., Naidoo, S.R., Allen, C.S., Sawada, H., van Aken, P.A., Kirkland, A.I., 2018. Imaging the atomic structure and local chemistry of platelets in natural type Ia diamond. Nat. Mater. 17, 243–248.

Ozawa, H., Ishiwata, H., Hatano, M., Iwasaki, T., 2018. Thermal stability of perfectly aligned nitrogen-vacancy centers for high sensitive magnetometers. Phys. Status Solidi A 215, 1800342.

Pakes, C.I., Garrido, J.A., Kawarada, H., 2014. Diamond surface conductivity: properties, devices, and sensors. MRS Bull. 39, 542.

Park, J.S., Sinclair, R., Rowcliffe, D., Stern, M., Davidson, H., 2007. Orientation relationship in diamond and silicon carbide composites. Diamond Relat. Mater. 16, 562–565.

Pécz, B., Tóth, L., Tsiakatouras, G., Adikimenakis, A., Kovács, A., Duchamp, M., Dunin-Borkowski, R.E., Yakimova, R., Neumann, P.L., Behmenburg, H., Foltynski, B., Giesen, C., Heuken, M., Georgakilas, A., 2015. GaN heterostructures with diamond and graphene. Semicond. Sci. Technol. 30, 114001.

Pfender, M., Aslam, N., Simon, P., Antonov, D., Thiering, G., Burk, S., Fávaro de Oliveira, F., Denisenko, A., Fedder, H., Meijer, J., Garrido, J.A., Gali, A., Teraji, T., Isoya, J., Doherty, M.W., Alkauskas, A., Gallo, A., Grüneis, A., Neumann, P., Wrachtrup, J., 2017. Protecting a diamond quantum memory by charge state control. Nano Lett. 17, 5931–5937.

Pham, T.T., Piñero, J.C., Maréchal, A., Gutiérrez, M., Lloret, F., Eon, D., Gheeraert, E., Rouger, N., Araújo, D., Pernot, J., 2018. Impact of nonhomoepitaxial defects in depleted diamond MOS capacitors. IEEE Trans. Electron Devices 65 (5), 1830–1837.

Piñero, J.C., Araujo, D., Traoré, A., Chicot, G., Maréchal, A., Muret, P., Alegre, M.P., Villar, M.P., Pernot, J., 2014. Metal–oxide–diamond interface investigation by TEM: toward MOS and Schottky power device behavior. Phys. Status Solidi A 211 (10), 2367–2371.

Pinero, J.C., Araújo, D., Fiori, A., Traoré, A., Villar, M.P., Eon, D., Muret, P., Pernot, J., Teraji, T., 2017. Atomic composition of WC/ and Zr/O-terminated diamond Schottky interfaces close to ideality. Appl. Surf. Sci. 395, 200–207.

Plitzko, J., Baumeister, W.P., 2019. Cryo-electron tomography. In: Hawkes, P.W., Spence, J.C.H. (Eds.), Springer Handbook of Microscopy. Springer Handbooks, Springer Nature Switzerland AG, p. 189. https://doi.org/10.1007/978-3-030-00069-1_4.

Radtke, M., Bernardi, E., Slablab, A., Nelz, R., Neu, E., 2019. Nanoscale sensing based on nitrogen vacancy centers in single crystal diamond and nanodiamonds: achievements and challenges. Nano Futures 3, 042004.

Ramaneti, R., Sankaran, K.J., Korneychuk, S., Yeh, C.J., Degutis, G., Leou, K.C., Verbeeck, J., Van Bael, M.K., Lin, I.N., Haenen, K., 2017. Vertically aligned diamond-graphite hybrid nanorod arrays with superior field electron emission properties. APL Mater. 5, 066102.

Rath, P., Ummethala, S., Nebel, C., Pernice, W.H.P., 2015. Diamond as a material for monolithically integrated optical and optomechanical devices. Phys. Status Solidi A 212 (11), 2385–2399.

Raymakers, J., Haenen, K., Maes, W., 2019. Diamond surface functionalization: from gemstone to photoelectrochemical applications. J. Mater. Chem. C 7, 10134.

Regan, B., Kim, S., Ly, A.T.H., Trycz, A., Bray, K., Ganesan, K., Toth, M., Aharonovich, I., 2019. Photonic devices fabricated from (111) oriented single crystal diamond. arXiv 1909.03728 (2019), InfoMat.

Regan, B., Aghajamali, A., Froech, J., Trong Tran, T., Scott, J., Bishop, J., Suarez-Martinez, I., Liu, Y., Cairney, J.M., Marks, N.A., Toth, M., Aharonovich, I., 2020. Plastic deformation of single-crystal diamond nanopillars. Adv. Mater. 32, 1906458.

Reimer, L., Kohl, H., 2008. Transmission Electron Microscopy: Physics of Image Formation, fifth ed. Springer.

Rose, H., 1977. Nonstandard imaging methods in electron microscopy. Ultramicroscopy 2, 251–267.

Rose, H., 1990. Outline of a spherically corrected semi-aplanatic medium-voltage transmission electron microscope. Optik 85, 19–24.

Rose, H., 2009. Historical aspects of aberration correction. J. Electron Microsc. (Tokyo) 58, 77–85.

Rose, H., 2012. Geometrical Charged-Particle Optics, second ed. Springer, Berlin.

Roth, T., Alianelli, L., Lengeler, D., Snigirev, A., Seiboth, F., 2017. Materials for x-ray refractive lenses minimizing wavefront distortions. MRS Bull. 42, 430.

Rudloff-Grund, J., Brenker, F.E., Marquardt, K., Howell, D., Schreiber, A., O'Reilly, S.Y., Griffin, W.L., Kaminsky, F.V., 2016. Nitrogen nanoinclusions in milky diamonds from Juina area, Mato Grosso State, Brazil. Lithos 265, 57–67.
Ruska, E., 1987. The development of the electron microscope and of electron microscopy (Nobel lecture). Angew. Chem. Int. Ed. Engl. 26, 595–608.
Salter, P.S., Booth, M.J., Courvoisier, A., Moran, D.A.J., MacLaren, D.A., 2017. High resolution structural characterisation of laser-induced defect clusters inside diamond. arXiv:1706.10137v2 [physics.app-ph] 11 Aug 2017, Appl. Phys. Lett. 111.
Sandstrom, R., Ke, L., Martin, A., Wang, Z., Kianinia, M., Green, B., Gao, W.-b., Aharonovich, I., 2017. Optical properties of Xe color centers in diamond. arXiv 1704.01636, Opt. Commun. 411, 182–186.
Sang, L., Wu, J., Xu, F., Qin, Z., Wang, M., Chen, D., Shen, B., 2019. Semiconductor diamond. In: Liao, M., Shen, B., Wang, Z. (Eds.), Ultra-wide Bandgap Semiconductor Materials. Materials Today, Elsevier Inc, pp. 111–261. Ch. 2.
Sankaran, K.J., Haenen, K., 2019. Nitrogen incorporated (ultra) nanocrystalline diamond films for field electron emission applications. In: Yang, N. (Ed.), Novel Aspects of Diamond. Topics in Applied Physics, vol. 121. Springer Nature, Switzerland AG, pp. 123–171. Ch. 5.
Sankaran, K.J., Deshmukh, S., Korneychuk, S., Yeh, C.-J., Palathinkal Thomas, J., Drijkoningen, S., Pobedinskas, P., Van Bael, M.K., Verbeeck, J., Leou, K.-C., Leung, K.-T., Roy, S.S., Lin, I.-N., Haenen, K., 2018. Fabrication, microstructure, and enhanced thermionic electron emission properties of vertically aligned nitrogen-doped nanocrystalline diamond nanorods. MRS Commun. 8, 1311–1320.
Sato, K., Iwasaki, T., Shimizu, M., Kato, H., Makino, T., Ogura, M., Takeuchi, D., Yamasaki, S., Hatano, M., 2015. Fabrication of diamond lateral p–n junction diodes on (111) substrates. Phys. Status Solidi A 212 (11), 2548–2552.
Sato, Y., Bugnet, M., Terauchi, M., Botton, G.A., Yoshiasa, A., 2016. Heterogeneous diamond phases in compressed graphite studied by electron energy-loss spectroscopy. Diamond Relat. Mater. 64, 190–196.
Sawabe, A., Fukuda, H., Suzuki, T., Ikuhara, Y., Suzuki, T., 2000. Interface between CVD diamond and iridium films. Surf. Sci. 467, L845–L849.
Schreck, M., Arnault, J.-C., 2018. Heteroepitaxy of diamond on Ir/metal-oxide/Si substrates. In: Koizumi, S., Umezawa, H., Pernot, J., Suzuki, M. (Eds.), Power Electronics Device Applications of Diamond Semiconductors. Elsevier Ltd, p. 58. Ch. 1.4.
Schreck, M., Hörmann, F., Roll, H., Lindner, J.K.N., Stritzker, B., 2001. Diamond nucleation on iridium buffer layers and subsequent textured growth: a route for the realization of single-crystal diamond films. Appl. Phys. Lett. 78 (2), 192.
Schreck, M., Gsell, S., Brescia, R., Fischer, M., Bernhard, P., Prunici, P., Hess, P., Stritzker, B., 2009. Diamond nucleation on iridium: local variations of structure and density within the BEN layer. Diamond Relat. Mater. 18, 107–112.
Schreck, M., Asmussen, J., Shikata, S., Arnault, J.-C., Fujimori, N., 2014. Large area high quality single crystal diamond. MRS Bull. 39, 504.
Schreck, M., Gsell, S., Brescia, R., Fischer, M., 2017. Ion bombardment induced buried lateral growth: the key mechanism for the synthesis of single crystal diamond wafers. Sci. Rep. 7, 44462.
Scott, E.A., Hattar, K., Braun, J.L., Rost, C.M., Gaskins, J.T., Bai, T., Wang, Y., Ganski, C., Goorsky, M., Hopkins, P.E., 2020. Orders of magnitude reduction in the thermal conductivity of polycrystalline diamond through carbon, nitrogen, and oxygen ion implantation. Carbon 157, 97–105.
Shechtman, D., Feldman, A., Hutchison, J., 1993a. High-order twin boundaries in CVD diamond films. Mater. Lett. 17, 211–216.

Shechtman, D., Hutchison, J.L., Robins, L.H., Farabaugh, E.N., Feldman, A., 1993b. Growth defects in diamond films. J. Mater. Res. 8 (3), 473–479.

Shenderova, O.A., Ciftan Hens, S.A., 2013. Nanodiamonds. In: Vajtai, R. (Ed.), Springer Handbook of Nanomaterials. Springer-Verlag, Berlin Heidelberg, p. 263. Ch. 8.

Shenderova, O.A., Gruen, D.M. (Eds.), 2012. Ultrananocrystalline Diamond—Synthesis, Properties, and Applications. Handbook 2nd Edition. Elsevier.

Shenderova, O.A., McGuire, G.E., 2015. Science and engineering of nanodiamond particle surfaces for biological applications. Biointerphases 10, 030802.

Shenderova, O.A., Shames, A.I., Nunn, N.A., Torelli, M.D., Vlasov, I., Zaitsev, A., 2019. Review article: synthesis, properties, and applications of fluorescent diamond particles. J. Vac. Sci. Technol. B 37, 030802.

Shibata, N., Findlay, S.D., Kohno, Y., Sawada, H., Kondo, Y., Ikuhara, Y., 2012. Differential phase-contrast microscopy at atomic resolution. Nat. Phys. 8, 611.

Shibata, N., Seki, T., Sánchez-Santolino, G., Findlay, S.D., Kohno, Y., Matsumoto, T., Ishikawa, R., Ikuhara, Y., 2017. Electric field imaging of single atoms. Nat. Commun. 8, 15631.

Shvyd'ko, Y., Blank, V., Terentyev, S., 2017. Diamond x-ray optics: transparent, resilient, high-resolution, and wavefront preserving. MRS Bull. 42, 437.

Sivazlian, F.R., Glass, J.T., Stoner, B.R., 1994. Investigation of the low angle grain boundaries in highly oriented diamond films via transmission electron microscopy. J. Mater. Res. 9, 2487–2489.

Smith, J.M., Meynell, S.A., Bleszynski Jayich, A.C., Meijer, J., 2019. Colour centre generation in diamond for quantum technologies. Nanophotonics 8 (11), 1889–1906.

Sola-Garcia, M., Meuret, S., Coenen, T., Polman, A., 2020. Electron-induced state conversion in diamond NV centers measured with pump – probe cathodoluminescence spectroscopy. ACS Photonics 7, 232–240.

Stammler, M., Stöckel, R., Ley, L., Albrecht, M., Strunk, H.P., 1997. Diamond nucleation on silicon during bias treatment in chemical vapour deposition as analysed by electron microscopy. Diamond Relat. Mater. 6, 747–751.

Steeds, J.W., Kohn, S., 2014. Annealing of electron radiation damage in a wide range of Ib and IIa diamond samples. Diamond Relat. Mater. 50, 110–122.

Steeds, J.W., Gilmore, A., Bussmann, K.M., Butler, J.E., Koidl, P., 1999. On the nature of grain boundary defects in high quality CVD diamond films and their influence on physical properties. Diamond Relat. Mater. 8, 996–1005.

Steeds, J.W., Charles, S.J., Davies, J., Griffin, I., 2000. Photoluminescence microscopy of TEM irradiated diamond. Diamond Relat. Mater. 9, 397–403.

Steeds, J.W., Mora, A.E., Butler, J.E., 2003. Observation of monolayer steps on {111}, $\Sigma = 3$ twin boundaries in chemically vapor-deposited polycrystalline diamond. Philos. Mag. Lett. 83 (5), 297–302.

Steeds, J.W., Sullivan, W., Wotherspoon, A., Hayes, J.M., 2009. Long-range migration of intrinsic defects during irradiation or implantation. J. Phys. Condens. Matter 21, 364219.

Stehl, C., Fischer, M., Gsell, S., Berdermann, E., Rahman, M.S., Traeger, M., Klein, O., Schreck, M., 2013. Efficiency of dislocation density reduction during heteroepitaxial growth of diamond for detector applications. Appl. Phys. Lett. 103, 151905.

Stehlik, S., Varga, M., Ledinsky, M., Miliaieva, D., Kozak, H., Skakalova, V., Mangler, C., Pennycook, T.J., Meyer, J.C., Kromka, A., Rezek, B., 2016. High-yield fabrication and properties of 1.4 nm nanodiamonds with narrow size distribution. Sci. Rep. 6, 38419.

Suenaga, K., 2014. Elemental analysis down to the single atom with electron beams. C. R. Phys. 15, 151–157.

Sumant, A.V., Auciello, O., Carpick, R.W., Srinivasan, S., Butler, J.E., 2014. Ultrananocrystalline and nanocrystalline diamond thin films for MEMS/NEMS applications. MRS Bull. 35, 281.
Sussmann, R.S. (Ed.), 2009. CVD Diamond for Electronic Devices and Sensors. John Wiley & Sons Ltd.
Suto, T., Yaita, J., Iwasaki, T., Hatano, M., 2017. Highly oriented diamond (111) films synthesized by pulse bias-enhanced nucleation and epitaxial grain selection on a 3C-SiC/Si (111) substrate. Appl. Phys. Lett. 110, 062102.
Szunerits, S., Nebel, C.E., Hamers, R.J., 2014. Surface functionalization and biological applications of CVD diamond. MRS Bull. 39 (6), 517.
Tabakoya, T., Kanada, S., Wakui, Y., Takamori, Y., Yamada, T., Nagai, M., Kojima, Y., Ariyada, O., Yamasaki, S., Nebel, C.E., Matsumoto, T., Inokuma, T., Tokuda, N., 2019. High-rate growth of single-crystalline diamond (100) films by hot-filament chemical vapor deposition with tantalum filaments at 3000C. Phys. Status Solidi A 216, 1900244.
Teraji, T., 2018. Homoepitaxial growth of ultrahigh purity diamond films. In: Koizumi, S., Umezawa, H., Pernot, J., Suzuki, M. (Eds.), Power Electronics Device Applications of Diamond Semiconductors. Elsevier Ltd, p. 1. Ch. 1.2.
Teraji, T., Yamamoto, T., Watanabe, K., Koide, Y., Isoya, J., Onoda, S., Ohshima, T., Rogers, L.J., Jelezko, F., Neumann, P., Wrachtrup, J., Koizumi, S., 2015. Homoepitaxial diamond film growth: high purity, high crystalline quality, isotopic enrichment, and single color center formation. Phys. Status Solidi A 212 (11), 2365–2384.
Tizei, L.H.G., Kociak, M., 2012. Spectrally and spatially resolved cathodoluminescence of nanodiamonds: local variations of the NV0 emission properties. Nanotechnology 23, 175702.
Tizei, L.H.G., Kociak, M., 2013. Spatially resolved quantum nano-optics of single photons using an electron microscope. Phys. Rev. Lett. 110, 153604.
Tokuda, N., 2019. Homoepitaxial diamond growth by plasma-enhanced chemical vapor deposition. In: Yang, N. (Ed.), Novel Aspects of Diamond—From Growth to Applications, second ed. Topics in Applied Physics, vol. 121. Springer, pp. 1–29. Ch. 1.
Tsao, J.Y., Chowdhury, S., Hollis, M.A., Jena, D., Johnson, N.M., Jones, K.A., Kaplar, R.J., Rajan, S., Van de Walle, C.G., Bellotti, E., Chua, C.L., Collazo, R., Coltrin, M.E., Cooper, J.A., Evans, K.R., Graham, S., Grotjohn, T.A., Heller, E.R., Higashiwaki, M., Islam, M.S., Juodawlkis, P.W., Khan, M.A., Koehler, A.D., Leach, J.H., Mishra, U.K., Nemanich, R.J., Pilawa-Podgurski, R.C.N., Shealy, J.B., Sitar, Z., Tadjer, M.J., Witulski, A.F., Wraback, M., Simmons, J.A., 2018. Ultrawide-bandgap semiconductors: research opportunities and challenges. Adv. Electron. Mater. 4, 1600501.
Tsubouchi, N., Mokuno, Y., 2016. Configuration of a single grown-in dislocation corresponding to one etch pit formed on the surface of CVD homoepitaxial diamond. J. Cryst. Growth 455, 71–75.
Tsubouchi, N., Mokuno, Y., 2017. Microstructures of threading dislocation bundles included in CVD homoepitaxial diamond plates. Diamond Relat. Mater. 78, 44–48.
Tsubouchi, N., Ogura, M., Makino, T., 2019. Microstructures of dome-shaped hillocks formed on B doped CVD homoepitaxial diamond films. Diamond Relat. Mater. 97, 10742.
Turner, S., Shenderova, O., Da Pieve, F., Lu, Y.-G., Yücelen, E., Verbeeck, J., Lamoen, D., Van Tendeloo, G., 2013. Aberration-corrected microscopy and spectroscopy analysis of pristine, nitrogen containing detonation nanodiamond. Phys. Status Solidi A 210 (10), 1976–1984.
Turner, S., Idrissi, H., Sartori, A.F., Korneychuck, S., Lu, Y.-G., Verbeeck, J., Schreck, M., Van Tendeloo, G., 2016. Direct imaging of boron segregation at dislocations in B: diamond heteroepitaxial films. Nanoscale 8, 2212.

Tzou, Y., Bruley, J., Ernst, F., Rühle, M., Raj, R., 1994. TEM study of the structure and chemistry of a diamond/silicon interface. J. Mater. Res. 9, 1566.

van Tendeloo, G., van Dyck, D., Pennycook, S.J. (Eds.), 2012. Handbook of Nanoscopy. John Wiley & Sons, Inc.

Verhoeven, H., Flöter, A., Reiß, H., Zachai, R., Wittorf, D., Jäger, W., 1997. Influence of the microstructure on the thermal properties of polycrystalline thin diamond films. Appl. Phys. Lett. 71, 1324–1331.

Vlasov, I.I., Barnard, A.S., Ralchenko, V.G., Lebedev, O.I., Kanzyuba, M.V., Saveliev, A.V., Konov, V.I., Goovaerts, E., 2009. Nanodiamond photoemitters based on strong narrow-band luminescence from silicon-vacancy defects. Adv. Mater. 21, 808–812.

Vlasov, I.I., Shenderova, O., Turner, S., Lebedev, O.I., Basov, A.A., Sildos, I., Rähn, M., Shiryaev, A.A., Van Tendeloo, G., 2010. Nitrogen and luminescent nitrogen-vacancy defects in detonation nanodiamond. Small 6 (5), 687–694.

Vlasov, I.I., Turner, S., Van Tendeloo, G., Shiryaev, A.A., 2012. Recent results on characterization of detonation nanodiamonds. In: Ultrananocrystalline Diamond. Elsevier Inc, p. 291. Ch. 9.

Vlasov, I.I., Shiryaev, A.A., Rendler, T., Steinert, S., Lee, S.-Y., Antonov, D., Vörös, M., Jelezko, F., Fisenko, A.V., Semjonova, L.F., Biskupek, J., Kaiser, U., Lebedev, O.I., Sildos, I., Hemmer, P.R., Konov, V.I., Gali, A., Wrachtrup, J., 2014. Molecular-sized fluorescent nanodiamonds. Nat. Nanotechnol. 9, 54.

Wade, T.C., Coffey, D.W., Ghosh, N., Wittig, J.E., Kang, W.P., Allard, L.F., Unocic, K.A., Davidson, J.L., Tolk, N.H., 2012. Nanostructure TEM analysis of diamond cold cathode field emitters. Diamond Relat. Mater. 22, 29–32.

Wang, J., He, Z., Tan, X., Wang, T., Liu, L., He, X., Liu, X.D., Zhang, L., Du, K., 2020. High-performance 2.6 V aqueous symmetric supercapacitor based on porous boron-doped diamond via regrowth of diamond nanoparticles. Carbon 160, 71–79.

Werner, M., Locher, R., 1998. Growth and application of undoped and doped diamond films. Rep. Prog. Phys. 61, 1665–1710.

Werner, M., Krötz, G., Möller, H., Eickhoff, M., Gluche, P., Adamschik, M., Johnston, C., Chalker, P.R., 1999. High-temperature sensors based on SiC and diamond technology. Sensors Update 5 (1), 141–190.

Widmann, C.J., Müller-Sebert, W., Lang, N., Nebel, C.E., 2016. Homoepitaxial growth of single crystalline CVD-diamond. Diamond Relat. Mater. 64, 1–7.

Wild, C., Kohl, R., Herres, N., Müller-Sebert, W., Koidl, P., 1994. Oriented CVD diamond films: twin formation, structure and morphology. Diamond Relat. Mater. 3, 373.

Williams, O.A., 2011. Nanocrystalline diamond. Diamond Relat. Mater. 20, 621–640.

Williams, O. (Ed.), 2014. Nanodiamond. In: RSC Nanoscience & Nanotechnology No. 31. The Royal Society of Chemistry.

Williams, O.A., Hees, J., Dieker, C., Jäger, W., Kirste, L., Nebel, C.E., 2010. Size-dependent reactivity of diamond nanoparticles. ACS Nano 4 (8), 4824.

Wittorf, D., 1999. Structural Characterization of Interfaces and Grain Boundaries in cVD Diamond Films on Si Substrates by Eans of High-Resolution Methods of Transmission Electron Microscopy (in German). Dissertation D 82 RWTH Aachen, Research Centre Jülich Report, Germany. JÜL-3637.

Wittorf, D., Jäger, W., Urban, K., Gutheit, T., Güttler, H., Schulz, G., Zachai, R., 1997a. Microstructure and growth of MWCVD diamond on $Si_{1-x}C_x$ buffer layers. Diamond Relat. Mater. 6, 649–653.

Wittorf, D., Jia, C.L., Jäger, W., Grushko, B., Urban, K., Jiang, X., Paul, M., Klages, C.-P., 1997b. Grain boundaries in diamond films on Si(100). Mater. Res. Soc. Symp. Proc. 466, 27–32.

Wittorf, D., Jäger, W., Jia, C.L., Urban, K., Flöter, A., Güttler, H., Zachai, R., 1997c. Interfaces of CVD diamond films on silicon(001). Inst. Phys. Conf. Ser. 157, 451.

Wittorf, D., Jäger, W., Dieker, C., Flöter, A., Güttler, H., 2000. Electron microscopy of interfaces in chemical vapour deposition diamond films on silicon. Diamond Relat. Mater. 9, 1696–1702.

Wrachtrup, J., Jelezko, F., Grotz, B., McGuinness, L., 2013. Nitrogen-vacancy centers close to surfaces. MRS Bull. 38, 149.

Wurzinger, P., Pongratz, P., Gerber, J., Ehrhardt, H., 1996. Bias-enhanced diamond nucleation on silicon—a TEM study. Diamond Relat. Mater. 5, 326.

Wurzinger, P., Pongratz, P., Hartmann, P., Haubner, R., Lux, B., 1997. Investigation of the boron incorporation in polycrystalline CVD diamond films by TEM, EELS and Raman spectroscopy. Diamond Relat. Mater. 6, 763–768.

Yaita, J., Iwasaki, T., Hatano, M., 2018. Heteroepitaxy of diamond on SiC. In: Koizumi, S., Umezawa, H., Pernot, J., Suzuki, M. (Eds.), Power Electronics Device Applications of Diamond Semiconductors. Elsevier Ltd, p. 1. Ch. 1.5.

Yamada, H., Chayahara, A., Umezawa, H., Tsubouchi, N., Mokuno, Y., Shikata, S., 2012. Fabrication and fundamental characterizations of tiled clones of single-crystal diamond with 1-inch size. Diamond Relat. Mater. 24, 29–33.

Yamasaki, S., Gheeraert, E., Koide, Y., 2014. Doping and interface of homoepitaxial diamond for electronic applications. MRS Bull. 39, 499.

Yang, N. (Ed.), 2019. Topics in applied physics. In: Novel Aspects of Diamond—From Growth to Applications, second ed. vol. 121. Springer.

Yang, N., Foord, J.S., Jiang, X., 2016. Diamond electrochemistry at the nanoscale: a review. Carbon 99, 90.

Yang, N., Yu, S., Macpherson, J.V., Einaga, Y., Zhao, H., Zhao, G., Swain, G.M., Jiang, X., 2019. Conductive diamond: synthesis, properties, and electrochemical applications. Chem. Soc. Rev. 48, 157.

Yoshikawa, T., Kodama, H., Kono, S., Suzuki, K., Sawabe, A., 2015. Wafer bowing control of free-standing heteroepitaxial diamond (100) films grown on Ir (100) substrates via patterned nucleation growth. Thin Solid Films 594, 120–128.

Yoshikawa, T., Herrling, D., Meyer, F., Burmeister, F., Nebel, C.E., Ambacher, O., Lebedev, V., 2019. Influence of substrate holder configurations on bias enhanced nucleation area for diamond heteroepitaxy: toward wafer-scale single-crystalline diamond synthesis. J. Vac. Sci. Technol. B 37, 021207.

Yu, Y., Zhi, J., 2019. Diamond nanowires: theoretical simulation and experiments. In: Yang, N. (Ed.), Novel Aspects of Diamond. Topics in Applied Physics, vol. 121. Springer Nature, Switzerland AG, pp. 313–362. Ch. 10.

Zhai, Z., Huang, N., Yang, B., Liu, L., Li, H., Chen, J., Zhang, B., Jiang, X., 2020. In situ construction of hierarchical diamond supported on carbon nanowalls/diamond for enhanced electron field emission. ACS Appl. Mater. Interfaces 12, 8522–8532.

Zhang, L., Erni, R., Verbeeck, J., Van Tendeloo, G., 2008. Retrieving the dielectric function of diamond from valence electron energy-loss spectroscopy. Phys. Rev. B 77, 195119.

Zhang, H., Glenn, D.R., Schalek, R., Lichtman, J.W., Walsworth, R.L., 2017. Efficiency of cathodoluminescence emission by nitrogen-vacancy color centers in nanodiamonds. Small 13, 1700543.

Zhuang, H., Yang, N., Fu, H., Zhang, L., Wang, C., Huang, N., Jiang, X., 2015. Diamond network: template-free fabrication and properties. ACS Appl. Mater. Interfaces 7, 5384–5390.

CHAPTER THREE

Fundaments of photoelectric readout of spin states in diamond

Emilie Bourgeois[a,b], Michal Gulka[a,b,c], Daniel Wirtitsch[d,e], Petr Siyushev[f], Huijie Zheng[g,h], Jaroslav Hruby[a,b], Arne Wickenbrock[g,h], Dmitry Budker[g,h,i], Adam Gali[j,k], Michael Trupke[d,e], Fedor Jelezko[f], and Milos Nesladek[a,b,*]

[a]Hasselt University, Hasselt, Belgium
[b]IMOMEC Division, IMEC, Leuven, Belgium
[c]Faculty of Biomedical Engineering, Czech Technical University in Prague, Kladno, Czech Republic
[d]Faculty of Physics, University of Vienna, VCQ, Vienna, Austria
[e]Institute for Atomic and Subatomic Physics, Vienna University of Technology, VCQ, Vienna, Austria
[f]Institute for Quantum Optics, Ulm University, Ulm, Germany
[g]Helmholtz Institute Mainz, GSI Helmholtzzentrum für Schwerionenforschung, Mainz, Germany
[h]Johannes Gutenberg University Mainz, Mainz, Germany
[i]Department of Physics, University of California at Berkeley, Berkeley, CA, United States
[j]Wigner Research Centre for Physics, Budapest, Hungary
[k]Department of Atomic Physics, Budapest University of Technology and Economics, Budapest, Hungary
*Corresponding author: e-mail address: milos.nesladek@uhasselt.be

Contents

1. Introduction: Photoelectric detection techniques for quantum physics 106
2. Photoelectric readout of NV spin 108
   2.1 Photoelectric detection of NV magnetic resonances (PDMR): Physical principle 108
   2.2 Continuous-wave and pulsed PDMR experiments 113
   2.3 Influence of diamond material properties on PDMR performance 121
   2.4 Single NV spin photoelectric detection and applications 129
   2.5 Photoelectric readout of nuclear spins 132
3. PDMR for magnetic field sensing applications 134
   3.1 PDMR NV magnetometry 134
   3.2 Microwave-free magnetometry 138
4. Conclusions 141
Acknowledgments 141
References 141

Semiconductors and Semimetals, Volume 104
ISSN 0080-8784
https://doi.org/10.1016/bs.semsem.2020.08.001

1. Introduction: Photoelectric detection techniques for quantum physics

The photoelectric quantum effect, discovered by Einstein (1905) initiated research on photons as the fundamental quantum mechanical particles. Photons became also the essential component of photonics and optoelectronic devices. For this reason, photon detection techniques are at the heart of quantum technologies. The detection of photons, for example for spin-state readout, is carried out by photodetectors, such as avalanche photodiodes (APD). For applications of solid-state qubits in wide-bandgap semiconductors, such as nitrogen-vacancy (NV) centers in diamond, the photon detection provides information about the spin transitions, using the so-called optical detection of magnetic resonances (ODMR) technique. However, in many instances, the solid-state semiconducting material itself can be actively used as a photon detector. This brought us to the idea (Bourgeois et al., 2015) of constructing quantum devices in which the spin readout is performed through the detection of free charge carriers resulting from optically promoted electronic transitions. These free charges are transported in the semiconductor device and are finally collected on electrodes. This idea gave birth to a technique, used in parallel with ODMR, called photoelectric detection of magnetic resonances (PDMR) (Bourgeois et al., 2015, 2017, 2020; Gulka et al., 2017; Hrubesch et al., 2017; Siyushev et al., 2019). Compared to ODMR, PDMR brings potential benefits for combining quantum devices with electronic peripheries. Such solid-state devices have a significant application potential in quantum sensing but also in quantum information science.

While the first realizations of quantum computing devices, for example, by Google (Arute et al., 2019) or IBM (MIT Technology Review, 2020), have been based on superconducting qubits or trapped ions, solid-state device technology is expected to take over this trend and realize quantum devices in an electronics-compatible manner and on a large scale. The use of point defects, made of impurity atoms or complex of atoms in a solid-state matrix, constitutes the ultimate step in the ongoing miniaturization of electronic and optoelectronic semiconductor devices (Bassett et al., 2019). Point defects may introduce levels in the fundamental bandgap of semiconductors or insulators that radically change the optical and magnetic properties of the host material. In particular, these point defects can be paramagnetic, i.e., have an electron spin greater than zero, which makes them potential solid-state qubits. In this scenario, the hosting material can take up an active role to interface the qubits to classical electronic devices, for example, by electronic charge transport. Several important material systems such as P donors in silicon or quantum dots (Zheng et al., 2019a; Zwanenburg et al., 2013) are being studied as qubits; however, their operation usually requires cooling to mK temperatures.

One of the best studied solid-state materials in which single atomic impurities are incorporated to serve as qubits is synthetic diamond (Awschalom et al., 2018). Contrary to other materials, diamond permits operation at room temperature, and even higher (Maurer et al., 2012). A prototypical example of diamond point defect qubit is the NV color center. This point defect consists of a nitrogen atom substituting a carbon atom adjacent to a missing carbon atom (a vacancy) in the diamond crystal. The optical readout of NV centers' spin state led to breakthrough discoveries such as the verification of a loophole-free Bell-inequality violation (Hensen et al., 2015), time crystals (Choi et al., 2017; Ho et al., 2017), quantum NMR (Aslam et al., 2017; Mamin et al., 2013), quantum magnetometry (Barry et al., 2020; Rondin et al., 2014), nanoscale imaging in spintronics (Chang et al., 2017) or entanglement in solid-state (Dolde et al., 2013; Reiserer et al., 2016), etc. The advantage of diamond as host material with respect to other solid-state quantum systems is that it enables vast electronic spin coherence time, above 2 ms at room temperature (Balasubramanian et al., 2009) and approaching 1 s at 77 K (Bar-Gill et al., 2013). These values must be compared to superconducting qubits, whose coherence time does not exceed several microseconds at temperatures below 25 mK (Wendin, 2017). Diamond lattice can provide naturally occurring nuclear spins such as ^{13}C, that have been used for fabrication of small quantum processors. Up to 20 nuclear spins can be coupled to a single electron spin, forming a quantum node (Abobeih et al., 2019). Apart from the NV center, also other promising defects are being studied, such as the group IV-(di)vacancy complexes (SiV, GeV, SnV, PbV, etc.) or the NiV in diamond, the divacancy in silicon carbide, rare-earth impurities in complex oxides, and others (Bassett et al., 2019).

One of the limiting factors for the optical detection of NV spin is the low collection efficiency of photons emitted by NV centers in bulk diamond, due to the limitations of objective optics and to the high index of refraction of diamond. Additionally, the use of NV centers for quantum computing is hindered by the magnetic dipole interaction strength, leading to a maximal entanglement distance between the spins of adjacent NV centers of the order of $\sim$30 nm (Dolde et al., 2013). The individual optical readout of such proximal NV centers requires a resolution below the diffraction limit, which can only be achieved by complex optical techniques with a very limited

Fig. 1 Prototypical sketch of a diamond NV center computer based on photoelectric readout of NV magnetic resonance (PDMR), in which spin qubits driven by electric and microwave gates interact by magnetic dipolar coupling. Electric charges are collected by a nanoscale diamond electronic device.

scalability. The alternative PDMR technique that we have recently developed leads to significantly (~250 times) higher detection rates than ODMR (Siyushev et al., 2019). A related electrical spin readout method was also described in Niethammer et al. (2019) where it was applied to the Si vacancy in SiC. By fabricating nanoscale devices of a size corresponding to the dipolar coupling length, entanglement of proximal NV centers might be achievable, bringing a perspective for scaling up and integrating the qubits with classical electronics (Fig. 1). The NV and SiV centers are mainly used in quantum applications due to their large photoluminescence (PL) yield. However, for PDMR, the radiative transitions are not needed and, on the contrary, reduce the quantum efficiency of the electron photoionization. Purely non-radiative electronic transitions showing spin polarization would thus be ideal for PDMR. Although till now demonstrated only for NV centers, photoelectric readout approach can be extended to non-conventional spin centers such as SiV, GeV, NiV in diamond and others.

2. Photoelectric readout of NV spin

2.1 Photoelectric detection of NV magnetic resonances (PDMR): Physical principle

2.1.1 NV ionization mechanisms

In PDMR, the spin state of NV^- ground state is read out through a detection of the electric signal resulting from NV^- photoionization (Fig. 2A).

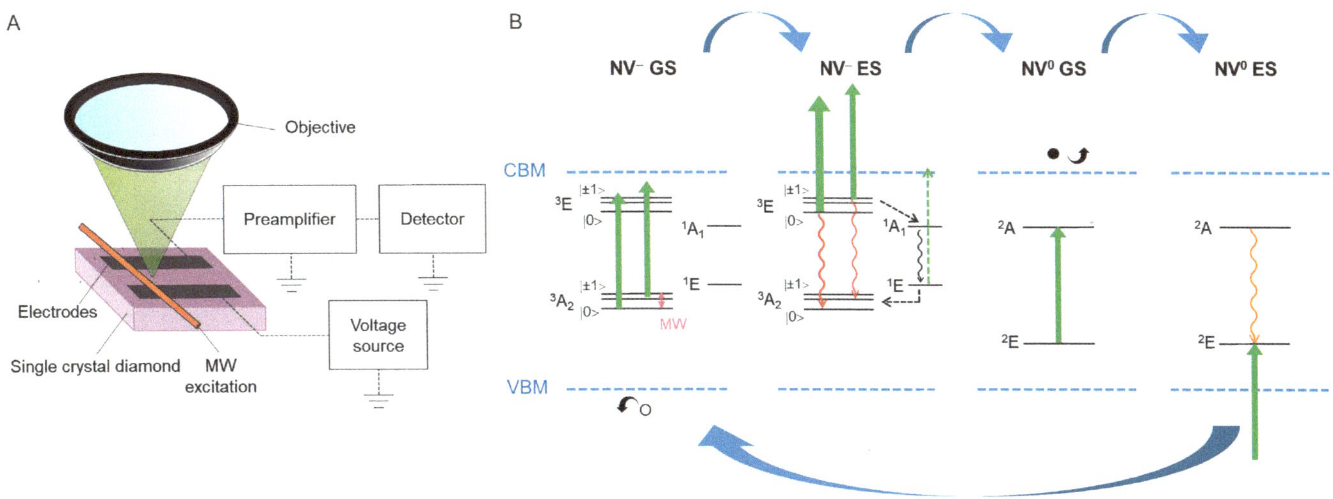

Fig. 2 (A) Simplified representation of a PDMR setup. A laser beam is focused on a single-crystal diamond plate containing ensembles of NV centers or single NV centers. Coplanar electrodes are deposited on one of the faces of the diamond plate. Under the effect of an external bias electric field, free charge carriers resulting from ionization of NV centers are collected in an external circuit enabling the photocurrent amplification and detection. A microwave wire is used for application of the microwave field. (B) Schematic representation of NV^- electronic structure and of the ODMR and PDMR mechanisms, under green excitation leading to NV two-photon ionization. Pink double-sided arrow: resonant microwave field used to manipulate the spin sublevels of NV^- triplet ground state (GS) $^3A_2$. Full green arrows: absorption of photons, promoting electrons first from $^3A_2$ to the triplet excited state (ES) ^3E of NV^- and then from this ES to above the conduction-band minimum (CBM). This second step leads to the ionization of NV^- and to the generation of free electrons detected in the PDMR technique. Red wavy arrows: photoluminescent transitions between the excited and ground triplet states ^3E and $^3A_2$, detected in ODMR. Dashed black arrows: spin-dependent ISC transitions from ^3E to the upper singlet state $^1A_1$, and from the lower metastable singlet state ^1E to $^3A_2$. Black wavy arrow: decay between the upper and lower singlet states. The spin dependence of these ISC transitions enables the spin contrast observed in photoluminescence intensity (ODMR) and in photocurrent (PDMR). The physical principles of ODMR and PDMR are further discussed in the text. Dashed green arrow: potential ionization from the lower singlet state of NV^-, discussed in the main text. The back-conversion from NV^0 to NV^- also occurs by a two-photon process (illustrated in the last two sections of the figure), in which an electron is first excited from the doublet ground state ^2E to the doublet excited state ^2A of NV^0 while a second photon promotes an electron from below the valence band maximum (VBM) to the vacated orbital of NV^0. This back conversion leads to formation of a free hole in the valence band, which contributes to the detected photoelectric signal. Wavy orange arrow: photoluminescent decay from the ES to the GS of NV^0.

The ionization of NV^- leads to its conversion into NV neutral state (NV^0). The back-conversion of NV^0 to NV^- is therefore required to ensure the stability of the process. The mechanisms leading to NV^- ionization and to the back-conversion from NV^0 to NV^- are described in this section.

Photochromism between NV^0 and NV^-: In its neutral state, the NV center contains five electrons (two coming from the nitrogen doublet and three from the dangling bonds of carbon atoms around the vacancy). The repartition of these electrons in the different molecular orbitals of NV^0 gives rise to a ground and an excited spin doublet states noted ^2E and ^2A, respectively (Fig. 2B) (Gali, 2009). Acquisition of an additional electron leads to the formation of NV^-, containing six electrons. The main properties of NV^- can be described by considering four multi-electron states: two spin-triplet states (ground state $^3A_2$ and excited state ^3E) and two spin-singlet states (lower state ^1E and a higher-lying state $^1A_1$) (Fig. 2B). Spin-conserving optical transitions (photoluminescence and absorption) occur between states of identical spin multiplicity. Photochromism between NV^0 and NV^- has been demonstrated, both on NV ensembles (Iakoubovskii et al., 1999, 2000; Manson and Harrison, 2005; Steeds et al., 2000; Waldermann et al., 2007) and on single NV centers (Gaebel et al., 2006). This phenomenon was first attributed to the ionization of dominant donor (for example, the substitutional nitrogen N_s) or acceptor defects, followed by the capture of free electrons or holes by NV centers (Gaebel et al., 2006; Iakoubovskii et al., 2000). However, the existence of direct photo-induced conversion between NV^- and NV^0 was later established.

NV^- ionization mechanism: Depending on the light wavelength, NV^- photoionization can occur via a one-photon or a two-photon process (Aslam et al., 2013; Chen et al., 2013). Ab-initio calculation, as well as experimental studies of the wavelength-dependence of NV ionization dynamics, demonstrate that a minimum photon energy around 2.7 eV is required to directly transfer an electron from the triplet ground state of NV^- to the diamond conduction band (Aslam et al., 2013; Bourgeois et al., 2017; Deák et al., 2014).

Photons with an energy below this threshold can still induce ionization of NV^- but the absorption of two photons is then necessary (Beha et al., 2012; Waldermann et al., 2007), as precisely characterized by the experimental study of single-charge transition events on single NV centers (Aslam et al., 2013; Siyushev et al., 2013). In this case, absorption of the first photon promotes an electron from the ground triplet state $^3A_2$ of NV^- to its triplet excited state ^3E (Fig. 2B). The decay of electrons from the excited to the ground triplet state leads to emission of photoluminescence light (characterized by a zero-phonon line—ZPL—at 636 nm (Walker, 1979) and a broad phonon side-band at longer wavelengths (Santori et al., 2010)) used for optical readout of NV spins (Gruber et al., 2012; van Oort et al., 1988). However, if a second photon is absorbed while NV^- is still in the triplet excited state, the electron is transferred to the conduction band, which results in the appearance of a free electron and leaves the NV center in its neutral form. A precise description of NV^- two-photon ionization mechanism in terms of molecular orbitals can be found in Siyushev et al. (2013).

The minimum photon energy required to induce two-photon ionization of NV^- corresponds to NV^- ZPL (1.95 eV). However, at room temperature, the proportion of the absorption occurring at the ZPL is low and the maximum of NV^- absorption is located in the absorption phonon sideband (around 2.2 eV). In most PDMR experiments, photoionization of NV^- is therefore performed using off-resonant green light (wavelength of about 532 nm, photon energy of about 2.33 eV)—close to the maximum of NV^- absorption (Beha et al., 2012). While two photons of the same energy can be used to ionize NV^-, both ab-initio modeling (Siyushev et al., 2013) and experiments (Chen et al., 2017c; Hacquebard and Childress, 2018; Meirzada et al., 2018; Shields et al., 2015) demonstrated that the second step of NV^- ionization (transition from the triplet excited state of NV^- to the conduction band) could be induced by photons of lower energy. NV^- can thus be ionized using a two-color scheme, in which off-resonant light is used to excite NV^- whereas red (Shields et al., 2015) or near-infrared (Chen et al., 2017c; Hacquebard and Childress, 2018; Meirzada et al., 2018) photons are used to induce the second ionization step.

Back-conversion from NV^0 to NV^-: In the same way, the recovery of NV negative charge state can be induced by a one-photon or a two-photon process. It was theoretically shown that an energy of approximately 2.8 eV enables direct transfer of an electron from the valence band edge to an empty molecular orbital of NV^0 (Bourgeois et al., 2017). For the two-photon back-conversion from NV^0 to NV^-, NV^0 is excited by absorption of a first photon, while a second photon promotes an electron from the valence band to the vacated orbital of NV^0 in the bandgap. This process leads to reformation of NV^- and creation of a free hole in the valence band (Siyushev et al., 2013). NV^- two-photon charge recovery requires a minimum photon energy corresponding to the ZPL of NV^0 at 2.16 eV. However, like in the case of NV^- ionization, at room temperature it is more efficiently induced by off-resonant light with wavelength close to the maximum of

NV^0 absorption phonon side-band. Several experimental studies have demonstrated that the second step of the back-conversion from NV^0 to NV^- (promotion of an electron from the valence band to NV^0 excited state) could be induced by near infrared light (Hopper et al., 2016; Ji and Dutt, 2016; Meirzada et al., 2018).

2.1.2 Spin dependence of NV center ionization dynamics

Spin non-conserving intersystem-crossing (ISC) transitions occur from the excited triplet state of NV^- ^3E to the higher singlet state $^1A_1$ and from the lower singlet state ^1E to the ground triplet state $^3A_2$ (Goldman et al., 2015; Thiering and Gali, 2018). Spin-selectivity of these transitions enables the optical polarization of the spin into the $m_s=0$ sublevel of NV^- ground triplet state, as well as the optical and photoelectric readout of NV^- spin. Electrons in the $m_s=\pm1$ spin sublevels of NV^- triplet excited state ^3E have a probability of approximately one third to decay non-radiatively to the higher singlet level $^1A_1$ (Doherty et al., 2013), while this probability is only a few percent in the case of electrons with an $m_s=0$ spin state (Goldman et al., 2015; Robledo et al., 2011; Tetienne et al., 2012). From $^1A_1$ (relaxation time <100 ps at room temperature (Ulbricht and Loh, 2018)), electrons rapidly decay to the singlet metastable state ^1E. They remain in this state for an average time of approximately 200 ns at room temperature (Acosta et al., 2010; Robledo et al., 2011), before undergoing a transition to the triplet ground state $^3A_2$ (see Fig. 2B). This non-radiative decay path competes with radiative transitions between the triplet excited and ground states. The $m_s=0$ state thus appears brighter in photoluminescence than the $m_s=\pm1$ states. This spin-dependence of the intensity of the photoluminescent light emitted by NV^- under illumination is used to optically determine NV^- spin state (for example, in ODMR).

By affecting the population of the triplet excited state, the spin-selective ISC transitions to the singlet state also impact the rate of electronic transitions from the ^3E level to the conduction band (Chen et al., 2013; Hopper et al., 2016; Shields et al., 2015). The rate of NV^- two-photon ionization is therefore lower for electrons in the $m_s=\pm1$ spin sublevel than for electrons in the $m_s=0$ sublevel. The spin dependence of NV^- ionization dynamics can be detected by a purely optical technique called spin-to-charge conversion (SCC) (Hopper et al., 2016, 2018a,b; Shields et al., 2015). In PDMR, it is sensed through direct detection of the spin-dependent photocurrent resulting from ionization. In this technique, the diamond itself is used as a detector, avoiding the necessity of conversion from an optical to an electric signal.

An alternative mechanism for the spin-dependent ionization of NV^-, based on transition of an electron from NV^- lower singlet state ^1E to the conduction band (represented by a dashed green arrow in Fig. 2B), was described in reference (Hopper et al., 2016). The 2.33 eV green excitation used in most continuous-wave PDMR (CW-PDMR) experiments could in principle induce one-photon transitions from ^1E to the conduction band, since this state was theoretically predicted to be located approximately 0.4 eV above the ground triplet state of NV^-, i.e., approximately 2.3 eV below the bottom of the conduction band (Thiering and Gali, 2018). However, the rate of NV^- ionization via this mechanism is expected to be higher for electrons in the $m_s = \pm 1$ spin state than for electrons in the $m_s = 0$ spin state (Hopper et al., 2016) and would therefore lead to an increase in the photoelectric signal resulting from NV ionization under application of a resonant microwave field. Since the magnetic resonances experimentally detected in photocurrent are negative on most samples (see Section 2.2.3), it seems that, for green light, the cross-section of transitions from the triplet excited state strongly exceeds the cross-section of transitions from the lower singlet state, and that the detected photocurrent predominantly results from the first type of transitions.

ISC transitions from the lower singlet state ^1E to the triplet ground state $^3A_2$ are also spin-dependent (higher transition rate to the $m_s = 0$ spin sublevel of $^3A_2$ than to the $m_s = \pm 1$ sublevels) (Kalb et al., 2018; Thiering and Gali, 2018), contributing to the initialization of NV^- spin into the $m_s = 0$ spin sublevel of the ground state under illumination.

2.2 Continuous-wave and pulsed PDMR experiments

2.2.1 Experimental implementation of PDMR

A simplified representation of a PDMR set-up is shown in Fig. 2A. PDMR is based on detection of the photocurrent resulting from NV centers' ionization. For this detection, electrodes have to be deposited on one of the faces of a single-crystal diamond plate, by standard lift-off optical lithography (Bourgeois et al., 2015; Siyushev et al., 2019) or by electron-beam lithography (Morishita et al., 2020) enabling to reach inter-electrode distances below 10 nm (Manfrinato et al., 2013). Before electrode deposition, the sample surface is cleaned and oxidized using an acidic mixture (Hrubesch et al., 2017; Siyushev et al., 2019) or an oxygen plasma, to avoid surface conduction. The contacts consist of a metal stack made of an interfacial layer (typically a thin sputtered titanium layer), topped by a protecting layer (aluminium or gold) also used for wire-bonding purposes. Annealing can be used to form a nm-thin titanium carbide (TiC) layer at the titanium/diamond interface, enabling formation of ohmic contacts. Electrodes are then wire-bonded to the tracks of a printed circuit board (PCB) or connected to the external circuitry using a flip-chip technique. To apply the microwave field utilized for spin manipulations, the sample can be equipped with a metal wire (~25–50 μm diameter, insulated from the readout electrodes by a polymer coating) spanned over the diamond surface. Alternatively, a strip-line or resonant microwave structures (printed on the PCB or directly on the diamond surface) can be used for this purpose. Light is focused in between electrodes to induce the ionization of NV centers. Since in PDMR the photon collection is not important, a simpler light focusing geometry can be used than in ODMR [classical lens instead of a high-numerical aperture (N.A.) objective], in particular for spin ensembles. However, for single NV measurements, a high N.A. objective is still required to locate the NV centers.

To drive the photo-generated charge carriers toward the contacts, a bias voltage is applied between the electrodes. The range of the applied voltage depends on the inter-electrode gap and on the current-voltage (I-V) characteristic of the contacts. Electric field magnitudes in the range from 2×10^4 to $5 \times 10^4\,\text{V cm}^{-1}$ are typically used to bias the photocurrent flow. The amplitude of the detected photocurrent depends on the number of NV centers in the focus, laser power, type of objective, applied voltage, I-V characteristic of the sample, etc. In general, the current from a single NV can range from 100 fA to about 2 pA (Gulka et al., 2017; Siyushev et al., 2019). Therefore, signal preamplification is necessary. A typical gain of 10^{11}–10^{12} V/A (Siyushev et al., 2019) is required for single NV, while for NV ensembles lower gains of 10^7–10^9 V/A can be sufficient (Bourgeois et al., 2015, 2017; Gulka et al., 2017; Hrubesch et al., 2017). The amplified DC signal can be monitored directly with a suitable ampere meter or data acquisition card. Alternatively, the NV-related photocurrent can be measured by lock-in detection. In this case, the signal is modulated (typically, at frequencies of tens of Hz to kHz) by pulsing the laser or the microwave field. This frequency is then provided to the lock-in amplifier as a reference. The lock-in technique has the advantage of filtering out asynchronous (including out-of-phase) noise (Gulka et al., 2017; Mandelis, 1994). Issues related to the fast detection of photocurrent in pulsed PDMR experiments are discussed in Section 2.2.4.

2.2.2 Photocurrent detection

Free electrons resulting from photoionization of NV^- and free holes resulting from back-conversion from NV^0 to NV^- can produce a photocurrent measurable in an external circuit, in case the diamond is equipped with electrodes and a DC electric field is applied between these contacts (see Section 2.2.1). Considering the simple case in which the electric field and the density of charge carriers in the material are uniform, the photocurrent density j ($\mathrm{A\,cm^{-2}}$) at the steady-state can be expressed as (Holovsky, 2011):

$$j = e(\mu_h \tau_h G_h + \mu_e \tau_e G_e)F, \tag{1}$$

where μ is the charge carrier mobility ($\mathrm{cm^2\,V^{-1}\,s^{-1}}$), τ is the lifetime of charge carriers (s), G is the charge-carrier generation rate per unit volume ($\mathrm{cm^{-3}\,s^{-1}}$), e is the elementary charge (C), and F is the applied electric field ($\mathrm{V\,cm^{-1}}$). The subscript h refers to holes while e refers to electrons. The detected photocurrent depends therefore on one hand on the charge-carrier generation rate, defined by the density of color centers in the material, the volume of the material addressed by the photoelectric detection, the illumination power and the optical cross-sections of the different electronic transitions involved in the ionization process at the used illumination wavelength. It is, on the other hand, affected by the electronic properties of the material, influencing the charge-carrier transport and recombination (see Section 2.3.2).

In the case of dispersed single NV centers separated by a distance above the diameter of the laser spot, photoelectric mapping enables to isolate the signal resulting from ionization of a single defect (see Section 2.4). However, when PDMR measurements are performed on a diamond sample presenting more than one defect in the fraction of the illumination cone contributing to the photocurrent, the individual contribution of NV centers to the measured signal cannot be simply separated from the background signal resulting from the ionization of defects other than NV centers in the diamond crystal. The nature of these defects and their role in the PDMR process are discussed in Section 2.3.1.

2.2.3 Continuous-wave PDMR

Due to electron spin-spin interactions, the $m_s = 0$ and $m_s = \pm 1$ spin sublevels of NV^- triplet ground state $^3A_2$ are separated by a zero-field splitting of 2.87 GHz. The spin state of NV^- ground state can, therefore, be coherently manipulated by the application of a microwave field at the resonant frequency. CW-PDMR is used to photoelectrically detect the electron spin resonances of NV centers. In these experiments, the sample is illuminated by continuous laser light—or by low-frequency pulsed light in the case of lock-in detection of the photocurrent—while applying a CW microwave field to the sample. The photocurrent resulting from NV ionization is measured as a function of the microwave frequency. At resonant microwave frequencies (leading to repopulation of the $m_s = \pm 1$ spin sublevels of NV^- ground state), dips are observed both in photoluminescence and in the photocurrent signal (Fig. 3). The lower photocurrent detected under resonant microwave field reflects the lower rate of NV^- ionization for electrons in the $m_s = \pm 1$ spin state compared to electrons in the $m_s = 0$ spin state (see Section 2.1.2).

The CW-PDMR technique was demonstrated on a variety of samples containing NV ensembles, ranging from large ensembles of NV centers in electron-irradiated type-Ib and type-IIa diamond (Bourgeois et al., 2015) or in as-grown CVD diamond, to small ensembles of 5–10 shallow NV centers implanted close to the surface of electronic grade diamond (Bourgeois et al., 2017; Gulka et al., 2017) (Fig. 3). It was also downscaled to a single NV center in HPHT electronic grade diamond (see Section 2.4). As in the case of CW-ODMR, the linewidth of photoelectrically detected magnetic resonances is determined by the spin-coherence properties of the NV-diamond material and—in the case of NV ensembles—by broadening originating from variations in the local environment of the different NV centers contributing to the signal in terms of strain, intrinsic electric fields and paramagnetic spin bath (Jensen et al., 2013; Manson et al., 2018; Mittiga et al., 2018). For a single NV defect or a thin layer of NV centers, i.e., in situations where the same spatially located NV ensembles contribute to the photoelectric and to the photoluminescence signals, no significant difference is observed between the linewidths extracted from CW-ODMR and CW-PDMR spectra (Gulka et al., 2017; Siyushev et al., 2019).

The electric field required to drive the free charge carriers resulting from NV ionization toward electrodes in PDMR experiments affects the fine structure of NV^- ground state spin sublevels, due to the Stark effect (Doherty et al., 2012). In the case of bulk NV ensembles in a diamond equipped with electrodes separated by a large distance, inhomogeneities in the electric field across the volume contributing to the photocurrent may lead to a broadening of PDMR resonances (Bourgeois et al., 2015). However, in the case of single NV centers or NV ensembles forming a narrow layer close to the diamond surface (i.e., located within the region of the material in which the electric field is uniform), the Stark effect should not limit NV sensing performance provided a highly stable electric field is applied.

We note that with NV ensembles, the CW-PDMR contrast is generally lower than the CW-ODMR contrast measured in the same conditions, and

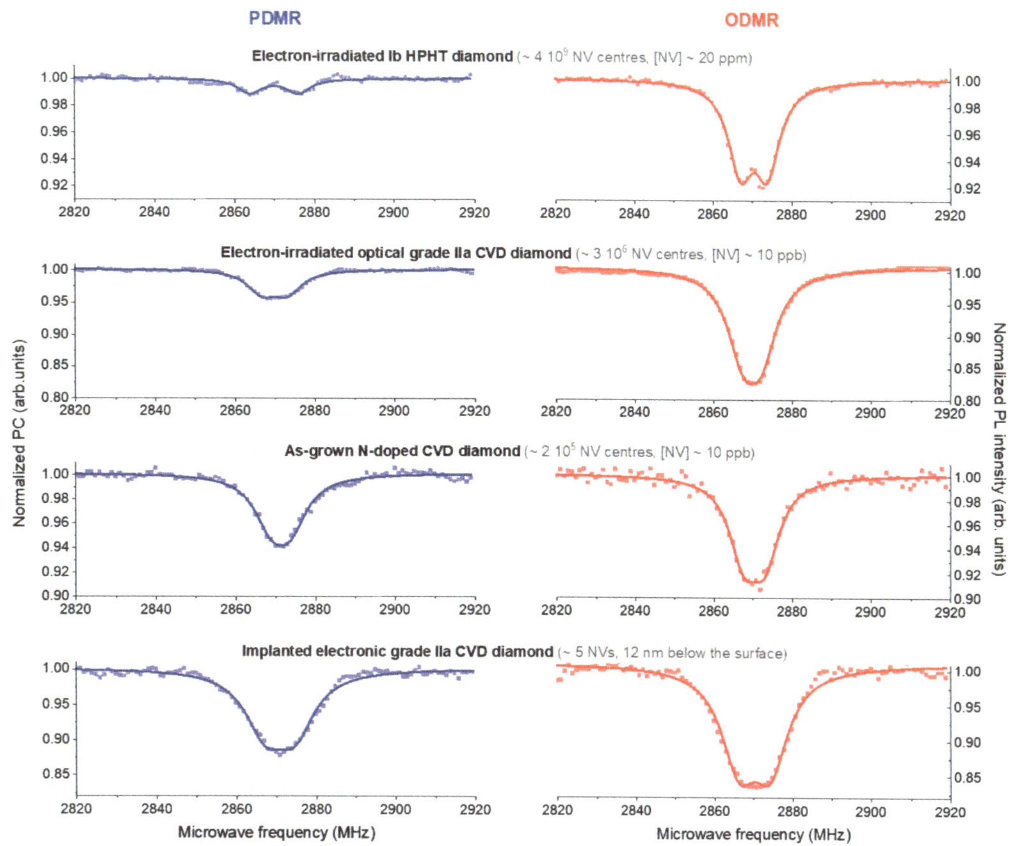

Fig. 3 Normalized CW-PDMR and CW-ODMR spectra measured simultaneously on various samples containing NV ensembles, under 532 nm green illumination and in the absence of an external magnetic field. Photocurrent detected by lock-in amplification, referenced to the green light pulsing frequency. Objective N.A.: 0.95. Electron-irradiated and annealed type-Ib HPHT diamond: electric field: 5×10^4 V cm^{-1}, inter-electrode distance: 15 μm, light pulsing frequency: 131 Hz, light power: 3 mW. Electron-irradiated and annealed type-IIa optical grade CVD diamond: electric field: 3.3×10^4 V cm^{-1}, inter-electrode distance: 15 μm, light pulsing frequency: 131 Hz, light power: 4 mW. As-grown N-doped CVD diamond: electric field: 3×10^4 V cm^{-1}, inter-electrode distance: 5 μm, light pulsing frequency: 131 Hz, light power: 10 mW. Implanted electronic grade type-IIa CVD diamond (implantation fluence $<10^{10}$ cm^2): electric field: 2×10^4 V cm^{-1}, inter-electrode distance: 50 μm, light pulsing frequency: 532 Hz, light power: 3.4 mW. Symbols: experimental data, solid lines: fitting with the sum of two Lorentzian functions.

that the difference between the ODMR and PDMR contrast is sample-dependent (Fig. 3). The origin of this difference in contrast, as well as possible technical and physical solutions to maximize the CW-PDMR contrast, are presented in Section 2.3.1.

2.2.4 Pulsed PDMR

The use of PDMR for quantum sensing and quantum computing applications implies coherent control of NV^- spins using pulsed protocols, combined with a pulsed electrical readout of the spin state. These experiments require application of sequences of laser pulses—used for spin initialization and readout—and microwave pulses—used for spin manipulations.

In pulsed ODMR, the existence of fast and sensitive techniques for photons detection enables determination of the spin projection by selectively collecting photons in the fraction of the readout laser pulses during which the spin contrast is maximal (typically for a duration of a few hundreds of ns) (Gupta et al., 2015), while the total duration of the laser pulse is usually much longer (up to a few μs) in order to re-initialize fully the NV^- spin back to the $m_s = 0$ state. However, the time selectivity (single electron counting) of the signal detection has not been realized in PDMR yet, due to the limited bandwidth of ultra-sensitive amplifiers used for detection of currents in the fA to pA range. For this reason, in pulsed PDMR experiments the laser pulses should be kept as short as possible—not to decrease the measurement contrast, but long enough to re-initialize the spin (Gulka et al., 2017; Hrubesch et al., 2017; Siyushev et al., 2019). An improvement could be achieved by performing electron counting, as it is done for electric spin-state readout in quantum dots (Vink et al., 2007), enabling in this way selective readout of the part of the signal with the highest spin contrast. However, amplifiers used for such detection need to be cooled down to cryogenic temperatures, which makes their use difficult for room temperature NV experiments. Some room temperature solutions however exist (Lany et al., 2008).

Photoelectric detection of coherently controlled NV spins was so far achieved using a DC detection—i.e., a continuous detection of the photocurrent while repeating the pulse sequences (Hrubesch et al., 2017), or lock-in detection of the photocurrent (Gulka et al., 2017; Siyushev et al., 2019). The pulsed PDMR experiments performed by Hrubesch et al. on NV ensembles (Hrubesch et al., 2017) demonstrated the suitability of DC detection for electrical readout of coherently manipulated NV^- spin ensembles. A direct DC readout of the photocurrent induced by coherently driven NV^- spins enabled to perform pulsed PDMR measurements, Rabi-oscillation detection and spin-echo experiments on NV spin ensembles. Single NV pulsed PDMR has however not been accomplished as yet using DC detection of photocurrent.

To achieve the sensitivity needed for a single NV spin detection, we used lock-in readout that integrates fast pulses in a low-frequency envelope. In this case, a low-frequency envelope (tens of Hz to kHz) is used as a modulation for the laser or for the microwave pulses. The frequency of the envelope is used as a reference for the lock-in amplifier. Pulse sequences are then repeated many times under one low-frequency pulse of the envelope. Due to the low bandwidth of the preamplifier, lock-in detection does not resolve the high-frequency (close to MHz range) current oscillations from laser pulses and records only the photocurrent signal modulated by the slow envelope. Using this method, combined with microwave-triggered detection of the PDMR signal (see Section 2.3.1), we achieved the detection of pulsed PDMR spectra (obtained by alternating laser pulses and microwave π-pulses while sweeping the microwave frequency) on small ensembles of shallow NV centers implanted in electronic grade diamond (Fig. 4A). The detection of Rabi oscillations, performed using the envelope technique on an ensemble of ~1000 NV centers implanted in the same sample, is shown in Fig. 3B. For these measurements, laser pulses are applied between resonant microwave pulses of variable durations. The oscillations observed when plotting the photocurrent as a function of the microwave-pulse duration reflect the coherent driving of NV spins between the $m_s = 0$ and $m_s = 1$ sublevels of NV ground state. In principle, most of the standard pulse sequences can be encoded into such low-frequency envelope. By this technique, we were able to achieve the detection of a single coherently manipulated NV center spin (Siyushev et al., 2019), as described in Section 2.4.

One of the advantages of pulsed PDMR compared to CW-PDMR is noise reduction. This reduction mostly concerns the microwave cross-talk noise occurring at the chip level. It has been shown that when the microwave excitation is pulsed (therefore not applied at the same time as the current readout), the microwave correlated noise can be reduced by more than one order of magnitude (Gulka et al., 2017). Additionally, low-noise pulsed schemes reducing the coupling to microwave [or radiofrequency (RF)] field used for spin manipulation can be employed (Dreher et al., 2015; Hoehne et al., 2012, 2013, 2017; Morishita et al., 2020). Like in ODMR (Dréau et al., 2011; Jensen et al., 2013), additional benefits of pulsed PDMR include avoiding broadening of magnetic resonance lines (induced by the simultaneous application of the microwave field and laser illumination in CW experiments) and contrast enhancement enabled by the application of laser

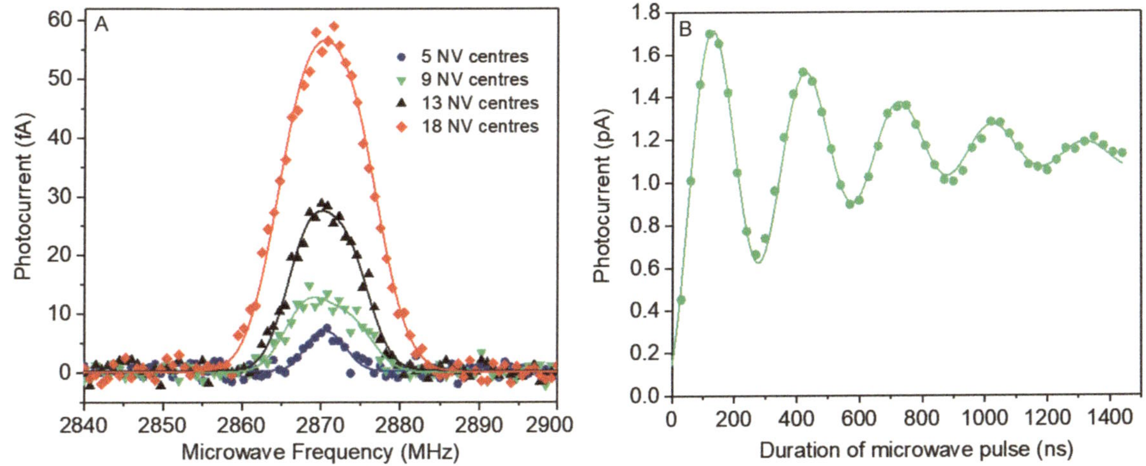

Fig. 4 Pulsed PDMR measurements on ensembles of shallow NV centers (12 ± 4 nm below the diamond surface) implanted in type-IIa single-crystal electronic grade CVD diamond (8-keV $^{14}N^{4+}$ ions, annealing at 900 °C). Green illumination: 9 mW, 532 nm, objective N.A.: 0.95, envelope frequency: 37 Hz, inter-electrode distance: 50 µm, electric field: 2×10^4 V cm^{-1}. Microwave-triggered lock-in detection of the photocurrent (see explanation in Section 2.3.1), using the envelope technique described in the text. Symbols: experimental data. Solid line: fitting using the functions described in Gulka et al., (2017). (A) Pulsed PDMR on small ensembles of 5–18 NV centers. (B) Photoelectric detection of Rabi oscillations on an ensemble of about 1000 NV centers. *Adapted from Gulka, M., Bourgeois, E., Hruby, J., Siyushev, P., Wachter, G., Aumayr, F., Hemmer, P. R., Gali, A., Jelezko, F., Trupke, M., Nesladek, M., 2017. Phys. Rev. Appl. 7, 044032. doi:10.1103/PhysRevApplied.7.044032. Copyright © 2017 American Physical Society.*

and microwave pulses separated in time. Further, the study of pulsed PDMR contrast (Hrubesch et al., 2017) on NV ensembles suggests that additional enhancement of the contrast can be obtained using two laser pulses, with one high-power laser pulse used for NV ionization and one low-power laser pulse for spin initialization.

2.3 Influence of diamond material properties on PDMR performance

2.3.1 Limitation of PDMR contrast by background photocurrent

Influence of the background photocurrent on the PDMR contrast: The presence of optoelectrically active defects other than NV centers limits the PDMR contrast, by resulting in the formation of a background photocurrent. While the photoluminescence detection enables the spectral selection of the signal resulting from ionization of NV^-, in the photoelectric detection such selection is not possible and all the defects ionized at the used laser wavelength contribute to free charge-carrier generation (Bourgeois et al., 2015; Chen et al., 2017b). The existence of this background photocurrent, usually non-affected by the microwave field, tends to reduce the PDMR contrast compared to the ODMR contrast, as can be observed in Fig. 3. The ratio $R_{PDMR/ODMR}$ between the maximal CW-PDMR and CW-ODMR contrasts obtainable under green illumination varies from sample to sample, indicating that this parameter strongly depends on the diamond composition.

To quantify this phenomenon, for different samples containing NV ensembles (prepared by irradiation and annealing of N-doped diamond or by nitrogen-ion implantation) the ratio $R_{PDMR/ODMR}$ was plotted as a function of the ratio $R_{Quadratic/Linear}$ (Fig. 5A). $R_{Quadratic/Linear}$ is defined as the ratio between the quadratic and linear fitting prefactors extracted from the fitting of the green light power dependence of the photocurrent with a parabolic function. In first approximation, the quadratic fraction of the photocurrent corresponds to the signal resulting from the two-photon ionization of NV, while the linear fraction corresponds to the background signal (Bourgeois et al., 2015). The resulting graph confirms that the CW-PDMR contrast tends to become closer to the CW-ODMR contrast when the contribution of the background photocurrent to the total signal decreases. A ratio $R_{PDMR/ODMR}$ of approximately 0.1 was reported on a sample containing high densities of NV centers prepared by high-energy irradiation of nitrogen-rich HPHT diamonds (Fig. 3, top panel). On the contrary, for a small ensemble of approximately five NV centers implanted in ultra-pure electronic grade diamond, i.e., in a sample containing a low proportion of background defects, a ratio $R_{PDMR/ODMR}$

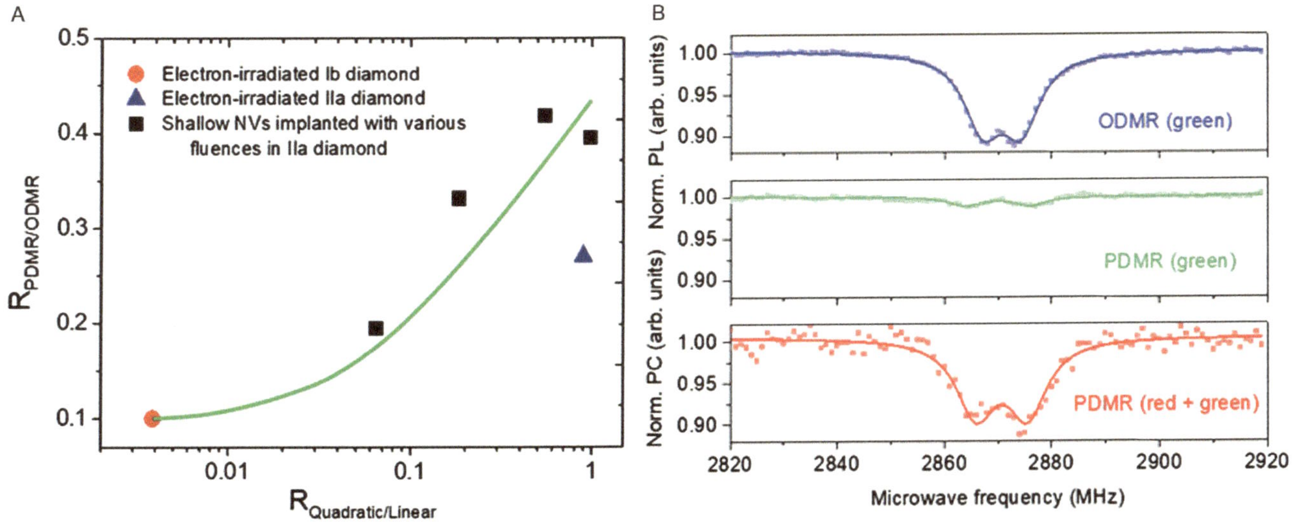

Fig. 5 (A) $R_{PDMR/ODMR}$ ratio between the maximal CW-PDMR and CW-ODMR contrasts obtained under green illumination on different samples. This parameter is plotted against the ratio $R_{Quadratic/Linear}$ between the prefactors a and b obtained by fitting the photocurrent I_{pc} measured as a function of the green light power P by the function: $I_{pc} = a\ P^2 + b\ P$. Here, in first approximation, NV-associated photocurrent is considered to follow a quadratic power dependence (see Bourgeois et al., 2015), while the power dependence of the background photocurrent is considered as linear. Line: guide for the eyes. Data published in Bourgeois et al., (2020). (B) PDMR spectrum measured under combined pulsed red excitation (660 nm to 2 mW) and CW green bias illumination (532 nm to 1.5 mW) (red curve, bottom part). This spectrum is compared to ODMR (blue curve, top part) and PDMR (green curve, middle part) spectra measured under 532 nm green illumination (with a power of 0.4 mW for ODMR and 3 mW for PDMR). Spectra obtained under the light powers leading to the maximal magnetic resonance contrast for each of the techniques are represented here. Measurements performed on large NV ensembles in electron-irradiated and annealed type-Ib diamond ($[NV^-] = 20$ ppm). Photocurrent detected by lock-in amplification, referenced to the red light pulsing frequency in case of two-color excitation and to the green light pulsing frequency in case of green excitation (objective N.A.: 0.95, light pulsing frequency: 131 Hz, inter-electrode distance: 15 μm, electric field: 5×10^4 V cm^{-1}). Symbols: experimental data, lines: fit to the sum of two Lorentzian functions. Data published in Bourgeois et al., (2020).

of 0.58 was measured (Fig. 3, bottom panel). In this case, a CW-PDMR contrast of approximately 11% could be obtained.

It was additionally shown that in some samples—usually prepared by high-energy irradiation of N-doped diamond—a further limitation of the PDMR contrast resulted from an increase in the hole photocurrent under application of a resonant microwave field (Bourgeois et al., 2020). This phenomenon is attributed to the existence of charge exchanges occurring through the conduction band between NV centers and other defects (capture of free electrons resulting from NV^- ionization by acceptor defects). These exchanges can lead to an increase in the photocurrent resulting from acceptor defects ionization at resonance, and depending on the material composition and on experimental conditions, can result in an inversion of the sign of PDMR resonances (Bourgeois et al., 2020).

Diamond materials used for NV quantum applications contain various nitrogen-associated defects, as well as other defects that can potentially be optoelectrically active and therefore affect the photoelectric detection of NV^- spin. The defects usually present in as-grown and irradiated N-doped diamond are for example reviewed in Deák et al. (2014) and Barry et al., (2020). The dominant defect in these samples is systematically the substitutional nitrogen, N_S (Acosta et al., 2009; Edmonds et al., 2012; Khan et al., 2013). In its neutral state N_S^0 this defect can be detected in EPR spectra, where it is known as the P1 center. In as-grown CVD diamond the ratio between $[N_S]$ and [NV] is typically around 50 to 300 (Edmonds et al., 2012), while in irradiated and annealed diamond $[N_S]$ usually still exceeds [NV] by more than a factor 10 (Acosta et al., 2009). N_S^0 can act as an electron donor, the absorption of a photon with an energy above ~2.2 eV leading to the formation of N_S^+ by the promotion of an electron to the conduction band (Nesládek et al., 1996, 1998; Rosa et al., 1999; Van Enckevort and Versteegen, 1992). The ionization of N_S^0 can thus be induced by the green excitation commonly used in PDMR experiments and is therefore in large part responsible for the background photocurrent observed on most samples (Bourgeois et al., 2015; Chen et al., 2017b). Other optoelectrically active defects are likely to be present in lower concentrations in N-doped diamond samples containing NV centers and can contribute to charge-carrier generation under illumination. Single vacancies are for example found in irradiated diamond (Collins and Dahwich, 2003; Kiflawi et al., 2007), while vacancy clusters can be formed during CVD growth of diamond (Hounsome et al., 2006) or during annealing of irradiated diamond (Twitchen et al., 1999). The calculated charge transition energy levels for the single vacancy (V) and the divacancy (V_2) indicate that for both of these defects ionization or recombination can be induced by visible light (Deák et al., 2014). Hydrogen-related defects, such as VH or NVH are also observed in CVD diamond (Khan et al., 2013). Two bands with thresholds at 1.1 and 2.2 eV detected in the optical absorption spectra of CVD diamond are attributed to ionization of NVH (Khan et al., 2013)—a defect present in approximately 30 times higher concentration than NV in as-grown CVD diamond (Edmonds et al., 2012). In the case of NV formation by nitrogen-ion implantation, the nitrogen to NV conversion remains incomplete, leading to an excess of N_s defects (Pezzagna et al., 2010). Ion implantation results, in addition, in the formation of point defects such as interstitials and vacancies (Lehtinen et al., 2016).

Technical solution to overcome the PDMR contrast limitation: A first solution to remove the unwanted background contribution to the photocurrent signal consists in detecting the photocurrent by lock-in amplification, referenced to the microwave pulsing frequency (Gulka et al., 2017). In this situation, the laser is set to the CW mode while the microwave field is pulsed at a frequency used to reference the lock-in amplifier. Only the variations in the photocurrent signal related to $m_s = 0 \leftrightarrow m_s = \pm 1$ spin transitions are then detected, meaning that a non-zero photocurrent is only measured if a resonant microwave field is applied. The non-modulated fraction of the photocurrent, and in particular the background signal resulting from ionization of defects other than NV centers, is effectively filtered out. A positive contrast is observed in microwave-triggered PDMR spectra, where the signal is the AC fraction of photocurrent (see microwave-triggered pulsed PDMR spectra presented in Fig. 4A). Because the AC off-resonant signal is close to the noise level, the PDMR detection contrast is close to 100% with the microwave-triggering technique. Although this detection method effectively avoids the detection of any background photocurrent, the maximum spin-projection contrast, defined by the spin-conserving and ISC transitions, is still limited to ~30%. It was demonstrated that the microwave-triggered PDMR scheme enables an enhancement of the signal-to-noise ratio—SNR (7.4 times improvement in SNR compared to laser-triggered PDMR in Gulka et al., (2017)).

Optimization of PDMR performance by using alternative illumination schemes: The PDMR contrast and the photoelectric detection rate can be controlled through the excitation wavelength. Indeed, the light wavelength affects NV excitation and photoionization dynamics, as well as the balance between the background photocurrent and NV-related signal. Green excitation, close to the maximum absorption of NV^-, induces both ionization of NV^- and back-conversion from NV^0 to NV^- by two-photon processes; it has thus been used in most PDMR experiments so far (Bourgeois et al., 2015; Gulka et al., 2017; Hrubesch et al., 2017; Siyushev et al., 2019). It involves however limitations in the PDMR contrast, as discussed above, and in the photoelectric detection rate—since under a given excitation power a two-photon ionization process leads to a lower charge-carrier generation rate than a one-photon process.

Multi-color light excitation schemes were developed to overcome these limitations. A first scheme is based on one-photon ionization of NV centers under combined blue and green excitation (Bourgeois et al., 2017). Another possibility is to improve the photoelectrically detected spin contrast by using photons of two different energies to induce, on the one hand, NV^- and NV^0 excitation and, on the other hand, the second step of NV^- photoionization or NV^0 recombination (see Section 2.1.1). The advantage of using longer-wavelength light (with photon energy below 2.2 eV) for the second step of NV^- photoionization is that the ionization of N_s^0 can be avoided, thus reducing the contribution of background photocurrent to the detected signal and increasing the CW-PDMR contrast. This two-color PDMR scheme was implemented (Bourgeois et al., 2020) using pulsed red light with a photon energy of 1.88 eV, above the minimum required to transfer an electron from NV^- excited state to the conduction band, and from the valence band to NV^0. Red light was combined with CW bias green light, used to facilitate the excitation of NV^- (threshold around 1.95 eV) and the first step of the two-photon back-conversion from NV^0 to NV^- (threshold around 2.16 eV). Experiments were performed on an irradiated and annealed type-Ib diamond containing dense NV ensembles, on which the maximal CW-ODMR contrast exceeds by a factor 8.5 the maximal CW-PDMR contrast obtainable under green illumination. To filter out the signal induced by the green light, the photocurrent was detected by lock-in amplification referenced to the red-light pulsing frequency. In this configuration, a maximal CW-PDMR contrast of 10% was obtained, similar to the maximal CW-ODMR contrast observed on the same sample and increased by a factor 8.5 compared to the maximal CW-PDMR contrast measured under only green excitation (Fig. 5B) (Bourgeois et al., 2020).

2.3.2 *Influence of diamond material properties on the photoelectric detection rate*

Optical readout of NV^- spin presents limitations in terms of detection rate. Saturation of the photoluminescence rate under a high illumination power (see Fig. 6A) results from saturation of the ^3E excited state (maximum rate of

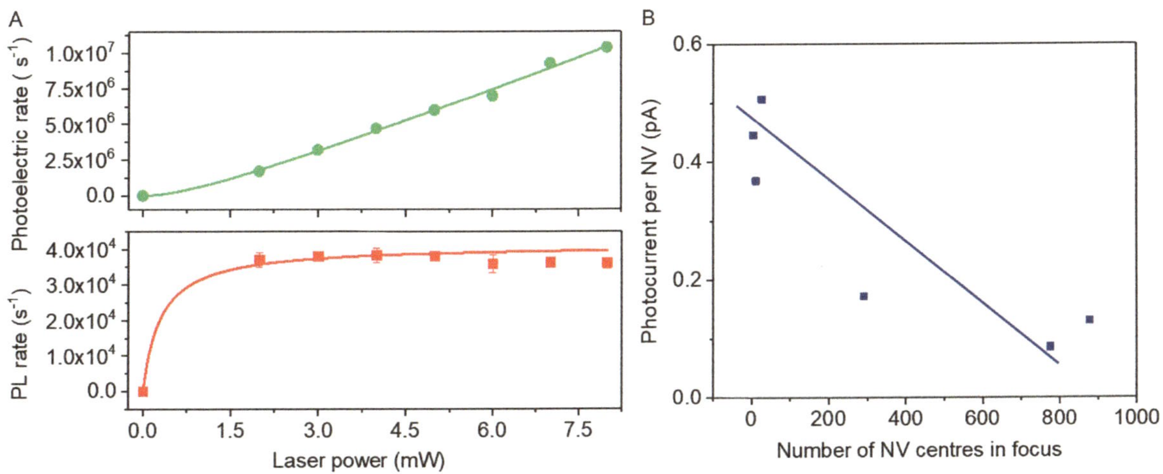

Fig. 6 (A) Light power dependence of the photoelectric and photoluminescence detection rates measured on a native single NV center in electronic grade CVD diamond (laser illumination: 532 nm, objective N.A.: 0.95, electric field: 1.6×10^4 V cm^{-1}, inter-electrode distance: 50 μm). Detection of DC photocurrent under CW green illumination. Symbols: experimental data. Solid lines: fitting with Eq. (2) for the photoelectric rate, and with the function presented in Bourgeois et al., (2020) for the photoluminescence rate. (B) Detected photocurrent per NV center as a function of the number of NV centers in focus of the objective, measured on an electronic grade type-IIa CVD diamond sample implanted with 8 keV $^{14}N^{4+}$ ions at fluences varying between 10^{10} and 10^{14} cm^{-2}. After annealing at 900 °C, implantation led to the formation of an NV layer with a thickness of 12 nm, located approximately 12 nm below the diamond surface. Photocurrent detected by lock-in amplification, referenced to the green light pulsing frequency. Green illumination: 532 nm, 6.8 mW, objective N.A.: 0.95, envelope frequency: 532 Hz, inter-electrode distance: 50 μm, electric field: 2×10^4 V cm^{-1}. Symbols: experimental data. Line: guide for the eyes. *Panel (A): Adapted from Siyushev, P., Nesladek, M., Bourgeois, E., Gulka, M., Hruby, J., Yamamoto, T., Trupke, M., Teraji, T., Isoya, J., Jelezko, F., 2019. Science 363, 728. Copyright © 2019, The Authors, some rights reserved; exclusive licensee American Association for the Advancement of Science.*

photons emission corresponding to the inverse of the photoluminescence lifetime). Due to the high refractive index of diamond and to limitations of collections optics, in case of a single NV center located under a flat diamond surface only a few percent of total emitted photons (Hadden et al., 2010; Momenzadeh et al., 2015) are effectively collected, resulting in typical detection rates of the order of 10^5 counts/s using a high-N.A. objective. Fabrication of nanostructures on the diamond surface enables obtaining detection rates above 10^6 counts/s (Hadden et al., 2010; Li et al., 2015; Momenzadeh et al., 2015; Wan et al., 2018). However, most of these methods require complex microfabrication techniques. Due to the low number of photons detected during the 200–300 ns detection time interval in which the spin contrast is maximal, averaging a large number of data is required to determine the spin projection with a sufficient SNR (Hopper et al., 2018a).

In case of single NV centers (Siyushev et al., 2019) or small ensembles of NV centers (Bourgeois et al., 2017) in ultra-pure diamond, the use of PDMR enables to enhance the detection rate compared to ODMR. The green light power dependence of the photoelectric and optical detection rates observed on a native single NV center in electronic grade CVD diamond is shown in Fig. 6A. Under green illumination, photoionization of NV^- and back-conversion from NV^0 to NV^- can only occur by two-photon processes. In this case, the photocurrent depends quadratically on the light power below the saturation of NV^- triplet excited state, and linearly beyond this value. The photocurrent I_{pc} can thus be expressed as:

$$I_{pc} = \frac{a\,b\,P^2}{1 + b\,P}, \tag{2}$$

where P is the laser power while a and b are constants (Siyushev et al., 2013, 2019). It can be observed that contrary to the photoluminescence rate, the photoelectric detection rate does not saturate in the range of excitation power considered in Fig. 6A. The photocurrent is indeed limited by the charge recovery dynamics and the electron lifetime, and not by NV^- photoluminescence lifetime (saturation of NV^- triplet excited state). Higher detection rates can thus be achieved in photocurrent than in photoluminescence. Under green excitation of 8 mW, a photoelectric detection rate of 10^7 charge carriers/s was obtained on a single NV (corresponding to a photocurrent of about 1.6 pA), 250 times higher than the optical detection rate observed in identical experimental conditions (Siyushev et al., 2019).

However, in samples containing a higher density of defects, the photoelectric detection rate can be limited by the charge-carrier drift mobility and recombination lifetime, controlling the carriers' transport and recombination. As shown by Eq. (1), electronic properties of the diamond strongly affect the photoelectric detection rate. In case of bulk ensembles of NV centers, it can also directly influence the volume of the material from which charge carriers are collected (Bourgeois et al., 2015, 2020). High drift mobilities (up to 4500 $cm^2\ V^{-1}\ s^{-1}$ for holes) and recombination lifetimes exceeding 2 μs were reported on ultra-pure CVD diamond layers at room temperature (Isberg et al., 2002). However, the lifetime of charge carriers is observed to drop when the concentration of nitrogen increases (Malinauskas et al., 2008; Ščajev et al., 2011), indicating that defects related to nitrogen act as recombination centers for charge carriers. The charge-carrier mobility is also affected by the presence of nitrogen, due to charge scattering by charged impurities and to trapping of free charge carriers by single nitrogen defects (Heremans et al., 2009). As an example, charge-carrier lifetime down to 170 ps and a mobility around 200 $cm^2\ V^{-1}\ s^{-1}$ were reported for CVD diamond containing 5–8 ppm of nitrogen (Malinauskas et al., 2008). Electron- or proton-irradiation treatments of diamond, used to increase the conversion from substitutional nitrogen to NV centers, are likely to further decrease the charge-carrier lifetime and mobility due to the creation of additional defects in the diamond lattice. A variation exceeding five orders of magnitude in the mobility-lifetime product can, therefore, be expected between ultra-high-quality diamond crystals and highly defective diamond material.

To illustrate the effect of diamond purity on photoelectric detection rate, photocurrent measured per NV center on different areas of an electronic grade CVD diamond implanted with various nitrogen-ion fluences is shown in Fig. 6B. The diameter of the laser spot being kept constant in this experiment, variations in the number of NV centers in focus of the objective correspond here to different NV densities. The detected photocurrent per NV center appears to drop with the defect concentration. This decrease can at least in part be attributed to a degradation of the charge-carrier mobility and lifetime, due to the presence of implanted nitrogen not converted into NV centers and of defects created in the material during ion implantation. It was also observed that a considerably higher detection rate per NV was obtained on electron-irradiated (dose of $10^{16}\ cm^{-2}$) type-IIa diamond with $[N_s^0] < 1$ ppm and [NV] $\approx$ 10 ppb or on as-grown N-doped diamond presenting the same NV concentration than on electron-irradiated (dose of 10^{18} cm^{-2}) type-Ib diamond with $[N_s^0] \approx 200$ ppm and $[NV] \sim 20$ ppm (Bourgeois et al., 2020). This result confirms that the photoelectric detection rate per NV decreases when the density of NV centers and other defects in the material increases.

2.4 Single NV spin photoelectric detection and applications

Through improvement of diamond material quality, electrode properties and insulation of the photocurrent detection circuit from external noise, the photoelectric readout of NV spins initially demonstrated on large NV ensembles (Bourgeois et al., 2015) could be downscaled to a single NV center (Siyushev et al., 2019).

Photoluminescence and photocurrent maps simultaneously recorded on an electronic grade diamond sample containing native single NV centers approximately 15 μm below the surface are shown in Fig. 7A (Siyushev et al., 2019). The photocurrent resulting from the ionization of a single NV center can be clearly distinguished in the photoelectric map. Compared to confocal photoluminescence imagining, photoelectric mapping leads to higher lateral resolution (full-width-half-maximum of the single defect spot reduced by 34%) and to higher imaging contrast (96% for photocurrent mapping, versus 70% for photoluminescence mapping) (Fig. 7B). This improved imaging performance can be explained by the two-photon ionization process inducing NV-related photocurrent, leading to a quadratic dependence of NV-related photocurrent under low illumination power (see Section 2.3.2.) and enabling thus a better discrimination from the background signal, which usually depends linearly on the light power.

Both CW and pulsed-PDMR measurements were successfully performed on a single NV center in HPHT electronic grade diamond. The corresponding data are presented in Fig. 7C and D and compared to optically detected signals. The use of the envelope scheme presented in Section 2.2.4 (with laser-triggering) enabled in particular to detect pulsed PDMR spectra (not shown here) (Siyushev et al., 2019) and Rabi oscillations (Fig. 7D) on a single NV spin. It should be noted that the laser powers used for both optical and photoelectric measurements were selected to optimize the spin contrast in the photoelectric signal and do not lead to the highest ODMR contrast.

The photoelectric readout of a coherently manipulated single NV^- spin opens the path toward the development of photoelectric quantum chips, with applications both in the field of nanoscale quantum sensing and

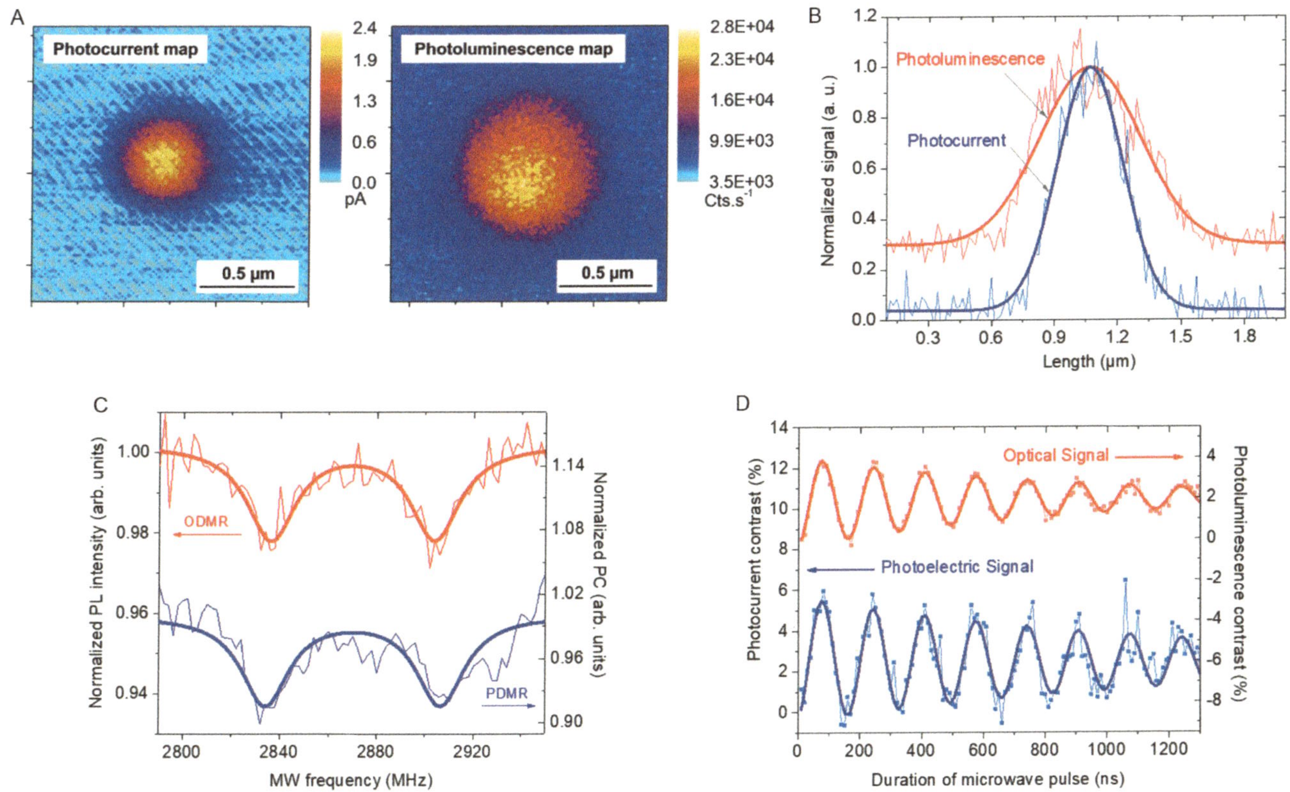

Fig. 7 See legend on opposite page.

quantum computing. Compared to optical detection, the higher photoelectric detection rates obtained on single NV centers (see Section 2.3.2) directly translate into potential improvements of detection sensitivities for magnetic field (Barry et al., 2020; Rondin et al., 2014), electric field (Chen et al., 2017a; Dolde et al., 2011) or temperature (Kucsko et al., 2013; Neumann et al., 2013), as detailed in Section 3.1. For the same reason, it can lead to progress in miniaturization of quantum NMR devices (Aslam et al., 2017; Schmitt et al., 2017; Shi et al., 2015) (see Section 2.5) and enable imaging and structural analysis of complex molecules by reducing the required acquisition time.

The PDMR technique enables individual addressing of proximal NV centers separated by a distance below the optical diffraction limit, provided that sets of contacts with a short inter-electrode distance are fabricated on the diamond surface. The enhanced spatial resolution allowed by the PDMR method could be beneficial for instance to achieve readout of arrays of

Fig. 7 (A) Mapping of the photocurrent (left) and photoluminescence rate (right) on a native single NV center in electronic grade type-IIa CVD diamond (laser illumination: 532 nm, 9 mW, objective N.A.: 0.95, electric field: 1.6×10^4 V cm^{-1}, inter-electrode distance: 50 μm). For these measurements, the DC photocurrent induced by CW illumination was detected after amplification using an ultra-low-noise preamplifier. (B) Normalized photocurrent and photoluminescence intensity profiles extracted from the maps. Narrow lines: experimental data. Thick lines: Gaussian fit. The full-width-half-maximum extracted from the fit is 0.56 μm for the photoluminescence map and 0.37 μm for the photocurrent map. (C) Simultaneous CW-PDMR and ODMR measurements on a native single NV in electronic grade type-IIa HPHT diamond (yellow illumination: 561 nm, objective N.A.: 0.95, electric field: 1.7×10^4 V cm^{-1}, inter-electrode distance: 5 μm, light pulsing frequency: 16 Hz). Photocurrent detected by lock-in amplification, referenced to the yellow light pulsing frequency. Narrow line: experimental data, thick line: fitting with the sum of two Lorentzian functions. Resonances are split by the application of an external magnetic field of 1.2 mT along the axis of the NV center. The ODMR and PDMR spectra were measured under a similar light power of 5 mW, above the power leading to optimal ODMR contrast. (D) Comparison of the Rabi oscillations measured optically and photoelectrically on a native single NV center in type-IIa electronic grade HPHT diamond. The photocurrent was detected by lock-in amplification, using the envelope technique with light-triggering (yellow illumination: 8 mW, 561 nm, laser pulse duration: 1 μs, objective N.A.: 0.95, envelope frequency: 131 Hz, inter-electrode distance: 5 μm, electric field: 4.4×10^4 V cm^{-1}). *Panels (A), (B) and (D): Adapted from Siyushev, P., Nesladek, M., Bourgeois, E., Gulka, M., Hruby, J., Yamamoto, T., Trupke, M., Teraji, T., Isoya, J., Jelezko, F., 2019. Science 363, 728. Copyright © 2019, The Authors, some rights reserved; exclusive licensee American Association for the Advancement of Science. Panel (C): Adapted from Bourgeois, E., Gulka, M., Nesladek, M., 2020. Adv. Opt. Mater. 8, 1902132. doi:10.1002/adom.201902132.*

entangled solid-state qubits, which is an important ongoing issue for the development of quantum computing.

2.5 Photoelectric readout of nuclear spins

Another important development for the PDMR technique is the readout of an individual nuclear spin through the photoelectric detection of single NV center. In the last decade, the NV center in diamond has demonstrated its potential as a room temperature qubit for quantum information processing, communication and sensing. Many of these applications, however, rely on coherent control of the spin state for extended durations. Even though the NV center holds record electron spin coherence time among solid-state qubits (Balasubramanian et al., 2009), nuclear spins coherence exceeds electron spin coherence by orders of magnitude (Pfender et al., 2017a). Combination of NV center electron spin with a nuclear spin thus allows for a small quantum register (Pfender et al., 2017b). The prolonged coherence of such system is utilized as a local quantum bit and the electron spin is used for efficient readout and sensing. The nuclear spin readout has already been demonstrated optically (Jelezko et al., 2004); however, combination of the nuclear spin control with PDMR would facilitate construction of scalable quantum processor units or compact NMR sensors. Thus, in order to enable the utilization of PDMR and its advantages for quantum computation and sensing, establishing the photoelectric readout of individual nuclear spin is essential.

PDMR nuclear spin readout was first demonstrated on an ensemble of ^{14}N nuclear spins (Morishita et al., 2020). The electrical detection of nuclear spin coherence was demonstrated using a direct DC readout sequence (Fig. 8A), where the photocurrent is read continuously. Low magnetic field ($\approx 10\,G$) was applied approximately along the $<111>$ direction of the diamond crystal to split resonances corresponding to the four different NV directions, and the transition between $|+1, 0\rangle$ and $|+1, +1\rangle$ (where $|m_s, m_I\rangle$ are electron and nuclear spins, respectively) was probed to achieve the electrical detection of Rabi oscillations (Fig. 8B) and electron-nuclear double resonance. Since the nuclear spin has not been fully polarized and only one of the four NV alignments was addressed, the detected contrast was small ($\approx$1–2% for ODMR). This is true especially for PDMR, in which the photocurrent generated during the whole laser pulse is recorded. However, by using an NV ensemble on which the detected photocurrent is high, the nuclear spin coupling to NV centers

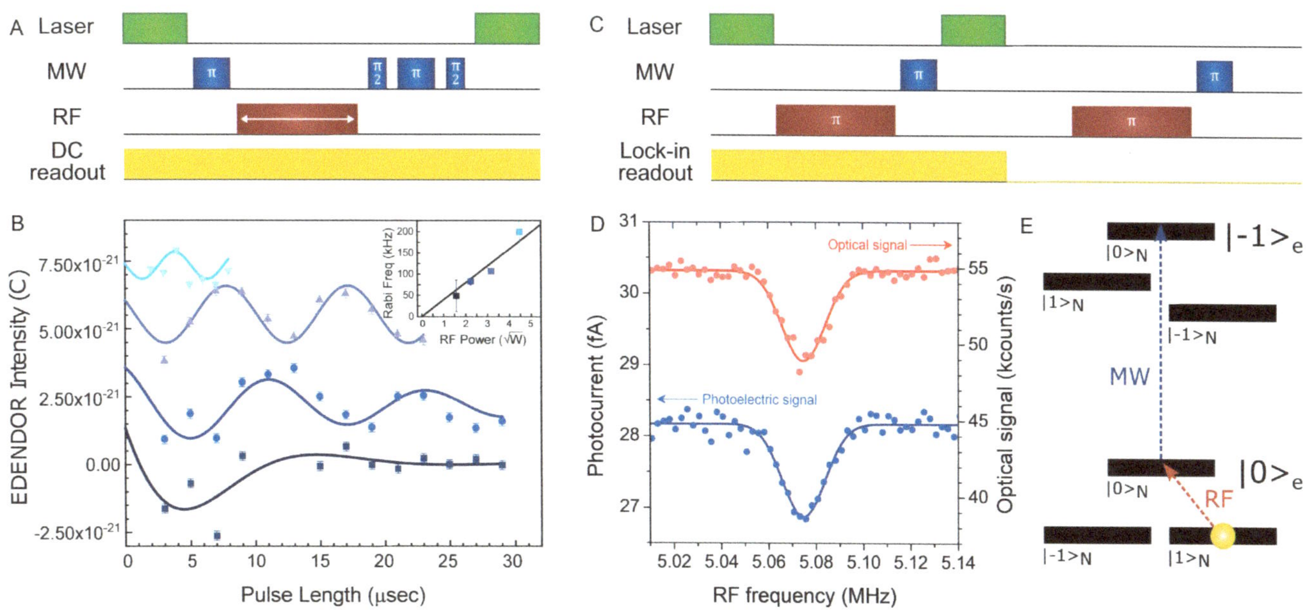

Fig. 8 (A) Pulsed sequence used for the direct DC readout of the electrical nuclear Rabi signal. (B) Electrically detected Rabi oscillations of nuclear spin ensembles measured for different RF driving powers. The inset shows the dependence of the Rabi frequency to the square root of RF power (laser illumination: 532 nm, 30 mW, objective N.A.: 0.8, electric field: 4×10^4 V cm^{-1}, inter-electrode distance: 2 μm). (C) Pulsed sequence used for the laser-triggered lock-in photoelectric detection of a single nuclear spin. (D) The single nuclear spin resonance measured at 440 G is obtained by sweeping the RF frequency. Electrical signal is shown in blue (scattered symbols), optical signal measured simultaneously is shown in red (scattered symbols), the solid lines are results of Gaussian fitting (yellow illumination: 6 mW, 561 nm, laser pulse duration: 4 μs, objective N.A.: 0.95, envelope frequency: 7 Hz, inter-electrode distance: 3.5 μm, electric field: 2.5×10^4 V cm^{-1}) [Unpublished data]. (E) The NV ground state hyperfine structure depicting the interaction of ^{14}N nuclear spin *N* with the NV electronic spin *e* and the resonant frequencies (arrows) used for the measurement in (D). *Panels (A) and (B): Adapted from Morishita, H., Kobayashi, S., Fujiwara, M., Kato, H., Makino, T., Yamasaki, S., Mizuochi, N., 2020. Sci. Rep. 10, 1.*

could be photoelectrically detected (Morishita et al., 2020). The situation is more challenging for single NV center, for which long nuclear spin sequences result in low photocurrent and excessive amount of DC-noise is accumulated during this period (especially for direct DC readout of photocurrent).

Recently, these obstacles were circumvented and coherent control and electrical readout of single nuclear spin was established using the lock-in detection (Gulka et al., n.d.). First, the ^{14}N nuclear spin was polarized to the $|0, +1\rangle$ state by application of a magnetic field aligned to NV axis and approaching the excited state level anticrossing (ESLAC, ≈440 G). After optical pumping close to ESLAC, the nuclear spin is polarized into the $|m_I\rangle = |+1\rangle$ state while, simultaneously, electron spin is initialized into $|m_s\rangle = |0\rangle$ (Fischer et al., 2013; Jacques et al., 2009; Steiner et al., 2010). The lock-in envelope readout technique (Gulka et al., 2017), was used to probe the transition between $|0, +1\rangle$ and $|0, 0\rangle$. A radiofrequency (RF) π-pulse was used to drive the nuclear spin from $|m_I\rangle = |+1\rangle$ state to $|m_I\rangle = |0\rangle$ state, followed by a MW π-pulse to excite the electron spin from $|m_s\rangle = |0\rangle$ state to $|m_s\rangle = |-1\rangle$ state (Fig. 8E). The whole sequence was encoded in the envelope frequency (see Section 2.2.4) used for the laser-triggering of the lock-in amplifier (Fig. 8C), which helped to filter out the DC and noise photocurrent. The nuclear spin resonance was obtained after sweeping the RF frequency around the resonant frequency (Fig. 8D). This is the first demonstration of electrically read single nuclear spin, which provides a basis for future PDMR applications in scalable quantum computing and nanoscale NMR.

3. PDMR for magnetic field sensing applications

3.1 PDMR NV magnetometry

The PDMR technique presents interesting perspective for the development of highly sensitive diamond magnetometers. Magnetometry based on optical readout of NV^- spin is a well-established technique, which has been widely studied, developed and applied in recent years (Barry et al., 2020; Rondin et al., 2014). The basic principle of CW NV magnetometry is the determination of Zeeman splitting between the $m_s = \pm 1$ spin sublevels of NV^- ground triplet state (Doherty et al., 2012), through detection of the difference between the magnetic resonance frequencies corresponding to transitions from $m_s = 0$ to $m_s = -1$, on the one hand, and from $m_s = 0$ to $m_s = +1$, on the other hand. The Zeeman splitting is proportional to the projection of the external magnetic field onto the axis of NV^- in the low-field regime

(Rondin et al., 2014). The CW detection of NV magnetic resonances constitutes therefore a simple magnetometry technique. In the case of magnetic resonance lines presenting a Lorentzian profile, the shot-noise limited sensitivity of an NV-based CW magnetometer, η, can be expressed as (Dréau et al., 2011):

$$\eta = \frac{4h}{3\sqrt{3}g\mu_B}\frac{\Delta\nu}{C\sqrt{R}} \tag{3}$$

where C is the magnetic resonance contrast, $\Delta\nu$ the magnetic resonance full-width-half-maximum and R the photon (for ODMR) or charge carrier (for PDMR) detection rate. While CW NV magnetometry is easy to implement, its sensitivity remains limited. The simultaneous application of the microwave field and laser illumination in CW experiments leads indeed to a broadening of magnetic resonance lines (Dréau et al., 2011). The optical and microwave fields need in addition to be balanced to avoid that the electron spin polarization rate becomes faster than the Rabi frequency associated with the microwave driving field (Jensen et al., 2013), reducing spin contrast. These drawbacks can be avoided by performing pulsed detection of magnetic resonances (in which laser pulses are applied alternately with π-microwave pulses) (Dréau et al., 2011) or by using the detection of Ramsey fringes (Nusran et al., 2012; Waldherr et al., 2012). These pulsed techniques also enable the implementation of advanced protocols, such as dynamic decoupling sequences (Bar-Gill et al., 2012; Naydenov et al., 2011), to prolong the NV spin-coherence time by decoupling the spin from its environment. In case of pulsed sensing using Ramsey sequences, the dependence of the sensitivity on the detection rate and spin contrast can be expressed with the same formula as in the case of CW magnetometry, with $\Delta\nu$ replaced by $1/\sqrt{(t_L T_2^*)}$, where t_L is the duration of the readout laser pulse and T_2^* the spin dephasing time (Rondin et al., 2014).

Considering Eq. (3), achieving high sensitivities for NV-based diamond sensors requires an optimization of the detection rate, the magnetic resonance linewidth (or equivalently of the spin-coherence time for pulsed measurements) and of the magnetic resonance contrast. Since the photoelectric readout of NV spins, in some conditions, provides higher detection rates than optical readout (see Section 2.3.2), it could lead to an improvement in the sensitivity of NV-based magnetometers. The PDMR technique can be used both for CW and pulsed NV magnetometry. Photoelectric detection of Ramsey oscillations has already been demonstrated on ensembles of approximately

1000 NV centers implanted in electronic grade diamond (Gulka et al., 2017). The photoelectric detection method also resolves some of the bottleneck issues for practical development of quantum sensors, by enabling fabrication of compact and scalable devices compatible with semiconductor-processing methods and easily interfaceable with electronics.

Here we focus our discussion on magnetometers based on NV ensembles. Indeed, although single NV magnetometry enables spatial resolutions below 10 nm (Hong et al., 2013; Rondin et al., 2012), thanks to the subnanometric size of the NV defect, its sensitivity remains above a few nT $Hz^{-1/2}$ (for AC magnetic field measurements) (Hong et al., 2013) since the maximal detection rates achievable using a single defect are limited. NV ensembles are therefore used for the development of ultra-sensitive diamond magnetometers. A record sensitivity of $\approx$900 fT $Hz^{-1/2}$ was obtained using pulsed magnetometry protocols on an ensemble of $\approx 10^{11}$ optically readout NV center spins, contained in electron-irradiated HPHT diamond ([NV] $\approx$ 0.9 ppm) (Wolf et al., 2015). NV-ensemble magnetometry constitutes in addition one of the possible solutions to perform vector magnetometry, by detecting the projection of the magnetic field on the axis of NV centers oriented along the four different $\langle 111 \rangle$ axes of the diamond crystal (Chipaux et al., 2015; Stürner et al., 2019).

Achieving high performance for PDMR on NV ensembles requires careful optimization of the diamond material properties. Similarly to what was reported for ODMR, the density of defects in the diamond affects PDMR performance by influencing NV charge stability and by creating magnetic noise limiting the spin-coherence time (Barry et al., 2020). However, due to the specifics of photoelectric readout, additional material properties impact PDMR performance, and optimal samples for PDMR-based magnetometry can thus differ from optimal samples for ODMR magnetometry. As highlighted in Section 2.3.1, the presence of optoelectrically active defects in diamond limits PDMR contrast by creating a background photocurrent and by potentially being involved in charge exchanges with NV centers. Since the ionization of N_S^0 is the main source of background photocurrent, its concentration should be minimized, as well as the presence of other optoelectrically active defects with ionization threshold below the energy of photons used for NV excitation. Maximizing CW-PDMR contrast involves therefore achieving a high conversion rate from N_S^0 to NV. It was observed, however, that the ratio between the CW-PDMR and CW-ODMR contrasts was higher for untreated CVD samples than for electron-irradiated and annealed CVD samples with the same NV concentration (see Fig. 3, second and third panels), although the latter type of samples most probably has a higher $[NV]/[N_S^0]$ ratio than the former one. This result seems to indicate that irradiation-induced defects could contribute to the background photocurrent in a significant way, which is in agreement with the observation of variations in the sign of PDMR resonances specifically for irradiated samples (Bourgeois et al., 2020). In addition, as-grown CVD diamond samples with $[NV] \approx 10\,\text{ppb}$ were shown to lead to a significantly higher photoelectric detection rate per NV center than an electron-irradiated HPHT diamond sample with $[NV] \approx 20\,\text{ppm}$, most probably due to the degradation of the charge-carrier mobility and recombination lifetime in highly defective diamond (see Section 2.3.2) (Bourgeois et al., 2020).

Considering these different factors, the optimal sample identified so far for NV-ensemble PDMR magnetometry is an untreated N-doped CVD diamond, containing approximately 10 ppb of NV centers. For this sample, the photoelectric detection rate, as well as the PDMR linewidth and contrast observed under green illumination are compared to the ODMR performance obtained under similar experimental conditions, see Table 1. These different parameters are used to calculate the parameter S defined as:

$$S = \frac{\Delta\nu}{C\sqrt{R}} \tag{4}$$

The CW magnetometry sensitivity presented in Eq. (3) is directly proportional to S. It has to be emphasized that the data presented in Table 1 were obtained without application of an external magnetic field to split the ODMR and PDMR resonances corresponding to the projection of the magnetic field on the NV centers of different orientations. For this reason,

Table 1 Comparison between the PDMR and ODMR performance obtained on as-grown CVD diamond with $[NV] \approx 10\,\text{ppb}$.

	ODMR	**PDMR**
R (s^{-1})	$3.7 \times 10^5 \pm 1.3 \times 10^3$	$3.1 \times 10^7 \pm 1.4 \times 10^5$
C (%)	10.8 ± 0.5	4.3 ± 0.9
FWHM (MHz)	7.4 ± 0.5	7.8 ± 0.6
S ($s^{-1/2}$)	1.1×10^5	3.3×10^4

Inter-electrode distance: 5 μm, electric field: $3\,\text{V}\,\mu\text{m}^{-1}$, objective NA: 0.95, green light: 532 nm, 12 mW. Lock-in detection of the photocurrent referenced to the green light pulsing frequency. Parameters extracted from the fitting of experimental PDMR and ODMR spectra with the sum of two Lorentzian functions. Uncertainties correspond to errors on the fitting parameters.

they do not allow to determine the absolute value of magnetic field sensitivity. These preliminary data indicate however an improvement in the sensitivity by a factor of 3.3 using PDMR compared to ODMR with similar excitation power. It can be noted that the PDMR contrast remains lower than the ODMR contrast, and that the resonance linewidths obtained by the two techniques are comparable. The observed improvement in sensitivity results therefore from an increase in the detection rate when using photoelectric detection. The PDMR performance could be further improved by implementing multi-color excitation schemes, as detailed in Section 2.3.1, and by performing pulsed experiments. The PDMR technique has thus a potential for development of sensitive vector magnetometers (Hruby et al., 2018), with a broad range of applications in aeronautics, automotive industry, robotics or biology.

3.2 Microwave-free magnetometry

It has been proposed and demonstrated that NV magnetometry can be performed without the microwave spin driving (Wickenbrock et al., 2016; Zheng et al., 2017, 2020b). Under a constant magnetic field of $\approx 102.4\,\text{mT}$, the so-called ground state level anticrossing (GSLAC) occurs (Wickenbrock et al., 2016). At this field, spin-population transfer can be observed between the Zeeman-sublevels of NV ground state ($m_s = -1$, $m_s = 0$), as depicted in Fig. 9. This results in a decrease of the intensity of NV-associated photoluminescence. The theoretical framework for these transitions was studied in Auzinsh et al. (2019), Ivády (2020), Ivády et al. (2015). In this arrangement, a minute change in the magnetic field around the GSLAC leads to a strong variation of photoluminescence intensity. This technique can be used for high-sensitivity CW magnetometry. Similarly to microwave-assisted magnetometry, the magnetic field sensitivity is given by the full-width-half-maximum of the PL peak at the GSLAC resonant field and by its contrast. A microwave-free magnetometer based on detection of absorption on the singlet transition $^1E \rightarrow {}^1A_1$ at GSLAC in place of the PL intensity was realized. This cavity-enhanced singlet-absorption GSLAC measurement method demonstrated an average noise floor of $0.45\,\text{nT}\,\text{Hz}^{-1/2}$, and a photon-shot-noise limit of $12.2\,\text{pT}\,\text{Hz}^{-1/2}$ (Zheng et al., 2019b). Detailed calculations of the processes near GSLAC revealed the importance of transition strengths between the ground state sublevels, and the coupling of NV centers to nuclear spins (N, ^{13}C). It should be stressed that additionally to the sharp decrease observed in the photoluminescence intensity at the GSLAC,

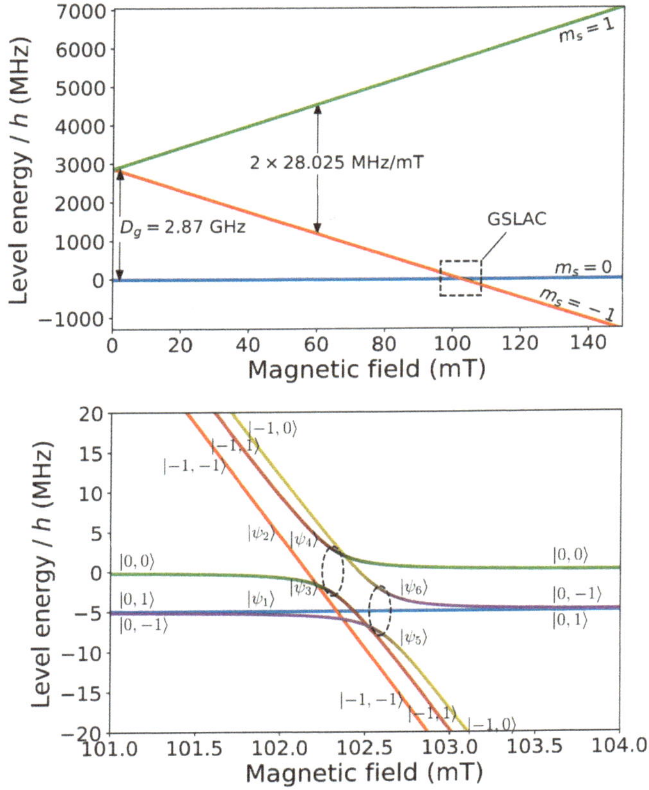

Fig. 9 Principle of the GSLAC spin mixing (upper panel). The Zeeman energies of NV centers show spin mixing for electron-spin magnetic sublevels. The scheme in the lower panel shows hyperfine energy levels in the vicinity of the GSLAC. The dashed ellipses indicate mixing near the GSLAC. *Reproduced from Auzinsh, M., Berzins, A., Budker, D., Busaite, L., Ferber, R., Gahbauer, F., Lazda, R., Wickenbrock, A., Zheng, H., 2019. Phys. Rev. B 100, 075204. https://doi.org/10.1103/PhysRevB.100.075204. Copyright © 2019 American Physical Society.*

additional resonances are detected and can be used for magnetometry. For example, a feature observed around 51.4 mT results from cross-relaxation between NV centers and single substitutional nitrogen centers (Wickenbrock et al., 2016).

Instead of optical detection, one can employ photoelectric detection of GSLAC transitions. In this case, minute changes in the magnetic field around the GSLAC induce variations in the photocurrent. The underlying spin-dependent ionization dynamics is responsible for the achieved contrast in the photocurrent and associated sensitivity.

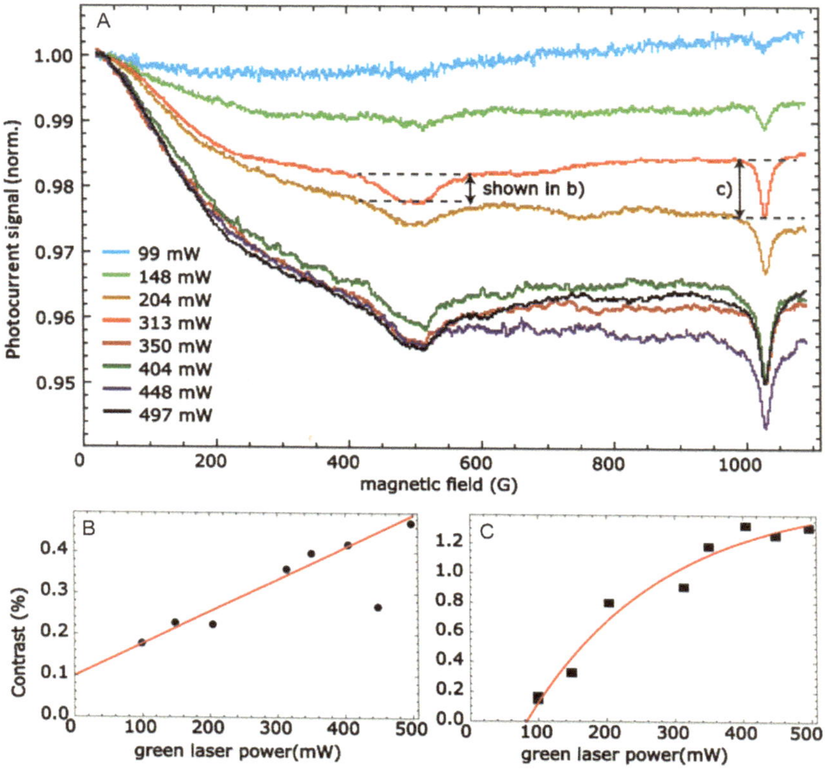

Fig. 10 (A) Photocurrent signal as a function of magnetic field. The green laser beam of various powers was optically modulated and the photocurrent was synchronously detected. (B) Contrast of photocurrent signal around 50 mT as a function of the green laser power. (C) GSLAC contrast of photocurrent signal as a function of the green laser power.

The preliminary data (Zheng et al., n.d.) presented in Fig. 10 show the decrease in photocurrent signal at the GSLAC and around 50 mT (Fig. 10A), as well as the corresponding measurement contrast (Fig. 10B and C). The contrast increases with the green light power, presenting similar tendencies to microwave-assisted PDMR. We attribute this increase to a higher proportion of NV-related photocurrent with respect to N_S^0-associated photocurrent under strong green excitation. Similarly to microwave-assisted PDMR, the photoelectric detection exhibits higher detection rates than optical detection, which would imply higher magnetic field sensitivities.

4. Conclusions

The aim of this chapter was to review the current state-of-the-art in photoelectric detection of NV spin magnetic resonance (PDMR), which we recently established. The PDMR method provides a viable alternative to optical detection of magnetic resonance (ODMR), with the particular advantages of high detection rates for single NV centers and small spin ensembles. In addition, this technique enables device scalability and integration with classical electronic peripheries, which is advantageous for novel applications in the field of quantum technology. We compared the performance of PDMR to ODMR and discussed CW and pulsed photoelectric measurements on NV ensembles. We also showed photoelectric detection of coherently driven single NV spins and demonstrated the photoelectric readout of individual nuclear spins. Coupling of nuclear and electronic spins could be used to realize small quantum processors. The possibility to read both nuclear and electron spins photoelectrically and in a scalable manner opens up applications for sensing and in the field of quantum information science. Finally, we demonstrated a microwave-free magnetometry technique which has perspectives for ultra-sensitive magnetic field detection in applications in which microwaves can be detrimental, for example, in biological sensing.

Acknowledgments

E.B., M.G, J. H. and M.N. acknowledge the support from the Grant Agency of the Czech Republic (Project Number 16-16336S), the FWO project DIAQUANT (S004018N), the FWO Project G0E7417N, the QUANTERA project Q-Magine (R-8843) and the EU project ASTERIQS (820394).

A.G. acknowledges the support from the National Office of Research, Development and Innovation in Hungary for Quantum Technology Program (Grant No. 2017-1.2.1-NKP-2017-00001) and National Excellence Program (Grant No. KKP129866), from the EU Commission (Asteriqs project, Grant No. 820394) and the EU QuantERA program (Q-Magine project, NKFIH Grant No. 127889).

P.S. acknowledges the Baden-Württemberg Stiftung for the financial support by the Eliteprogramme for Postdocs.

References

Abobeih, M.H., Randall, J., Bradley, C.E., Bartling, H.P., Bakker, M.A., Degen, M.J., Markham, M., Twitchen, D.J., Taminiau, T.H., 2019. Nature 576, 411.

Acosta, V.M., Bauch, E., Ledbetter, M.P., Santori, C., Fu, K.M.C., Barclay, P.E., Beausoleil, R.G., Linget, H., Roch, J.F., Treussart, F., Chemerisov, S., Gawlik, W.,

Budker, D., 2009. Phys. Rev. B Condens. Matter Mater. Phys. 80, 115202. https://doi.org/10.1103/PhysRevB.80.115202.
Acosta, V.M., Jarmola, A., Bauch, E., Budker, D., 2010. Phys. Rev. B 82, 201202.
Arute, F., Arya, K., Babbush, R., Bacon, D., Bardin, J.C., Barends, R., Biswas, R., Boixo, S., Brandao, F.G.S.L., Buell, D.A., Burkett, B., Chen, Y., Chen, Z., Chiaro, B., Collins, R., Courtney, W., Dunsworth, A., Farhi, E., Foxen, B., Fowler, A., Gidney, C., Giustina, M., Graff, R., Guerin, K., Habegger, S., Harrigan, M.P., Hartmann, M.J., Ho, A., Hoffmann, M., Huang, T., Humble, T.S., Isakov, S.V., Jeffrey, E., Jiang, Z., Kafri, D., Kechedzhi, K., Kelly, J., Klimov, P.V., Knysh, S., Korotkov, A., Kostritsa, F., Landhuis, D., Lindmark, M., Lucero, E., Lyakh, D., Mandrà, S., McClean, J.R., McEwen, M., Megrant, A., Mi, X., Michielsen, K., Mohseni, M., Mutus, J., Naaman, O., Neeley, M., Neill, C., Niu, M.Y., Ostby, E., Petukhov, A., Platt, J.C., Quintana, C., Rieffel, E.G., Roushan, P., Rubin, N.C., Sank, D., Satzinger, K.J., Smelyanskiy, V., Sung, K.J., Trevithick, M.D., Vainsencher, A., Villalonga, B., White, T., Yao, Z.J., Yeh, P., Zalcman, A., Neven, H., Martinis, J.M., 2019. Nature 574, 505.
Aslam, N., Waldherr, G., Neumann, P., Jelezko, F., Wrachtrup, J., 2013. New J. Phys. 15, 013064.
Aslam, N., Pfender, M., Neumann, P., Reuter, R., Zappe, A., Fávaro de Oliveira, F., Denisenko, A., Sumiya, H., Onoda, S., Isoya, J., Wrachtrup, J., 2017. Science 357, 67.
Auzinsh, M., Berzins, A., Budker, D., Busaite, L., Ferber, R., Gahbauer, F., Lazda, R., Wickenbrock, A., Zheng, H., 2019. Phys. Rev. B 100, 075204.
Awschalom, D.D., Hanson, R., Wrachtrup, J., Zhou, B.B., 2018. Nat. Photonics 12, 516.
Balasubramanian, G., Neumann, P., Twitchen, D., Markham, M., Kolesov, R., Mizuochi, N., Isoya, J., Achard, J., Beck, J., Tissler, J., Jacques, V., Hemmer, P.R., Jelezko, F., Wrachtrup, J., 2009. Nat. Mater. 8, 383.
Bar-Gill, N., Pham, L.M., Belthangady, C., Le Sage, D., Cappellaro, P., Maze, J.R., Lukin, M.D., Yacoby, A., Walsworth, R., 2012. Nat. Commun. 3, 858.
Bar-Gill, N., Pham, L.M., Jarmola, A., Budker, D., Walsworth, R.L., 2013. Nat. Commun. 4, 1743. https://doi.org/10.1038/ncomms2771.
Barry, J.F., Schloss, J.M., Bauch, E., Turner, M.J., Hart, C.A., Pham, L.M., Walsworth, R.L., 2020. Rev. Mod. Phys. 92 (1), 015004. https://doi.org/10.1103/RevModPhys.92.015004.
Bassett, L.C., Alkauskas, A., Exharos, A.L., Fu, K.-M.C., 2019. Micromachines 9, 437. https://doi.org/10.3390/mi9090437.
Beha, K., Batalov, A., Manson, N.B., Bratschitsch, R., Leitenstorfer, A., 2012. Phys. Rev. Lett. 109, 097404.
Bourgeois, E., Jarmola, A., Siyushev, P., Gulka, M., Hruby, J., Jelezko, F., Budker, D., Nesladek, M., 2015. Nat. Commun. 6, 8577. https://doi.org/10.1038/ncomms9577.
Bourgeois, E., Londero, E., Buczak, K., Hruby, J., Gulka, M., Balasubramaniam, Y., Wachter, G., Stursa, J., Dobes, K., Aumayr, F., Trupke, M., Gali, A., Nesladek, M., 2017. Phys. Rev. B 95, 041402(R). https://doi.org/10.1103/PhysRevB.95.041402.
Bourgeois, E., Gulka, M., Nesladek, M., 2020. Adv. Opt. Mater. 8, 1902132. https://doi.org/10.1002/adom.201902132.
Chang, K., Eichler, A., Rhensius, J., Lorenzelli, L., Degen, C.L., 2017. Nano Lett. 17, 2367.
Chen, X.D., Zou, C.L., Sun, F.W., Guo, G.C., 2013. Appl. Phys. Lett. 103, 013112. https://doi.org/10.1063/1.4813120.
Chen, E.H., Clevenson, H.A., Johnson, K.A., Pham, L.M., Englund, D.R., Hemmer, P.R., Braje, D.A., 2017a. Phys. Rev. A 95, 053417. https://doi.org/10.1103/PhysRevA.95.053417.
Chen, X.D., Li, S., Shen, A., Dong, Y., Dong, C.H., Guo, G.C., Sun, F.W., 2017. Phys. Rev. Appl. 7, 014008. https://doi.org/10.1103/PhysRevApplied.7.014008.

Chen, J., Lourette, S., Rezai, K., Hoelzer, T., Lake, M., Nesladek, M., Bouchard, L.-S., Hemmer, P., Budker, D., 2017b. Appl. Phys. Lett. 110, 011108.
Chipaux, M., Tallaire, A., Achard, J., Pezzagna, S., Meijer, J., Jacques, V., Roch, J.F., Debuisschert, T., 2015. Eur. Phys. J. D 69, 166. https://doi.org/10.1140/epjd/e2015-60080-1.
Choi, S., Choi, J., Landig, R., Kucsko, G., Zhou, H., Isoya, J., Jelezko, F., Onoda, S., Sumiya, H., Khemani, V., Von Keyserlingk, C., Yao, N.Y., Demler, E., Lukin, M.D., 2017. Nature 543, 221.
Collins, A.T., Dahwich, A., 2003. J. Phys. Condens. Matter 15 (37), L591–L596. https://doi.org/10.1088/0953-8984/15/37/L06.
Deák, P., Aradi, B., Kaviani, M., Frauenheim, T., Gali, A., 2014. Phys. Rev. B 89, 075203.
Doherty, M.W., Dolde, F., Fedder, H., Jelezko, F., Wrachtrup, J., Manson, N.B., Hollenberg, L.C.L., 2012. Phys. Rev. B 85, 205203.
Doherty, M.W., Manson, N.B., Delaney, P., Jelezko, F., Wrachtrup, J., Hollenberg, L.C.L., 2013. Phys. Rep. 528 (1), 1–45.
Dolde, F., Fedder, H., Doherty, M.W., Nöbauer, T., Rempp, F., Balasubramanian, G., Wolf, T., Reinhard, F., Hollenberg, L.C.L., Jelezko, F., Wrachtrup, J., 2011. Nat. Phys. 7, 459.
Dolde, F., Jakobi, I., Naydenov, B., Zhao, N., Pezzagna, S., Trautmann, C., Meijer, J., Neumann, P., Jelezko, F., Wrachtrup, J., 2013. Nat. Phys. 9, 139.
Dréau, A., Lesik, M., Rondin, L., Spinicelli, P., Arcizet, O., Roch, J.-F., Jacques, V., 2011. Phys. Rev. B 84, 195204.
Dreher, L., Hoehne, F., Morishita, H., Huebl, H., Stutzmann, M., Itoh, K.M., Brandt, M.S., 2015. Phys. Rev. B 91, 075314. https://doi.org/10.1103/PhysRevB.91.075314.
Edmonds, A.M., D'Haenens-Johansson, U.F.S., Cruddace, R.J., Newton, M.E., Fu, K.-M. C., Santori, C., Beausoleil, R.G., Twitchen, D.J., Markham, M.L., 2012. Phys. Rev. B 86, 035201.
Einstein, A., 1905. Ann. Phys. 17, 132.
Fischer, R., Jarmola, A., Kehayias, P., Budker, D., 2013. Phys. Rev. B 87, 125207.
Gaebel, T., Domhan, M., Wittmann, C., Popa, I., Jelezko, F., Rabeau, J., Greentree, A., Prawer, S., Trajkov, E., Hemmer, P.R., Wrachtrup, J., 2006. Appl. Phys. B 82, 243.
Gali, A., 2009. Phys. Rev. B 79, 235210.
Goldman, M.L., Doherty, M.W., Sipahigil, A., Yao, N.Y., Bennett, S.D., Manson, N.B., Kubanek, A., Lukin, M.D., 2015. Phys. Rev. B 91, 165201.
Gruber, A., Dräbenstedt, A., Tietz, C., Fleury, L., Wrachtrup, J., von Borczyskowski, C., 1997. Science 276, 2012–2014.
Gulka, M., Wirtitsch, D., Ivády, V., Vodnik, J., Hruby, J., Magchiels, G., Bourgeois, E., Gali, A., Trupke, M., Nesladek, M., n.d. To be determined. In preparation.
Gulka, M., Bourgeois, E., Hruby, J., Siyushev, P., Wachter, G., Aumayr, F., Hemmer, P.R., Gali, A., Jelezko, F., Trupke, M., Nesladek, M., 2017. Phys. Rev. Appl. 7, 044032. https://doi.org/10.1103/PhysRevApplied.7.044032.
Gupta, A., Hacquebard, L., Childress, L., 2015. J. Opt. Soc. Am. B 33, B28–B34. https://doi.org/10.1364/JOSAB.33.000B28.
Hacquebard, L., Childress, L., 2018. Phys. Rev. A 97, 063408.
Hadden, J.P., Harrison, J.P., Stanley-Clarke, A.C., Marseglia, L., Ho, Y.-L.D., Patton, B.R., O'Brien, J.L., Rarity, J.G., 2010. Appl. Phys. Lett. 97, 241901.
Hensen, B., Bernien, H., Dréau, A.E., Reiserer, A., Kalb, N., Blok, M.S., Ruitenberg, J., Vermeulen, R.F.L., Schouten, R.N., Abellán, C., Amaya, W., Pruneri, V., Mitchell, M.W., Markham, M., Twitchen, D.J., Elkouss, D., Wehner, S., Taminiau, T.H., Hanson, R., 2015. Nature 526, 682.
Heremans, F.J., Fuchs, G.D., Wang, C.F., Hanson, R., Awschalom, D.D., 2009. Appl. Phys. Lett. 94, 152102. https://doi.org/10.1063/1.3120225.

Ho, W.W., Choi, S., Lukin, M.D., Abanin, D.A., 2017. Phys. Rev. Lett. 119, 010602.

Hoehne, F., Dreher, L., Behrends, J., Fehr, M., Huebl, H., Lips, K., Schnegg, A., Suckert, M., Stutzmann, M., Brandt, M.S., 2012. Rev. Sci. Instrum. 83, 043907. https://doi.org/10.1063/1.4704837.

Hoehne, F., Dreher, L., Suckert, M., Franke, D.P., Stutzmann, M., Brandt, M.S., 2013. Phys. Rev. B 88, 155301. https://doi.org/10.1103/PhysRevB.88.155301.

Holovsky, J., 2011. Fourier transform photocurrent spectroscopy on non-crystalline semiconductors. In: StrategFourier Transforms—New Analytical Approaches and FTIR Strategies. InTech, Rijeka, Croatia, p. 257.

Hong, S., Grinolds, M.S., Pham, L.M., Le Sage, D., Luan, L., Walsworth, R.L., Yacoby, A., 2013. MRS Bull. 38, 155.

Hopper, D.A., Grote, R.R., Exarhos, A.L., Bassett, L.C., 2016. Phys. Rev. B 94, 241201 (R). https://doi.org/10.1103/PhysRevB.94.241201.

Hopper, D., Shulevitz, H., Bassett, L., 2018a. Micromachines 9, 437.

Hopper, D.A., Grote, R.R., Parks, S.M., Bassett, L.C., 2018b. ACS Nano 12, 4678.

Hounsome, L.S., Jones, R., Martineau, P.M., Fisher, D., Shaw, M.J., Briddon, P.R., Öberg, S., 2006. Phys. Rev. B 73, 125203. https://doi.org/10.1103/PhysRevB.73.125203.

Hrubesch, F.M., Braunbeck, G., Stutzmann, M., Reinhard, F., Brandt, M.S., 2017. Phys. Rev. Lett. 118, 037601.

Hruby, J., Magchiels, G., Gybels, D., Mladenov, T., Winter, K., Vodnik, J., Prooth, J., Gulka, M., Bourgeois, E., Nesladek, M., 2018. Proc. 69th Int. Astronautical Congress (IAC 2018), International Astronautical Federation (IAF), Bremen, Germany, 9401–9405.

Iakoubovskii, K., Adriaenssens, G.J., Nesládek, M., Stals, L.M., 1999. Diamond Relat. Mater. 8, 717.

Iakoubovskii, K., Adriaenssens, G.J., Nesladek, M., 2000. J. Phys. Condens. Matter 12, 189.

Isberg, J., Hammersberg, J., Johansson, E., Wikström, T., Twitchen, D.J., Whitehead, A.J., Coe, S.E., Scarsbrook, G.A., 2002. Science 297, 1670.

Ivády, V., 2020. Phys. Rev. B 101, 155203.

Ivády, V., Szász, K., Falk, A.L., Klimov, P.V., Christle, D.J., Janzén, E., Abrikosov, I.A., Awschalom, D.D., Gali, A., 2015. Phys. Rev. B 92, 115206.

Jacques, V., Neumann, P., Beck, J., Markham, M., Twitchen, D., Meijer, J., Kaiser, F., Balasubramanian, G., Jelezko, F., Wrachtrup, J., 2009. Phys. Rev. Lett. 102, 057403.

Jelezko, F., Gaebel, T., Popa, I., Domhan, M., Gruber, A., Wrachtrup, J., 2004. Phys. Rev. Lett. 93, 130501.

Jensen, K., Acosta, V.M., Jarmola, A., Budker, D., 2013. Phys. Rev. B 87. https://doi.org/10.1103/PhysRevB.87.014115.

Ji, P., Dutt, M.V.G., 2016. Phys. Rev. B 94, 024101. https://doi.org/10.1103/PhysRevB.94.024101.

Kalb, N., Humphreys, P.C., Slim, J.J., Hanson, R., 2018. Phys. Rev. A 97, 062330. https://doi.org/10.1103/PhysRevA.97.062330.

Khan, R.U.A., Cann, B.L., Martineau, P.M., Samartseva, J., Freeth, J.J.P., Sibley, S.J., Hartland, C.B., Newton, M.E., Dhillon, H.K., Twitchen, D.J., 2013. J. Phys. Condens. Matter 25, 275801.

Kiflawi, I., Collins, A.T., Iakoubovskii, K., Fisher, D., 2007. J. Phys. Condens. Matter 19, 046216. https://doi.org/10.1088/0953-8984/19/4/046216.

Kucsko, G., Maurer, P.C., Yao, N.Y., Kubo, M., Noh, H.J., Lo, P.K., Park, H., Lukin, M.D., 2013. Nature 500, 54.

Lany, M., Boero, G., Popovic, R., 2008. Appl. Phys. Lett. 92, 022111.

Lehtinen, O., Naydenov, B., Börner, P., Melentjevic, K., Müller, C., McGuinness, L.P., Pezzagna, S., Meijer, J., Kaiser, U., Jelezko, F., 2016. Phys. Rev. B 93, 035202.

Li, L., Chen, E.H., Zheng, J., Mouradian, S.L., Dolde, F., Schröder, T., Karaveli, S., Markham, M.L., Twitchen, D.J., Englund, D., 2015. Nano Lett. 15, 1493.

Malinauskas, T., Jarasiunas, K., Ivakin, E., Ralchenko, V., Gontar, A., Ivakhnenko, S., 2008. Diamond Relat. Mater. 17, 1212.
Mamin, H.J., Kim, M., Sherwood, M.H., Rettner, C.T., Ohno, K., Awschalom, D.D., Rugar, D., 2013. Science 339, 557.
Mandelis, A., 1994. Rev. Sci. Instrum. 65, 3309.
Manfrinato, V.R., Zhang, L., Su, D., Duan, H., Hobbs, R.G., Stach, E.A., Berggren, K.K., 2013. Nano Lett. 13, 1555.
Manson, N.B., Harrison, J.P., 2005. Diamond Relat. Mater. 14, 1705.
Manson, N.B., Hedges, M., Barson, M.S.J., Ahlefeldt, R., Doherty, M.W., Abe, H., Ohshima, T., Sellars, M.J., 2018. New J. Phys. 20, 113037. https://doi.org/10.1088/1367-2630/aaec58.
Maurer, P.C., Kucsko, G., Latta, C., Jiang, L., Yao, N.Y., Bennett, S.D., Pastawski, F., Hunger, D., Chisholm, N., Markham, M., Twitchen, D.J., Cirac, J.I., Lukin, M.D., 2012. Science 336, 1283.
Meirzada, I., Hovav, Y., Wolf, S.A., Bar-Gill, N., 2018. Phys. Rev. B 98, 245411.
MIT Technology Review. Inside the race to build the best quantum computer on Earth. 2020. (Accessed 27 August 2020).
Mittiga, T., Hsieh, S., Zu, C., Kobrin, B., Machado, F., Bhattacharyya, P., Rui, N.Z., Jarmola, A., Choi, S., Budker, D., Yao, N.Y., 2018. Phys. Rev. Lett. 121, 246402.
Momenzadeh, S.A., Stöhr, R.J., De Oliveira, F.F., Brunner, A., Denisenko, A., Yang, S., Reinhard, F., Wrachtrup, J., 2015. Nano Lett. 15, 165.
Morishita, H., Kobayashi, S., Fujiwara, M., Kato, H., Makino, T., Yamasaki, S., Mizuochi, N., 2020. Sci. Rep. 10, 1.
Naydenov, B., Dolde, F., Hall, L.T., Shin, C., Fedder, H., Hollenberg, L.C.L., Jelezko, F., Wrachtrup, J., 2011. Phys. Rev. B 83, 081201(R). https://doi.org/10.1103/PhysRevB.83.081201.
Nesládek, M., Meykens, K., Stals, L.M., Vaněček, M., Rosa, J., 1996. Phys. Rev. B 54, 5552.
Nesládek, M., Stals, L.M., Stesmans, A., Iakoubovskij, K., Adriaenssens, G.J., Rosa, J., Vaněček, M., 1998. Appl. Phys. Lett. 72, 3306.
Neumann, P., Jakobi, I., Dolde, F., Burk, C., Reuter, R., Waldherr, G., Honert, J., Wolf, T., Brunner, A., Shim, J.H., Suter, D., Sumiya, H., Isoya, J., Wrachtrup, J., 2013. Nano Lett. 13, 2738.
Niethammer, M., Widmann, M., Rendler, T., Morioka, N., Chen, Y.-C., Stöhr, R., Hassan, J.U., Onoda, S., Ohshima, T., Lee, S.-Y., Mukherjee, A., Isoya, J., Son, N.T., Wrachtrup, J., 2019. Nat. Commun. 101, 1.
Nusran, N.M., Momeen, M.U., Dutt, M.V.G., 2012. Nat. Nanotechnol. 7, 109.
Pezzagna, S., Naydenov, B., Jelezko, F., Wrachtrup, J., Meijer, J., 2010. New J. Phys. 12, 065017.
Pfender, M., Aslam, N., Sumiya, H., Onoda, S., Neumann, P., Isoya, J., Meriles, C.A., Wrachtrup, J., 2017a. Nat. Commun. 8, 1.
Pfender, M., Aslam, N., Simon, P., Antonov, D., Thiering, G., Burk, S., Fávaro de Oliveira, F., Denisenko, A., Fedder, H., Meijer, J., Garrido, J.A., Gali, A., Teraji, T., Isoya, J., Doherty, M.W., Alkauskas, A., Gallo, A., Grüneis, A., Neumann, P., Wrachtrup, J., 2017b. Nano Lett. 17, 5931.
Reiserer, A., Kalb, N., Blok, M.S., van Bemmelen, K.J.M., Taminiau, T.H., Hanson, R., Twitchen, D.J., Markham, M., 2016. Phys. Rev. X 6, 021040. https://doi.org/10.1103/PhysRevX.6.021040.
Robledo, L., Bernien, H., van der Sar, T., Hanson, R., 2011. New J. Phys. 13, 025013.
Rondin, L., Tetienne, J.P., Spinicelli, P., Dal Savio, C., Karrai, K., Dantelle, G., Thiaville, A., Rohart, S., Roch, J.F., Jacques, V., 2012. Appl. Phys. Lett. 100, 153118. https://doi.org/10.1063/1.3703128.
Rondin, L., Tetienne, J.P., Hingant, T., Roch, J.F., Maletinsky, P., Jacques, V., 2014. Rep. Prog. Phys. 77, 056503. https://doi.org/10.1088/0034-4885/77/5/056503.

Rosa, J., Vaněček, M., Nesládek, M., Stals, L.M., 1999. Phys. Status Solidi 172, 113.
Santori, C., Barclay, P.E., Fu, K.-M.C., Beausoleil, R.G., Spillane, S., Fisch, M., 2010. Nanotechnology 21, 274008.
Ščajev, P., Gudelis, V., Ivakin, E., Jarašiūnas, K., 2011. Phys. Status Solidi 208, 2067.
Schmitt, S., Gefen, T., Stürner, F.M., Unden, T., Wolff, G., Müller, C., Scheuer, J., Naydenov, B., Markham, M., Pezzagna, S., Meijer, J., Schwarz, I., Plenio, M., Retzker, A., McGuinness, L.P., Jelezko, F., 2017. Science 356, 832.
Shi, F., Zhang, Q., Wang, P., Sun, H., Wang, J., Rong, X., Chen, M., Ju, C., Reinhard, F., Chen, H., Wrachtrup, J., Wang, J., Du, J., 2015. Science 347, 1135.
Shields, B.J., Unterreithmeier, Q.P., de Leon, N.P., Park, H., Lukin, M.D., 2015. Phys. Rev. Lett. 114, 136402.
Siyushev, P., Pinto, H., Vörös, M., Gali, A., Jelezko, F., Wrachtrup, J., 2013. Phys. Rev. Lett. 110, 167402.
Siyushev, P., Nesladek, M., Bourgeois, E., Gulka, M., Hruby, J., Yamamoto, T., Trupke, M., Teraji, T., Isoya, J., Jelezko, F., 2019. Science 363, 728.
Steeds, J.W., Charles, S.J., Davies, J., Griffin, I., 2000. Diamond Relat. Mater. 9, 397.
Steiner, M., Neumann, P., Beck, J., Jelezko, F., Wrachtrup, J., 2010. Phys. Rev. B 81, 035205.
Stürner, F.M., Brenneis, A., Kassel, J., Wostradowski, U., Rölver, R., Fuchs, T., Nakamura, K., Sumiya, H., Onoda, S., Isoya, J., Jelezko, F., 2019. Diamond Relat. Mater. 93, 59.
Tetienne, J.-P., Rondin, L., Spinicelli, P., Chipaux, M., Debuisschert, T., Roch, J.-F., Jacques, V., 2012. New J. Phys. 14, 103033.
Thiering, G., Gali, A., 2018. Phys. Rev. B 98, 085207.
Twitchen, D.J., Newton, M.E., Baker, J.M., Anthony, T.R., Banholzer, W.F., 1999. Phys. Rev. B 59, 12900.
Ulbricht, R., Loh, Z.-H., 2018. Phys. Rev. B 98, 094309.
Van Enckevort, W.J.P., Versteegen, E.H., 1992. J. Phys. Condens. Matter 4, 2361.
van Oort, E., Manson, N.B., Glasbeek, M., 1988. J. Phys. C Solid State Phys. 21, 4385.
Vink, I.T., Nooitgedagt, T., Schouten, R.N., Vandersypen, L.M.K., Wegscheider, W., 2007. Appl. Phys. Lett. 91, 123512.
Waldermann, F.C., Olivero, P., Nunn, J., Surmacz, K., Wang, Z.Y., Jaksch, D., Taylor, R.A., Walmsley, I.A., Draganski, M., Reichart, P., Greentree, A.D., Jamieson, D.N., Prawer, S., 2007. Diamond Relat. Mater. 16, 1887.
Waldherr, G., Beck, J., Neumann, P., Said, R.S., Nitsche, M., Markham, M.L., Twitchen, D.J., Twamley, J., Jelezko, F., Wrachtrup, J., 2012. Nat. Nanotechnol. 7, 105.
Walker, J., 1979. Rep. Prog. Phys. 42, 1605.
Wan, N.H., Shields, B.J., Kim, D., Mouradian, S., Lienhard, B., Walsh, M., Bakhru, H., Schröder, T., Englund, D., 2018. Nano Lett. 18, 2787.
Wendin, G., 2017. Rep. Prog. Phys. 80, 106001.
Wickenbrock, A., Zheng, H., Bougas, L., Leefer, N., Afach, S., Jarmola, A., Acosta, V.M., Budker, D., 2016. Appl. Phys. Lett. 109, 053505.
Wolf, T., Neumann, P., Nakamura, K., Sumiya, H., Ohshima, T., Isoya, J., Wrachtrup, J., 2015. Phys. Rev. X 5, 041001.
Zheng, H., Chatzidrosos, G., Wickenbrock, A., Bougas, L., Lazda, R., Berzins, A., Gahbauer, F.H., Auzinsh, M., Ferber, R., Budker, D., 2017. in Slow Light. Fast Light. Opto-Atomic Precis. Metrol. X, SPIE, 101190X.
Zheng, H., Hruby, J., Bourgeois, E., Omar, M., Ribeiro, J., Siyushev, P., Jelezko, F., Nesladek, M., Wickenbrock, A., Budker, D., n.d. To be determined. In preparation.
Zheng, G., Samkharadze, N., Noordam, M.L., Kalhor, N., Brousse, D., Sammak, A., Scappucci, G., Vandersypen, L.M.K., 2019a. Nat. Nanotechnol. 14, 742.

Zheng, H., Xu, J., Iwata, G.Z., Lenz, T., Michl, J., Yavkin, B., Nakamura, K., Sumiya, H., Ohshima, T., Isoya, J., Wrachtrup, J., Wickenbrock, A., Budker, D., 2019b. Phys. Rev. Appl. 11, 064068.
Zheng, H., Sun, Z., Chatzidrosos, G., Zhang, C., Nakamura, K., Sumiya, H., Ohshima, T., Isoya, J., Wrachtrup, J., Wickenbrock, A., Budker, D., 2020b. Phys. Rev. Appl. 13, 044023.
Zwanenburg, F.A., Dzurak, A.S., Morello, A., Simmons, M.Y., Hollenberg, L.C.L., Klimeck, G., Rogge, S., Coppersmith, S.N., Eriksson, M.A., 2013. Rev. Mod. Phys. 85, 961.

CHAPTER FOUR

Integrated quantum photonic circuits made from diamond

Francesco Lenzini, Nico Gruhler, Nicolai Walter, and Wolfram H.P. Pernice*
Institute of Physics, University of Muenster, Muenster, Germany
*Corresponding author: e-mail address: wolfram.pernice@uni-muenster.de

Contents

1. Introduction 149
2. Fabrication of diamond-based integrated photonic circuits 152
   2.1 Nanophotonic component fabrication 152
   2.2 Passive photonic elements 154
3. Single photon emitters for waveguide devices 155
   3.1 Coupling to diamond photonic structures 157
   3.2 Hybrid approaches 159
4. Active components 161
5. Waveguide integrated single-photon detectors 163
6. Conclusion 167
References 167

1. Introduction

Photonic integrated circuits (PICs) have materialized into of the leading platforms for implementing photonic quantum technologies (O'Brien et al., 2009; Tanzilli et al., 2011) for applications in secure quantum communication (Martin et al., 2012), enhanced quantum sensing (Matthews et al., 2011), and quantum information processing (Politi et al., 2009). Linear optical circuits based on lithographically patterned waveguides provide a scalable approach for manipulating quantum states of light (Carolan et al., 2015), and guarantee high fidelity of quantum operation due to inherent stability and accurate control of interacting modes in on-chip interferometers (Laing et al., 2010). PICs also provide numerous ways to support the generation of non-classical light (Spring et al., 2017). Equally important, waveguide based photonic circuits allow for the characterization of single

Semiconductors and Semimetals, Volume 104
ISSN 0080-8784
https://doi.org/10.1016/bs.semsem.2020.08.004

photon states using integrated single-photon detectors (Pernice et al., 2012). Since the demonstration of active waveguide components individually, integration of all resources for functional quantum devices on a common platform has been a key technology driver to enable continuous increase in complexity of quantum photonic circuits (Khasminskaya et al., 2016; Wang et al., 2018), and enhance the stability and scalability of quantum photonic technologies. Despite significant effort, finding an optimal material platform for this endeavor has been challenging. Complementing leading silicon photonic approaches, diamond-based PICs offers significant advantages, which have contributed to the attractiveness of this material choice.

Diamond offers outstanding optical and mechanical properties, which are fundamental for realizing a platform for integrated photonic circuits and especially quantum information technologies. Waferscale applications in photonics have been enabled by the continuous progress in the synthesis and processing of chemical vapor deposition (CVD) diamond. CVD diamond in particular provides high quality optical material with reproducible properties and template sizes which are compatible with PIC fabrication (Nemanich et al., 2014). Particularly relevant are a large refractive index and a wide transparency window spanning visible and infrared wavelengths, which make diamond a prime candidate for realizing integrated photonic circuits. The refractive index of around 2.4 in the visible and near infrared (NIR) wavelength regime provides a large index contrast to optical buffer layers with lower refractive index and thus allows for tight confinement of light in diamond waveguides (Hiscocks et al., 2008; Phillip and Taft, 1964). The large electronic bandgap of 5.5 eV opens a transmission window from ultraviolet (UV) wavelengths and stretching to the very far infrared (Dore et al., 1998; Mildren and Rabeau, 2013; Mollart et al., 2003). Importantly, the large bandgap also prevents two-photon absorption even in the visible and NIR regime, which, for example, plague silicon photonic circuits at higher optical powers. The high thermal conductivity and low thermo-optic coefficient of diamond, in contrast, also allow guiding of modes with high optical power in waveguide devices (Nebel, 2003). By exploiting a relatively high nonlinear refractive index (Boyd, 2008; Levenson and Bloembergen, 1974) ($n_2 = 1.3*10^{-19} m^2 W^{-1}$ for visible wavelengths) diamond furthermore provides attractive platform for integrated non-linear optics (Hausmann et al., 2014; Latawiec et al., 2015; Mildren et al., 2008).

While the properties mentioned above find numerous applications in classical integrated optics, in view of quantum photonic devices diamond offers additional benefits. In particular, optically active defect centers have attracted intense interest. Currently over 500 so-called color centers are known in diamond, which cover a wide range of emission wavelengths from the UV to the NIR (Zaitsev, 2001). Such color centers can be used as a stable source of single photons and, in some cases, also provide a controllable coherent electron spin (Jelezko et al., 2004; Kurtsiefer et al., 2000). Prominent color centers in diamond, such as the nitrogen- or silicon-vacancy center (NV, SiV), are of particular interest as they are long-term stable. In addition, a low density of phonon states (diamond provides the highest Debye temperature of ~2000 K) causes low electron-phonon coupling, which implies that they can be used for single photon generation even under ambient conditions (Becker & Becher, 2019; Lee et al., 2012; Leifgen et al., 2014). Thus, by integrating these color centers directly in monolythic waveguides, diamond can provide a single material platform for quantum photonic applications.

For implementing reconfigurable photonic building blocks with low dissipation, diamond also offers outstanding mechanical properties, including an exceptionally high Young's modulus of 1100 GPa. Together with low thermo-elastic dissipation due to its high thermal conductivity (Najar et al., 2014), diamond has therefore emerged as a prime material for implementing mechanical resonators operating at high frequencies without suffering from significant damping (Burek et al., 2016; Mitchell et al., 2016; Rath et al., 2013a). Such devices are particularly attractive for applications in precision sensing (Hanay et al., 2012; Tao et al., 2014). Going beyond sensing devices, combining optical and mechanical degrees of freedom in waveguide-based components allows for reconfiguring integrated photonic circuits with optomechanical components. These devices modulate the effective refractive index of a guided mode due to the displacement of a mechanically moveable part placed next to a waveguide, which in turn induces a phase shift to the propagating light field (Poot and Tang, 2014a; Rath et al., 2014). Integrated phase shifters form key elements for reprogramming diamond-based quantum photonic circuits (Ma et al., 2011; O'Brien et al., 2009).

Last, as a counterpart to waveguide integrated single photon sources, single photon detectors can be seamlessly integrated with diamond PICs. Superconducting films can be deposited on top of diamond waveguides and used for the realization of integrated superconducting nanowire single-photon detectors (SNSPDs) (Rath et al., 2015). In combination with the above mentioned devices, these components provide a complete toolbox for realizing a diamond-based integrated quantum photonic architecture.

2. Fabrication of diamond-based integrated photonic circuits

To date, predominantly two complementary diamond platforms have been used for realizing nanophotonic circuits. First, bulk single crystalline diamond (SCD) offers high material quality at the cost of more challenging fabrication procedures. Second, polycrystalline diamond (PCD) provides wafer-scale thin films for nanofabrication processes closer to traditional PIC manufacture. Since SCD grows on existing SCD template, waferscale substrates with thin SCD films are challenging to prepare for PIC preparation. Instead, PICs can be prepared from thick slabs of SCD with a limited size of typically below 1 cm^2. Following structuring, transfer and etching methods have to be applied for the fabrication of integrated circuits. Important for diamond quantum photonic devices, the high material quality of SCD provides a low concentration of color centers and a low background fluorescence (Mizuochi et al., 2012; Schreck et al., 2014).

While many of the outstanding properties of diamond are preserved in PCD material, the transparency range and the thermal conductivity are reduced because of grain boundaries, which incorporate sp^2-carbon and additional dopants (Nebel, 2003; Zaitsev, 2001). Grain boundaries also induce additional scattering losses. Impurities and strain in addition deteriorate the optical properties of color center (Mildren and Rabeau, 2013). In contrast to bulk SCD, thin PCD films can be grown by chemical vapor deposition (CVD) on various substrates suitable for PICs including in particular SiO_2 optical buffer layers and thereby enable the realization of large area photonic circuits (Rath et al., 2013b).

Integrated optical devices and circuits are realized in diamond using a wide range of different methods. Especially the fabrication of nanophotonic devices in bulk SCD slabs seen a rise of different strategies varying in their complexity.

2.1 Nanophotonic component fabrication

Two approaches for structuring photonic components from bulk diamond have emerged successfully: (i) Thin diamond films or membranes are obtained from a thick diamond piece and subsequently processed; (ii) Photonic structures are carved directly into bulk diamond samples.

To fabricate SCD thin films various methods can be used. In the *thin down approach* dry etching is employed to reduce the thickness of a bulk diamond sample from tens of microns to some hundreds of nanometers required to fabricate photonic waveguides. With this approach often the size is limited and thickness variations of the bulk slab are maintained and therefore necessitates further precautions (Jung et al., 2016; Li et al., 2013; Tao and Degen, 2013). Thickness variations can be alleviated with a *lift-off procedure* in which a damaged layer is implanted in diamond by ion irradiation. This layer is subsequently graphitized using an annealing step. The damaged layer can afterwards be selectively removed via wet etching (Li et al., 2010; Olivero et al., 2008; Wang et al., 2007). Because the crystal structure of the membrane is damaged, additional diamond overgrowth is used to recover high quality material, while the damaged layer is removed in a further dry etching step (Aharonovich et al., 2012; Gaathon et al., 2013; Magyar et al., 2011). Alternatively, *heteroepitaxial growth* of diamond removes polycrystalline character within a few to tens microns on specific substrates (Gsell et al., 2004; Teraji et al., 2015). A diamond thin film is then obtained by removing the substrate and buffer layers, followed by dry etching the PCD backside. Such thin SCD membranes can then be used for PIC nanostructuring with electron beam lithography (EBL) or ion milling methods. In order to avoid handling of very small samples, a prepatterned transferable silicon mask can be directly used as an etching mask for structuring thin as well as bulk diamond samples (Bayn et al., 2014; Li et al., 2015a,b). In the *angle etching technique* an angled anisotropic dry etching step is used to realize freestanding photonic components with triangular cross-section (Atikian et al., 2017; Burek et al., 2012). Rectangular cross-sections with vertical side walls can be realized using the *undercut etching procedure*. First, an anisotropic plasma etching step is used followed by a Si_3N_4 protecting coating and a quasi-isotropic reactive ion etching (Khanaliloo et al., 2015a,b). A racetrack resonator and a nanobeam (including its cross-section) made with this fabrication method are shown in Fig. 5A–C. Maskless processing based on *focused ion beam (FIB) milling* using typically Ga^+-ions allows for directly cutting and structuring (Bayn et al., 2011). This process is limited in terms of scalability. Moreover, the material undergoing FIB experiences ion damage during structuring. Direct patterning by electron-beam induced etching (EBIE) might circumvent this issue (Zachreson et al., 2014). A part of a photonic crystal cavity fabricated by ion milling is presented in Fig. 5D). D*irect laser writing (DLW)* with ultrashort laser pulses can further be used to structure diamond. With this method, the diamond lattice is damaged which leads to a slightly decreased refractive index. This way buried 3D waveguides can be created in bulk SCD (see also Fig. 5E) (Bharadwaj et al., 2017) which can be coupled to color centers directly.

The availability of polished PCD thin films on a waferscale (Füner et al., 1998) allows for creating photonic devices with larger area coverage. Such templates can be processed with standard lithography, dry and wet etching commonly used in PIC manufacture. Recent progress in the deposition of high purity diamond can improve the quality of PCD thin films with respect to transparency and may eventually allow for the incorporation of color centers.

2.2 Passive photonic elements

To build quantum photonic circuits a library of photonic elements is required from which complex circuits can be assembled. A key building block is a *waveguide,* which is used to route light across a chip. Waveguides are also used to interconnect individual components into photonic circuits. Second, coupling interfaces or *waveguide couplers* are necessary to guide light into and out of the on-chip waveguides. Such interfaces are for example required to combine off-chip optical elements such as light sources or detectors with on-chip circuits. Alternatively, out-of-plane coupling can be achieved via reflection at angled total internal reflection mirrors (Hiscocks et al., 2008) or via tapered fiber coupling. A contact-free optical coupling method, which provides fast out-of-plane access to many different circuits, is based on grating couplers (see Fig. 5F) (Gao et al., 2016). In-plane signal routing can be achieved using beam splitters in the form of y-splitters and directional couplers (see Fig. 5G). Photonic beam splitters are essential to realize on-chip Mach-Zehnder interferometers frequently used in quantum photonic circuits.

For optical signal processing in PICs *optical resonators or cavities* are needed. Cavities are essential for Purcell enhancement of single photon emission, but are also used as narrow-band filters, for signal routing and waveguide characterization. Different cavity geometries are employed, which can be categorized by their optical quality factor (Q-factor). Typical designs include one-dimensional photonic crystal cavities (highest Q up to 9900 (18300) in the visible (NIR) regime for SCD) (Burek et al., 2014; Li et al., 2015a), two-dimensional PhC (highest Q up to 4700 in the visible regime for SCD and up to 600 (6500) for PCD in the visible (NIR) regime) (Li et al., 2015b; Wang et al., 2007), disk resonators (highest Q up to 3000 (115000) in the visible (NIR) regime for SCD) (Khanaliloo et al., 2015b), as well as ring and racetrack resonators (highest Q up to 59,000 (1000000) in the visible (NIR) regime for SCD and up to 30,000 for PCD in the NIR) (Atikian et al., 2017; Hausmann et al., 2014). An overview of these elements is presented in Fig. 1, including a racetrack resonator in (A) and part of a photonic crystal cavity in (D).

Diamond PICs vary in terms their respective propagation losses. Waveguides fabricated in PCD by direct laser writing feature propagation losses of 8–16 dB/cm in the visible regime (Courvoisier et al., 2016). Very low losses in the range of 0.3–0.6 dB/cm were demonstrated in the near-IR regime using a thin down approach of bulk SCD slabs to thin membranes (Gao et al., 2016). In PCD waveguides with near-IR propagation losses of 40 dB/cm were reported using standard EBL and dry etching procedures. The circuit quality was further improved by reactive ion beam angled etching (Atikian et al., 2017). Looking beyond wavelength regimes used for traditional PICs, the wide transparency window of diamond also enables waveguide devices operating at mid- and long-wavelengths for sensing applications in the fingerprint region of molecules.

3. Single photon emitters for waveguide devices

While passive photonic devices provide the means for signal routing and processing, active elements are needed for optical driving. Diamond naturally provides attractive single photon emitters via color centers. Thus, the integration of these emitters with waveguide devices is a natural route for quantum PICs. Coupling of color centers to photonic structures allows to enhance the collection efficiency for later collection with free-space optics. Without suitable measures, the coupling efficiency from defect centers in bulk diamond is limited by the high refractive index of diamond, which imposes a total internal reflection angle of $\simeq 25°$ of the diamond-air interface. Different photonic structures, including solid immersion lenses (Hadden et al., 2010; Robledo et al., 2011; Schröder et al., 2011), nanopillars (Babinec et al., 2010) and circular bullseye gratings (Li et al., 2015c) have thus far been used as a mitigation strategy, providing up to a 15-fold enhancement in the emission rate.

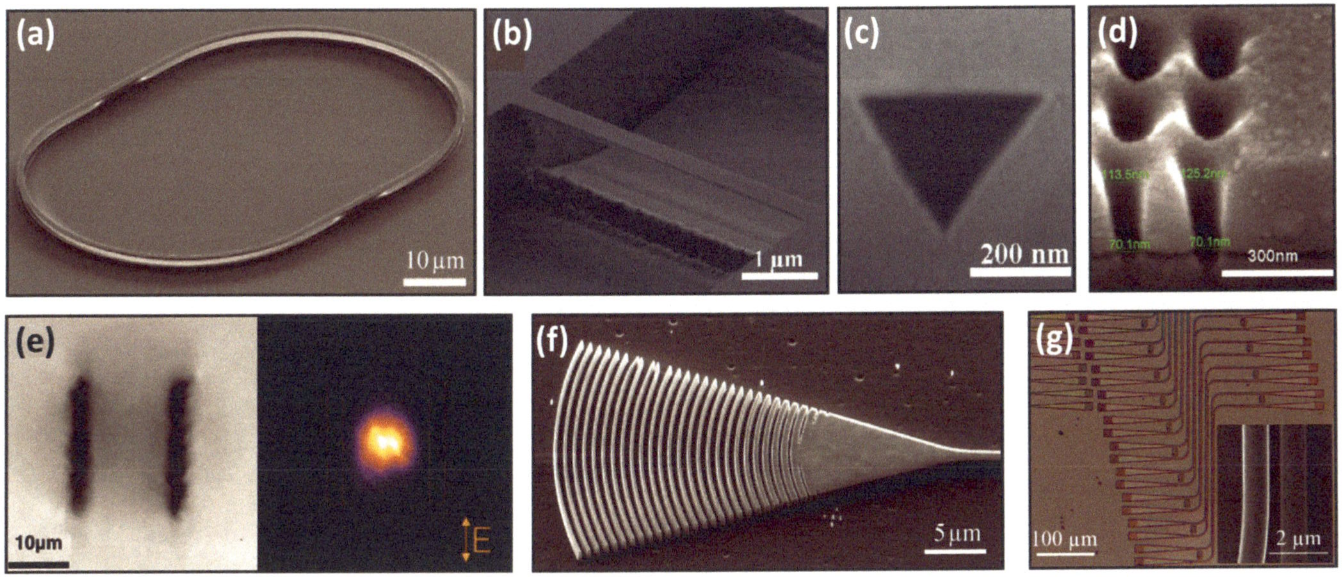

Fig. 1 Diamond photonic devices. (A)–(C) Selection of devices fabricated by angled reactive ion etching: (A) Racetrack resonator; (B) 1D nanobeam with triangular cross-section (C). (D) Photonic crystal cavity prepared from SCD by FIB milling. (E) Type II waveguide made by direct laser writing w in a bulk SCD. (F) Grating coupler made from PCD. (G) Directional couplers with the SEM inset showing the start of the coupler region.

Of particular interest for this review is the coupling of color centers to optical cavities and waveguides monolithically integrated in diamond, whose fabrication method will be described in detail in Section 4. Hybrid architectures, where color centers are integrated with non-diamond photonic structures, are briefly discussed in Section 3.2.

3.1 Coupling to diamond photonic structures

In order to enhance the emission rate into the ZPL via the Purcell effect, coupling of NV centers to optical cavities can be used. With high-Q resonators, NV centers can potentially achieve pure ZPL emission with a near-unity efficiency and an excited state lifetime smaller than 100 ps. A useful figure of merit is the Purcell factor F_P, defined as the enhancement in emission rate induced by the cavity structure. Planar photonic crystal cavities, thanks to their small volume, have enabled the highest Purcell enhancement reported to date for NV centers (see Fig. 2C and D). A record Purcell factor $F_p = 70$, which implies that $\simeq 70\%$ of the total photons are emitted into the ZPL, was reported in Faraon et al. (2012) using a planar cavity fabricated around a native defect. Optical cavities based on waveguides as for example microring resonators (Faraon et al., 2011) and photonic crystal nanobeam cavities (Hausmann et al., 2013) have been demonstrated, as illustrated in Fig. 2A,B, and E. Waveguide-based resonators lead to Purcell enhancement with $F_P = 62$ (Li et al., 2015a), corresponding to emission of more than 50% of the radiated field into the ZPL. These values are on par with the best values reported for planar cavities. Even higher Purcell factors are currently limited by misalignment between the emission dipole of NV centers and the cavity mode field. All the experiments reported so far were performed in (100)-oriented diamond, the most commonly used crystal axis, where NV centers display a random orientation. A promising approach is the fabrication of photonic structures in CVD grown (111)-oriented diamond, where NV centers have shown almost perfect (>95%) preferential orientation (Michl et al., 2014). These experiments made use of photonic cavities fabricated around native defects, Alternatively, ion implanted NV centers can be aligned to photonic crystal cavity. To achieve precise positioning of the defect in the center of the cavity, the same lithography mask for targeted nitrogen ion implantation and dry etching of photonic structures can be used. In this case, 5-fold enhancement of the intensity in the ZPL was measured, which is close to the values obtained using native NV centers. Deterministic positioning of a laser-written NV centers inside a buried

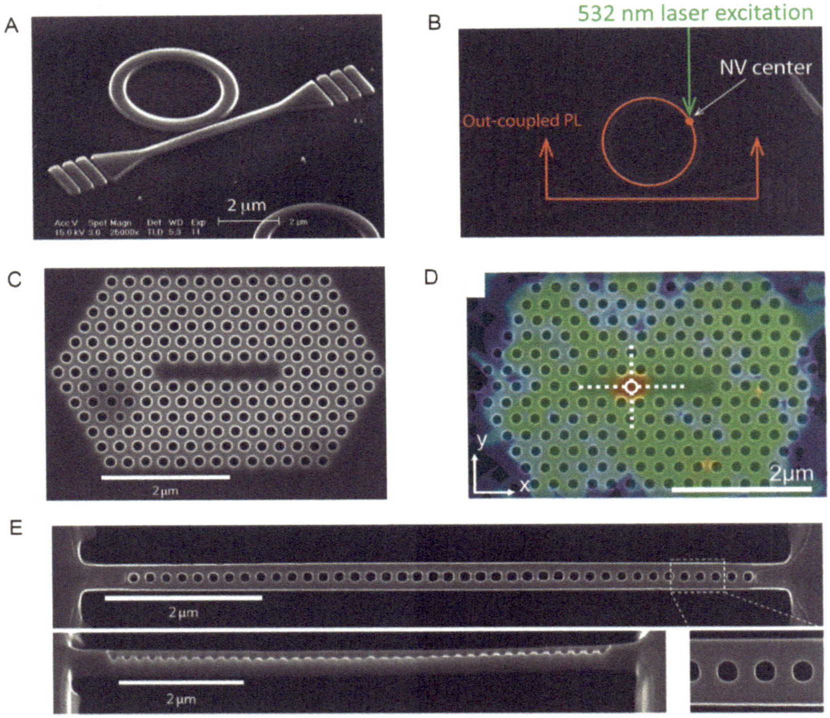

Fig. 2 Single photon emitters based on color centers coupled to different photonic structures. (A) A microdisk resonator for Purcell enhancement of NV centers and (B) scheme for laser excitation and fluorescence collection. (C) A planar photonic crystal cavity for Purcell enhancement of SiV centers. (D) SEM image of a planar photonic crystal cavity overlapped with the fluorescence scan of a SiV centers. (E) Nanobeam cavity used for Purcell enhancement of SiV centers.

waveguides similarly overcomes alignment challenges between emitter and guided modes (Hadden et al., 2018). In this case, however, mismatch between the waveguide mode and the focused laser excitation spot limited the measured count rate to $\simeq$10 counts/s.

Besides the nitrogen vacancy center, silicon vacancy defect centers are promising candidates for waveguide integrated single photon sources. In contrast to NV, SiV centers naturally emit the radiated light predominantly into the ZPL. As with NV, coupling to optical cavities can be employed to further boost their quantum efficiency by enhancing their emission rate and reducing the probability of non-radiative decays. Purcell enhancement has been shown with photonic structures fabricated around defects generated by in situ doping (Riedrich-Möller et al., 2012), as well as with SiV centers generated by focused ion beams in the center of the cavity (Sipahigil et al., 2016). Using a 1D photonic crystal cavity, a Purcell factor $F_P \simeq 26$ leading to a 42-fold enhancement in emission intensity, has been recently reported. (Zhang et al., 2018).

For yet another class of defect centers, Germanium vacancy (GeV) centers, ion implanted defects were used for coupling to photonic structures in nanoscale diamond waveguide with a partially reflective Bragg mirror (Bhaskar et al., 2017). The authors predict a 10% probability of emitting one photon per excitation into the waveguide. This is enabled by the large overlap of the cross-section of the defect with the waveguide mode, as well as the high quantum efficiency of this single emitter.

3.2 Hybrid approaches

In order to overcome the challenges of nanoprocessing in a monolithic all-diamond platforms, hybrid systems joining diamond with different materials provide attractive routes for implementing integrated quantum photonic circuits. In hybrid systems diamond emitters are used for single photon, while established photonic platforms are employed for waveguiding and optical signal processing. This approach was pursued with gallium phosphide bonded to a diamond substrate (GaP-on -diamond). GaP offers a higher refractive index than diamond and therefore allows for waveguiding in the GaP layer. GaP is an electro-optically active material and hence can be used for on-chip modulation. In GaP-on-diamond waveguides coupling to NV centers has been achieved (Fu et al., 2008), as well as coupling to optical resonators (Englund et al., 2010) has been demonstrated. Further, waveguide-coupled microdisk resonators attached to grating couplers demonstrating the potential to collect photons from NV on a hybrid platform for quantum communication applications (see Fig. 3A and B) (Gould et al., 2016).

The less exotic material Silicon nitride (SiN) was similarly chosen for hybrid integration. SiN provides a large bandgap and supports waveguiding in the emission wavelength of color centers. Silicon nitride photonic circuitry can be realized with established fabrication process compatible with silicon nanofabrication. Deterministic coupling of a single nanodiamond to an integrated SiN cavity was first demonstrated using a pick-and-place

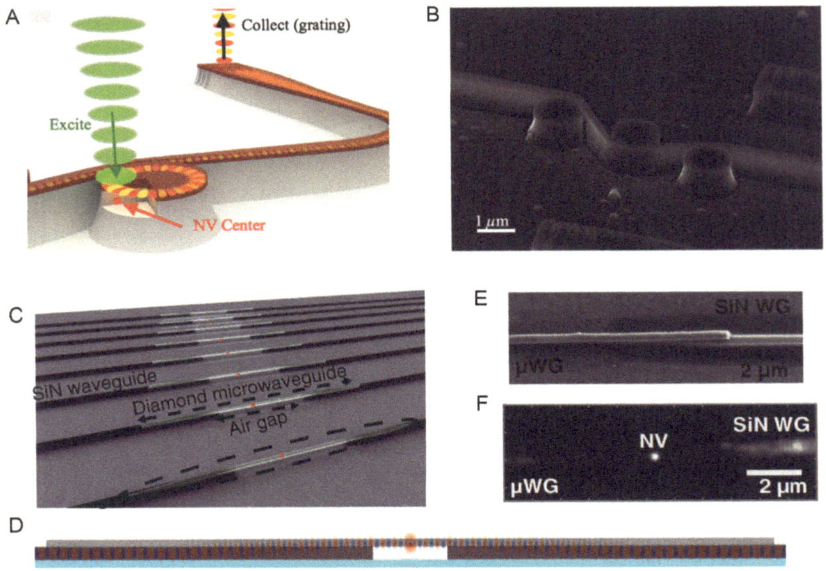

Fig. 3 Hybrid diamond-based photonic architectures. (A and B) GaP-on-Diamond devices. (A) Schematic for the collection of the ZPL of one NV center using a grating coupler and a waveguide coupled microdisk resonator. (B) Waveguide coupled microdisk resonators. (C–F) Silicon Nitride–Diamond hybrid devices. (C) A diamond waveguide containing a single NV center bridging a SiN waveguide. (D) FDTD simulation of the field transferred from the diamond waveguide mode to the SiN waveguide. (E) A diamond waveguide bridging a silicon nitride waveguide. (F) Photoluminescence map of the diamond waveguide showing the NV center emission.

technique (Barth et al., 2009). The coupling efficiency was subsequently improved using tapered diamond waveguide containing a single NV center placed over an integrated SiN cavity (Mouradian et al., 2015). A novel pick-and-place allowed pre-characterizing the color center, followed by deterministic integration into the desired structure (see Fig. 3C and D). While SiN is very attractive from a processing point of view, a drawback lies in the strong fluorescence emission in the visible wavelength range which results from optical pumping of the defect center.

This fluorescent background can be suppressed by employing aluminum nitride (AlN) as a wideband semiconductor. AlN provides optical transparency in the UV, as well as second order optical nonlinearity and piezo-electric properties. This enables to integration of active components with diamond devices (see Section 5) and thus makes this hybrid system an attractive alternative to all-diamond photonic platforms.

4. Active components

Besides the passive photonic components described above, active components for reconfiguring the optical properties of quantum PICs after fabrication are required. For instance, in applications where quantum interference from different single-photon emitters is required, achieving high emission efficiency from each single photon emitter is needed while at the same time spectrally overlapping the emission wavelengths in order to render the photons indistinguishable. The emission efficiency can be increased using Purcell enhancement by coupling the emitter to an optical cavity. Yet, fabrication inhomogeneity across a chip makes it challenging to realize cavities with precisely the same resonance wavelength. To overcome spectral spread tuning techniques are required to actively align the resonances of individual cavities with respect to each other. The same challenge arises for aligning the single photon emitters. The emission wavelengths of photons emitted from different color centers differ because of variations in the electric environment of the host material. In order to boost the indistinguishability of the emitted photons each emitter needs to operate at exactly the same frequency. Active PIC tuning is also required to change the functionality of an integrated circuit, e.g. for controlling the routing of single photons through the chip or to perform reconfigurable linear optical operations on quantum states of light.

To actively control cavities or to route light across the chip, the effective refractive index of the propagating modes has to be locally altered in a controllable way. In established semiconducting materials such as silicon or GaAs, this change of the refractive index can be achieved by free carrier injection, by using elctro-optical modification or use the thermos-optical effect. In diamond these tuning mechanism are not available because of the large electronic bandgap. In addition, injection of free carriers would also perturb the charge state of color centers.

Instead, methods based on gas adsorption to shift the resonance to higher wavelengths can be used. Other approaches, albeit non-volatile, use the deposition and etching of an additional material, e.g., HfO_2, to shift the resonance frequency. Even though this way individual cavities can be tuned, both methods affect the complete chip and are therefore not site-specific. A locally addressable approach relies on resistive microheaters to exploit the temperature dependence of the refractive index for local tuning. This approach is used in particular on silicon-based material platforms. Microheaters suffer however from thermal crosstalk and large power dissipation. These side effects limit the integration density. Importantly, heavy thermal load is incompatible with cryogenic operation needed for integrated superconducting single photon detectors.

An alternative way to actively tune the effective index of diamond waveguide uses electro-mechanical tuning (Poot and Tang, 2014a). Using mechanically movable waveguides, the effective refractive index of a desired device can be locally altered by bringing the free-standing part closer to the waveguide (see Fig. 5). Depending on the material system, the actuation of the moving arm can be achieved by electrostatic forces, or by the piezoelectric effect for hybrid AlN/diamond systems.

In monolithic diamond PICs, electro-mechanical tuning enables the realization of cryo-compatible variable phase shifters. In Mach Zehnder interferometer configuration, phase shifters allow switching between constructive interference (transmission) and destructive interference (reflection), and thus enable the routing of light inside a waveguide network (Deotare et al., 2012). In photonic quantum circuits, variable phase shifters are essential elements for the implementation of arbitrary linear optics transformations (Carolan et al., 2015). In Fig. 4, the experimental implementation of an opto-mechanical phase-shifter on a diamond on insulator is shown (Rath et al., 2014). This component delivers an additional important building block for a diamond-based integrated quantum photonics architecture. Because of the high Young's modulus of diamond, the device can be operated at frequencies up to 115 MHz, demonstrating the potential for fast routing applications (Poot and Tang, 2014b).

With electro-mechanical tuning the resonance wavelength of optical cavities can also be changed to a desired value. An electro-mechanically tuneable optical cavity was recently proposed (Mouradian and Englund, 2017) for the hybrid AlN/diamond platform (see Fig. 5A–C). Electro-static tuning can be employed to increase the indistinguishability of single photon emitters with different spectral properties (Tamarat et al., 2006). The patterning of electrodes around a single emitter makes it possible to compensate

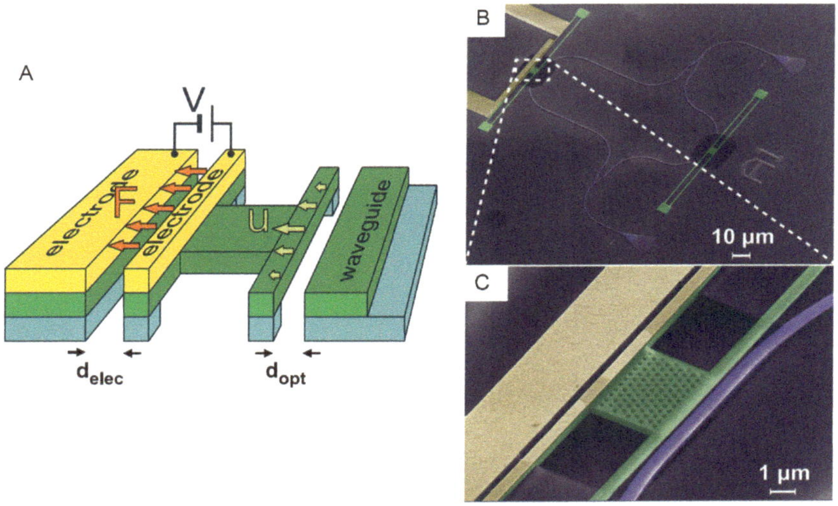

Fig. 4 Electro-mechanical phase shifter. (A) Operation principle of an evanescently coupled opto-mechanical phase shifter made from silicon nitride. (B) Electro-mechanical phase shifter on a diamond on insulator platform in a Mach Zehnder interferometer configuration. (C) Detailed view of the free-standing region of the phase shifter. The distance between the movable arm (green) and the waveguide (purple) can tuned via the voltage applied between the electrodes (yellow).

for local charge inhomogeneities in the vicinity of the color center by tuning its emission wavelength via the DC Stark effect. Using this technique, dynamic stabilization of the emission wavelength of NV centers in the range of minutes was shown (Acosta et al., 2012) and two-photon interference from two different NV centers (Sipahigil et al., 2012) was achieved. Electrical tuning and stabilization of NV centers in an integrated optical circuit has also been recently demonstrated on the hybrid GaP-on-diamond photonic platform (see Fig. 5D and E).

5. Waveguide integrated single-photon detectors

Having discussed passive and active photonic elements, a last ingredient for functional PICs are co-integrated single photon detectors.

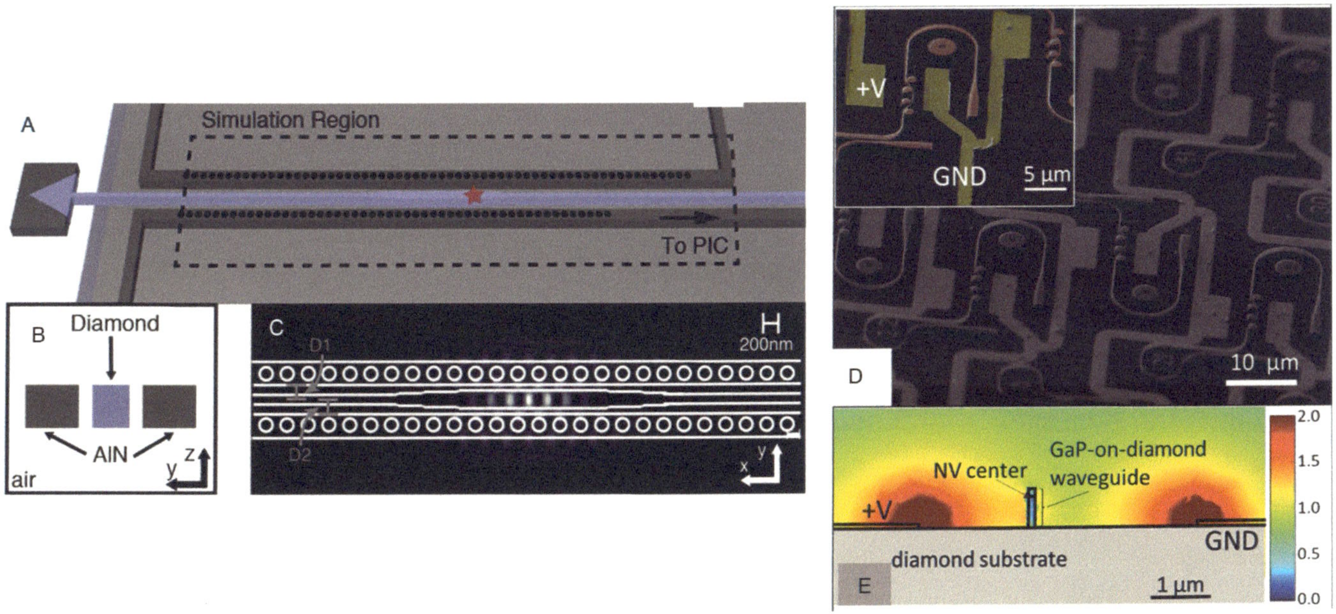

Fig. 5 Active photonic elements in hybrid architectures. (A) Electro-mechanical displacement of the AlN arms is used to tune the resonance frequency of the cavity. (B) Sideview of the center of the cavity. (C) Simulated cavity mode profile. (D) SEM image showing several integrated devices. Electrodes patterned on the side of microdisk resonators containing single NV centers are used to tune their emission wavelength via the Stark effect. (E) Device cross section with simulated electric field magnitude.

Efficient single-photon detectors are essential to further develop optical quantum technologies. Single-photon detectors are needed to measure the state of a photonic qubit in quantum computing and communication protocols. They are also needed to take advantage of measurement-induced nonlinearities for generation of two-photon entanglement (Knill et al., 2001). Typically, such measurements are performed by measuring light from the chip with external single-photon detectors, such as avalanche photodiodes (APDs). This approach comes with several disadvantages, such as the introduction of additional losses to the system and limited scalability. In order to circumvent these issues, the detectors would have to be integrated directly on chip. Following this methodology, superconducting nanowire single photon detectors (SNSPDs) (Gol'tsman et al., 2001) have been patterned on top of optical waveguides. SNSPDs feature detection efficiencies in excess of 90% (Pernice et al., 2012) and, importantly, provide scalability. Waveguide integrated SNSPDs have shown superior performance over all their competing platforms, including dark-count rates lower than 10 Hz (Schuck et al., 2013), and GHz count rates at low timing jitter ($\simeq$32 ps) (Vetter et al., 2016).

Typical SNSPDs are patterned into thin (width < 100 nm) nanowire prepared from a superconducting thin film (typical thickness ~ 5 nm). These nanowires become superconducting if cooled below the critical temperature of the film. For sufficiently narrow wires the absorption of one single photon leads to the breakdown of the superconducting state, resulting in a voltage spike which can be amplified and detected by counting electronics.

SNSPDs have been integrated on a diamond platform using different superconducting films. Niobium titanium nitride (NbTiN) thin films deposited on polished single crystalline diamond substrates have been shown similar performances to other substrate materials (Atikian et al., 2014). On top of PCD diamond waveguides (see Fig. 6) SNSPDs have been prepared from niobium nitride (NbN) thin films (Kahl et al., 2016; Rath et al., 2015). The detectors showed an on chip detection efficiency up to $\simeq$50%, count rates in the MHz regime, and a timing jitter <30 ps. The results show that SNSPDs are suitable for developing fully integrated quantum photonic circuits based on the diamond platform.

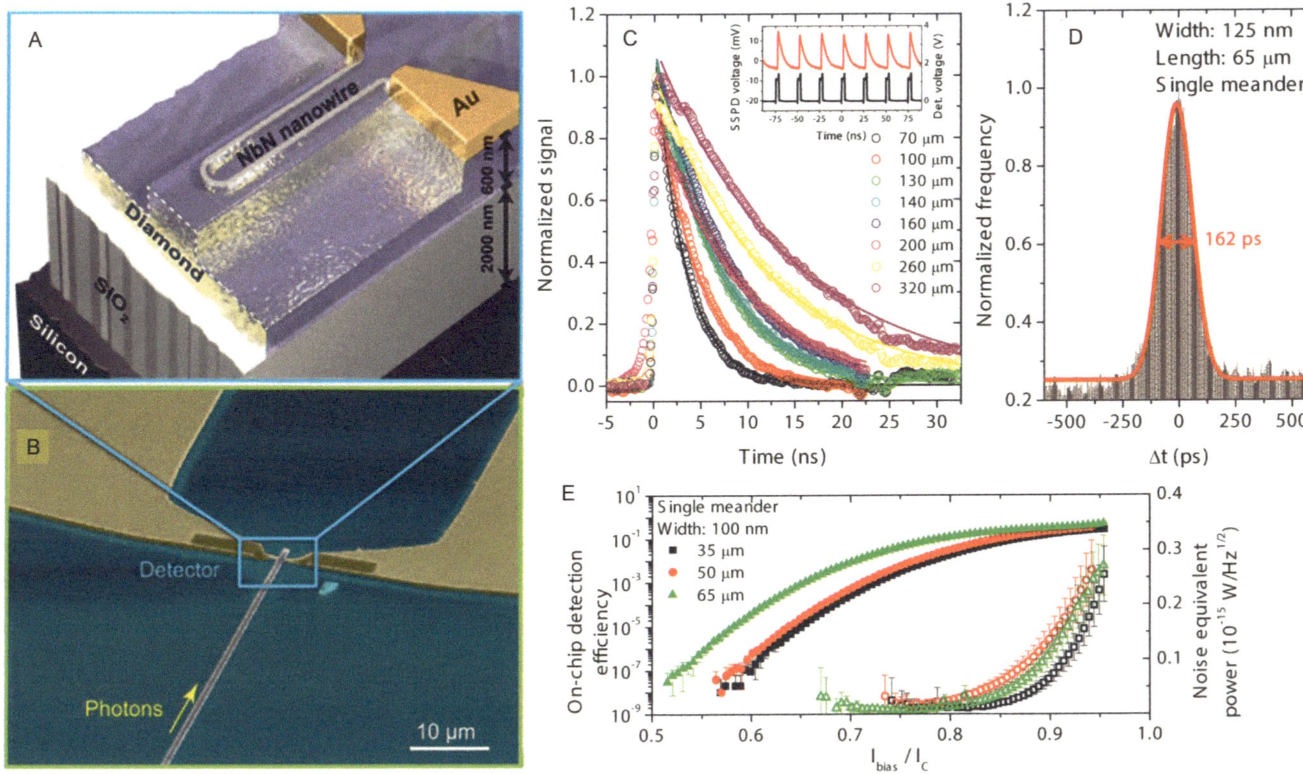

Fig. 6 SNSPDs patterned on top of diamond waveguides. (A) Rendered image of a waveguide SNSPD. (B) False color electron micrograph of detector. Photons guided in the waveguide get absorbed along the length of the detector. (C) Pulse traces recorded for different nanowire lengths. Smaller wires allow faster repetition rates. Inset: comparison of SNSPD traces (red) and signal from 12 GHz diode (black). (D) Exemplary timing jitter. (E) On-Chip detection efficiency and noise equivalent power measured versus normalized bias current for different nanowire lengths.

6. Conclusion

With recent progress in the realization of passive and active integrated optical devices, Diamond has been firmly established as a promising platform for quantum photonics. Current challenges lie in the co-integration of all elements required for functional devices in a single circuit. Besides demanding design questions, technical challenges lie in the optimization of photonic devices for low loss and large-scale integration. In addition, further efforts in the development of optical filter techniques for pump suppression of single photon emitters are needed, as well as improvement in the performance of single-photon detectors. Nevertheless, diamond-based integrated quantum photonics has matured to a technology level able to compete with other platforms, such as silicon quantum photonics, which received higher attention in past years. At a circuit level, realizing large numbers of identical elements requires further engineering, which, however, equally applies to alternative material platforms.

References

Acosta, V.M., Santori, C., Faraon, A., Huang, Z., Fu, K.M.C., Stacey, A., Simpson, D.A., Ganesan, K., Tomljenovic-Hanic, S., Greentree, A.D., Prawer, S., Beausoleil, R.G., 2012. Phys. Rev. Lett. 108, 6.

Aharonovich, I., Lee, J.C., Magyar, A.P., Buckley, B.B., Yale, C.G., Awschalom, D.D., Hu, E.L., 2012. Adv. Mater. 24, OP54.

Atikian, H.A., Eftekharian, A., Jafari Salim, A., Burek, M.J., Choy, J.T., Hamed Majedi, A., Lončar, M., 2014. Appl. Phys. Lett. 104, 122602.

Atikian, H.A., Latawiec, P., Burek, M.J., Sohn, Y.-I., Meesala, S., Gravel, N., Kouki, A.B., Lončar, M., 2017. APL Photonics 2, 051301.

Babinec, T.M., Hausmann, B.J.M., Khan, M., Zhang, Y., Maze, J.R., Hemmer, P.R., Lončar, M., 2010. Nat. Nanotechnol. 5, 195.

Barth, M., Nüsse, N., Löchel, B., Benson, O., 2009. Opt. Lett. 34, 1108.

Bayn, I., Meyler, B., Lahav, A., Salzman, J., Kalish, R., Fairchild, B.A., Prawer, S., Barth, M., Benson, O., Wolf, T., Siyushev, P., Jelezko, F., Wrachtrup, J., 2011. Diamond Relat. Mater. 20, 937.

Bayn, I., Mouradian, S., Li, L., Goldstein, J.A., Schröder, T., Zheng, J., Chen, E.H., Gaathon, O., Lu, M., Stein, A., Ruggiero, C.A., Salzman, J., Kalish, R., Englund, D., 2014. Appl. Phys. Lett. 105, 211101.

Becker, J.N., Becher, C., 2017. Phys. Status Solidi 214, 1700586.

Bharadwaj, V., Courvoisier, A., Fernandez, T.T., Ramponi, R., Galzerano, G., Nunn, J., Booth, M.J., Osellame, R., Eaton, S.M., Salter, P.S., 2017. Opt. Lett. 42, 3451.

Bhaskar, M.K., Sukachev, D.D., Sipahigil, A., Evans, R.E., Burek, M.J., Nguyen, C.T., Rogers, L.J., Siyushev, P., Metsch, M.H., Park, H., Jelezko, F., Lončar, M., Lukin, M.D., 2017. Phys. Rev. Lett. 118, 1.

Boyd, R.W., 2008. Nonlinear Optics. Academic Press.

Burek, M.J., De Leon, N.P., Shields, B.J., Hausmann, B.J.M., Chu, Y., Quan, Q., Zibrov, A.S., Park, H., Lukin, M.D., Lončar, M., 2012. Nano Lett. 12, 6084.

Burek, M.J., Chu, Y., Liddy, M.S.Z., Patel, P., Rochman, J., Meesala, S., Hong, W., Quan, Q., Lukin, M.D., Lončar, M., 2014. Nat. Commun. 5, 5718.
Burek, M.J., Cohen, J.D., Meenehan, S.M., El-Sawah, N., Chia, C., Ruelle, T., Meesala, S., Rochman, J., Atikian, H.A., Markham, M., Twitchen, D.J., Lukin, M.D., Painter, O., Lončar, M., 2016. Optica 3, 1404.
Carolan, J., Harrold, C., Sparrow, C., Martin-Lopez, E., Russell, N.J., Silverstone, J.W., Shadbolt, P.J., Matsuda, N., Oguma, M., Itoh, M., Marshall, G.D., Thompson, M.G., Matthews, J.C.F., Hashimoto, T., O'Brien, J.L., Laing, A., 2015. Science 349, 711.
Courvoisier, A., Booth, M.J., Salter, P.S., 2016. Appl. Phys. Lett. 109, 031109.
Deotare, P.B., Bulu, I., Frank, I.W., Quan, Q., Zhang, Y., Ilic, R., Loncar, M., 2012. Nat. Commun. 3, 846.
Dore, P., Nucara, A., Cannavò, D., De Marzi, G., Calvani, P., Marcelli, A., Sussmann, R.S., Whitehead, A.J., Dodge, C.N., Krehan, A.J., Peters, H.J., 1998. Appl. Optics 37, 5731.
Englund, D., Shields, B., Rivoire, K., Hatami, F., Vučković, J., Park, H., Lukin, M.D., 2010. Nano Lett. 10, 3922.
Fairchild, B.A., Olivero, P., Rubanov, S., Greentree, A.D., Waldermann, F., Taylor, R.A., Walmsley, I., Smith, J.M., Huntington, S., Gibson, B.C., Jamieson, D.N., Prawer, S., 2008. Adv. Mater. 20, 4793.
Faraon, A., Barclay, P.E., Santori, C., Fu, K.M.C., Beausoleil, R.G., 2011. Nat. Photonics 5, 301.
Faraon, A., Santori, C., Huang, Z., Acosta, V.M., Beausoleil, R.G., 2012. Phys. Rev. Lett. 109, 33604.
Fu, K.M.C., Santori, C., Barclay, P.E., Aharonovich, I., Prawer, S., Meyer, N., Holm, A.M., Beausoleil, R.G., 2008. Appl. Phys. Lett. 93, 234107.
Füner, M., Wild, C., Koidl, P., 1998. Appl. Phys. Lett. 72, 1149.
Gaathon, O., Hodges, J.S., Chen, E.H., Li, L., Bakhru, S., Bakhru, H., Englund, D., Osgood, R.M., 2013. Opt. Mater. (Amst) 35, 361.
Gao, F., Huang, Z., Feigel, B., Van Erps, J., Thienpont, H., Beausoleil, R.G., Vermeulen, N., 2016. J. Light. Technol. 34, 5576.
Gol'tsman, G.N., Okunev, O., Chulkova, G., Lipatov, A., Semenov, A., Smirnov, K., Voronov, B., Dzardanov, A., Williams, C., Sobolewski, R., 2001. Appl. Phys. Lett. 79, 705.
Gould, M., Schmidgall, E.R., Dadgostar, S., Hatami, F., Fu, K.M.C., 2016. Phys. Rev. Appl. 6, 011001.
Gsell, S., Bauer, T., Goldfuß, J., Schreck, M., Stritzker, B., 2004. Appl. Phys. Lett. 84, 4541.
Hadden, J.P., Harrison, J.P., Stanley-Clarke, A.C., Marseglia, L., Ho, Y.-L.D., Patton, B.R., O'Brien, J.L., Rarity, J.G., 2010. Appl. Phys. Lett. 97, 241901.
Hadden, J.P., Bharadwaj, V., Sotillo, B., Rampini, S., Osellame, R., Witmer, J.D., Jayakumar, H., Fernandez, T.T., Chiappini, A., Armellini, C., Ferrari, M., Ramponi, R., Barclay, P.E., Eaton, S.M., 2018. Opt. Lett. 43, 3586.
Hanay, M.S., Kelber, S., Naik, A.K., Chi, D., Hentz, S., Bullard, E.C., Colinet, E., Duraffourg, L., Roukes, M.L., 2012. Nat. Nanotechnol. 7, 602.
Hausmann, B.J.M., Shields, B.J., Quan, Q., Chu, Y., de Leon, N.P., Evans, R., Burek, M.J., Zibrov, A.S., Markham, M., Twitchen, D.J., Park, H., Lukin, M.D., Loncăr, M., 2013. Nano Lett. 13, 5791.
Hausmann, B.J.M., Bulu, I., Venkataraman, V., Deotare, P., Loncar, M., 2014. Nat. Photonics 8, 369.
Hiscocks, M.P., Ganesan, K., Gibson, B.C., Huntington, S.T., Ladouceur, F., Prawer, S., 2008. Opt. Express 16, 19512.
Jelezko, F., Gaebel, T., Popa, I., Domhan, M., Gruber, A., Wrachtrup, J., 2004. Phys. Rev. Lett. 93, 130501.

Jung, T., Kreiner, L., Pauly, C., Mücklich, F., Edmonds, A.M., Markham, M., Becher, C., 2016. Phys. Status Solidi Appl. Mater. Sci. 213, 3254.
Kahl, O., Ferrari, S., Rath, P., Vetter, A., Nebel, C., Pernice, W.H.P., 2016. J. Light. Technol. 34, 249.
Khanaliloo, B., Jayakumar, H., Hryciw, A.C., Lake, D.P., Kaviani, H., Barclay, P.E., 2015a. Phys. Rev. X 5, 041051.
Khanaliloo, B., Mitchell, M., Hryciw, A.C., Barclay, P.E., 2015b. Nano Lett. 15, 5131.
Khasminskaya, S., Pyatkov, F., Słowik, K., Ferrari, S., Kahl, O., Kovalyuk, V., Rath, P., Vetter, A., Hennrich, F., Kappes, M.M., Gol'tsman, G., Korneev, A., Rockstuhl, C., Krupke, R., Pernice, W.H.P., 2016. Nat. Photonics 10, 727.
Knill, E., Laflamme, R., Milburn, G.J., 2001. Nature 409, 46.
Kurtsiefer, C., Mayer, S., Zarda, P., Weinfurter, H., 2000. Phys. Rev. Lett. 85, 290.
Laing, A., Peruzzo, A., Politi, A., Verde, M.R., Halder, M., Ralph, T.C., Thompson, M.G., O'Brien, J.L., 2010. Appl. Phys. Lett. 97, 211109.
Latawiec, P., Venkataraman, V., Burek, M.J., Hausmann, B.J.M., Bulu, I., Lončar, M., 2015. Optica 2, 924.
Lee, K.C., Sussman, B.J., Sprague, M.R., Michelberger, P., Reim, K.F., Nunn, J., Langford, N.K., Bustard, P.J., Jaksch, D., Walmsley, I.A., 2012. Nat. Photonics 6, 41.
Leifgen, M., Schröder, T., Gädeke, F., Riemann, R., Métillon, V., Neu, E., Hepp, C., Arend, C., Becher, C., Lauritsen, K., Benson, O., 2014. New J. Phys. 16, 023021.
Levenson, M.D., Bloembergen, N., 1974. Phys. Rev. B 10, 4447.
Li, L., Trusheim, M., Gaathon, O., Kisslinger, K., Cheng, C.-J., Lu, M., Su, D., Yao, X., Huang, H.-C., Bayn, I., Wolcott, A., Osgood, R.M., Englund, D., 2013. J. Vac. Sci. Technol. B, Nanotechnol. Microelectron. Mater. Process. Meas. Phenom 31, 06FF01.
Li, L., Schröder, T., Chen, E.H., Walsh, M., Bayn, I., Goldstein, J., Gaathon, O., Trusheim, M.E., Lu, M., Mower, J., Cotlet, M., Markham, M.L., Twitchen, D.J., Englund, D., 2015a. Nat. Commun. 6, 6173.
Li, L., Bayn, I., Lu, M., Nam, C.-Y., Schröder, T., Stein, A., Harris, N.C., Englund, D., 2015b. Sci. Rep. 5, 7802.
Li, L., Chen, E.H., Zheng, J., Mouradian, S.L., Dolde, F., Schröder, T., Karaveli, S., Markham, M.L., Twitchen, D.J., Englund, D., 2015c. Nano Lett. 15, 1493.
Liao, M., Li, C., Hishita, S., Koide, Y., 2010. J. Micromech. Microeng. 20, 085002.
Ma, X., Zotter, S., Tetik, N., Qarry, A., Jennewein, T., Zeilinger, A., 2011. Opt. Express 19, 22723.
Magyar, A.P., Lee, J.C., Limarga, A.M., Aharonovich, I., Rol, F., Clarke, D.R., Huang, M., Hu, E.L., 2011. Appl. Phys. Lett. 99, 081913.
Martin, A., Alibart, O., De Micheli, M.P., Ostrowsky, D.B., Tanzilli, S., 2012. New J. Phys. 14, 25002.
Matthews, J.C.F., Politi, A., Bonneau, D., O'Brien, J.L., 2011. Phys. Rev. Lett. 107, 163602.
Michl, J., Teraji, T., Zaiser, S., Jakobi, I., Waldherr, G., Dolde, F., Neumann, P., Doherty, M.W., Manson, N.B., Isoya, J., Wrachtrup, J., 2014. Appl. Phys. Lett. 104, 102407.
Mildren, R.P., Rabeau, J., 2013. Optical Engineering of Diamond. Wiley-VCH Verlag GmbH & Co, KGaA.
Mildren, R.P., Butler, J.E., Rabeau, J.R., 2008. Opt. Express 16, 18950.
Mitchell, M., Khanaliloo, B., Lake, D.P., Masuda, T., Hadden, J.P., Barclay, P.E., 2016. Optica 3, 963.
Mizuochi, N., Makino, T., Kato, H., Takeuchi, D., Ogura, M., Okushi, H., Nothaft, M., Neumann, P., Gali, A., Jelezko, F., Wrachtrup, J., Yamasaki, S., 2012. Nat. Photonics 6, 299.

Mollart, T.P., Lewis, K.L., Pickles, C.S.J., Wort, C.J.H., 2003. Semicond. Sci. Technol., 18. https://doi.org/10.1088/0268-1242/18/3/317.
Mouradian, S.L., Englund, D., 2017. APL Photonics 2, 046103.
Mouradian, S.L., Schröder, T., Poitras, C.B., Li, L., Goldstein, J., Chen, E.H., MichaelWalsh, J., Cardenas, M., Markham, L., Twitchen, D.J., Lipson, M., Englund, D., 2015. Phys. Rev. X 5, 031009.
Najar, H., Chan, M.L., Yang, H.A., Lin, L., Cahill, D.G., Horsley, D.A., 2014. Appl. Phys. Lett. 104, 151903.
Nebel, C.E., 2003. J. Ristein, Semiconductors and Semimetals: Thin-Film Diamond I. Elsevier.
Nemanich, R.J., Carlisle, J.A., Hirata, A., Haenen, K., 2014. MRS Bull. 39, 490.
O'Brien, J.L., Furusawa, A., Vučković, J., 2009. Nat. Photonics 3, 687.
Pernice, W.H.P., Schuck, C., Minaeva, O., Li, M., Goltsman, G.N., Sergienko, A.V., Tang, H.X., 2012. Nat. Commun. 3, 1325.
Phillip, H.R., Taft, E.A., 1964. Phys. Rev. 136, A1445.
Politi, A., Matthews, J.C.F., O'Brien, J.L., 2009. Science 325, 1221.
Poot, M., Tang, H.X., 2014a. Appl. Phys. Lett. 104, 061101.
Poot, M., Tang, H.X., 2014b. Appl. Phys. Lett. 104, 061101.
Rath, P., Khasminskaya, S., Nebel, C., Wild, C., Pernice, W.H.P., 2013a. Nat. Commun. 4, 1690.
Rath, P., Gruhler, N., Khasminskaya, S., Nebel, C., Wild, C., Pernice, W.H.P., 2013b. Opt. Express 21, 11031.
Rath, P., Ummethala, S., Diewald, S., Lewes-Malandrakis, G., Brink, D., Heidrich, N., Nebel, C., Pernice, W.H.P., 2014. Appl. Phys. Lett., 105. https://doi.org/10.1063/1.4901105.
Rath, P., Kahl, O., Ferrari, S., Sproll, F., Lewes-Malandrakis, G., Brink, D., Ilin, K., Siegel, M., Nebel, C., Pernice, W., 2015. Light Sci. Appl. 4, 1.
Riedrich-Möller, J., Kipfstuhl, L., Hepp, C., Neu, E., Pauly, C., Mücklich, F., Baur, A., Wandt, M., Wolff, S., Fischer, M., Gsell, S., Schreck, M., Becher, C., 2012. Nat. Nanotechnol. 7, 69.
Robledo, L., Childress, L., Bernien, H., Hensen, B., Alkemade, P.F.A., Hanson, R., 2011. Nature 477, 574.
Schreck, M., Asmussen, J., Shikata, S., Arnault, J.C., Fujimori, N., 2014. MRS Bull. 39, 504.
Schröder, T., Gädeke, F., Banholzer, M.J., Benson, O., 2011. New J. Phys. 13, 55017.
Schuck, C., Pernice, W.H.P., Tang, H.X., 2013. Appl. Phys. Lett. 102, 051101.
Sipahigil, A., Goldman, M.L., Togan, E., Chu, Y., Markham, M., Twitchen, D.J., Zibrov, A.S., Kubanek, A., Lukin, M.D., 2012. Phys. Rev. Lett. 108, 1.
Sipahigil, A., Evans, R.E., Sukachev, D.D., Burek, M.J., Borregaard, J., Bhaskar, M.K., Nguyen, C.T., Pacheco, J.L., Atikian, H.A., Meuwly, C., Camacho, R.M., Jelezko, F., Bielejec, E., Park, H., Lončar, M., Lukin, M.D., 2016. Science 354, 847.
Spring, J.B., Mennea, P.L., Metcalf, B.J., Humphreys, P.C., Gates, J.C., Rogers, H.L., Söller, C., Smith, B.J., Kolthammer, W.S., Smith, P.G.R., Walmsley, I.A., 2017. Optica 4, 90.
Tamarat, P., Gaebel, T., Rabeau, J.R., Khan, M., Greentree, A.D., Wilson, H., Hollenberg, L.C.L., Prawer, S., Hemmer, P., Jelezko, F., Wrachtrup, J., 2006. Phys. Rev. Lett. 97, 1.
Tanzilli, S., Martin, A., Kaiser, F., De Micheli, M.P., Alibart, O., Ostrowsky, D.B., 2011. Laser Photon. Rev. 6, 115.
Tao, Y., Degen, C., 2013. Adv. Mater. 25, 3962.
Tao, Y., Boss, J.M., Moores, B.A., Degen, C.L., 2014. Nat. Commun. 5, 3638.

Teraji, T., Yamamoto, T., Watanabe, K., Koide, Y., Isoya, J., Onoda, S., Ohshima, T., Rogers, L.J., Jelezko, F., Neumann, P., Wrachtrup, J., Koizumi, S., 2015. Phys. status solidi 212, 2365.

Vetter, A., Ferrari, S., Rath, P., Alaee, R., Kahl, O., Kovalyuk, V., Diewald, S., Goltsman, G.N., Korneev, A., Rockstuhl, C., Pernice, W.H.P., 2016. Nano Lett. 16, 7085.

Wang, C.F., Hanson, R., Awschalom, D.D., Hu, E.L., Feygelson, T., Yang, J., Butler, J.E., 2007. Appl. Phys. Lett. 91, 201112.

Wang, J., Paesani, S., Ding, Y., Santagati, R., Skrzypczyk, P., Salavrakos, A., Tura, J., Augusiak, R., Mančinska, L., Bacco, D., Bonneau, D., Silverstone, J.W., Gong, Q., Acín, A., Rottwitt, K., Oxenløwe, L.K., O'Brien, J.L., Laing, A., Thompson, M.G., 2018. Science.

Zachreson, C., Martin, A.A., Aharonovich, I., Toth, M., 2014. ACS Appl. Mater. Interfaces 6, 10367.

Zaitsev, A.M., 2001. Optical Properties of Diamond: A Data Handbook. Springer.

Zhang, J.L., Sun, S., Burek, M.J., Dory, C., Tzeng, Y.-K., Fischer, K.A., Kelaita, Y., Lagoudakis, K.G., Radulaski, M., Shen, Z.-X., Melosh, N.A., Chu, S., Lončar, M., Vučković, J., 2018. Nano Lett. 18, 1360.

CHAPTER FIVE

Diamond membranes for photonic devices

Kerem Bray, Blake Regan, and Igor Aharonovich*
School of Mathematical and Physical Sciences, University of Technology Sydney, NSW, Australia
*Corresponding author: e-mail address: igor.aharonovich@uts.edu.au

Contents

1. Introduction 173
2. Diamond devices 176
3. Diamond membrane fabrication 178
   3.1 Fabrication of diamond membranes 179
   3.2 Diamond overgrowth 181
   3.3 Fabrication of p-type diamond membranes 181
   3.4 Flip and thin 182
   3.5 Optical characterization of diamond membranes 184
   3.6 Fabrication of diamond color centers 185
4. Fabrication of optical resonators from diamond membranes 186
   4.1 Photonic crystal cavities 188
5. Electroluminescence of diamond 191
6. Single crystal diamond membranes for nanoelectronics 197
   6.1 Characterization of p-type diamond membranes 199
   6.2 Fabrication of vertical nanoscale p-n junctions 200
   6.3 Characterization of vertical nanoscale p-n junctions 200
   6.4 Characterization of vertical p-i-n junctions 203
7. Diamond electroluminescence 210
References 211

1. Introduction

Decades of rapid advances in quantum information technology, especially lasers and light emitting diodes, have led to the development of devices that are high performance, affordable and reliable. New concepts in "top-down" or "bottom up" synthesis have driven the discovery and research of new candidates for uses in nano-optics and nanoelectronics (Aharonovich et al., 2013; Lu and Lieber, 2007).

Semiconductors and Semimetals, Volume 104
ISSN 0080-8784
https://doi.org/10.1016/bs.semsem.2020.08.002

Quantum technology spans numerous applications including quantum metrology, quantum key distribution for secured communications and most ambitiously the quantum computer (Awschalom et al., 2007; Gisin et al., 2002; Mouradian et al., 2015). The fast-growing need for stronger and faster computational devices has led to the investigation of using quantum mechanical devices as bits, or qubits, which would enable faster processing speeds when compared to classical bits (Kieling et al., 2007; Knill, 2010). Simply, qubits have the possibility of an additional, superimposed state of two bits forming a third state. This allows for simultaneous computations, an example of this was recently exploited by Google to achieve quantum supremacy where they created quantum states on 53 qubits. Quantum supremacy is the demonstration that a programmable quantum device can solve a problem that classical computers practically cannot (Arute et al., 2019). Furthermore, to fabricate these practical devices and components for quantum applications, the material would need to be efficient, reliable, scalable and allow manipulation of individual quantum states at a nanometer scale.

Recently, there has been a lot of interest in non-classical light sources, where an emitter can produce controllable, quantized packets of light. A fundamental building block for these emitters is single photon emitters which are very important for many scalable quantum technologies (Buckley et al., 2012; Lodahl et al., 2015; Northup and Blatt, 2014; O'Brien et al., 2009). Single photon emission is desirable for quantum applications because it travels at the speed of light, possesses low noise, and allows for free space communication (Alléaume et al., 2004; Lounis and Orrit, 2005; O'Brien et al., 2009).

Promising candidates for single photon applications are diamond color centers. They possess various outstanding properties such as room temperature operation, photostability and ease of fabrication. Diamond can host several single photon emitters for nanophotonic applications. The most prominently studied defect is the Nitrogen vacancy (NV) defect due to its interesting optical, read out and spin properties (Childress and Hanson, 2013; Jelezko and Wrachtrup, 2004). However, it possesses a few drawbacks such as a broad photoluminescence spectrum where the zero-phonon line comprises of $\sim$4%. This low emission leads to a diminished count rate of single photons being emitted from the NV defect. Silicon vacancy (SiV) centers have been suggested as an alternative to NV color centers, owing to their narrow zero-phonon line (ZPL) width of $\sim$5 nm at room temperature along with their small phonon-coupling (Wang et al., 2006).

Furthermore, its emission is in the near-infrared at 737 nm and has a short lifetime of 1.2 ns. However, they possess low single photon count rates (~1000 counts/s), a quantum yield of 0.05 and the fluorescence is quenched in the presence of graphite which leads to the need for high-quality diamond (Aharonovich, 2010; Neu et al., 2011a, 2012).

Single crystal diamond containing SiV color centers is imperative for a range of optoelectrical quantum applications because diamond possess a high breakdown voltage (~10^7 Vcm^{-1}) and relatively high carrier mobilities (0.2 m^{2}V^{-1} s^{-1}). They have recently been realized for optoelectronic applications, where electroluminescence signal has been observed from SiV color centers (Berhane et al., 2015; Liang et al., 2007). Electrical excitation of diamond is essential to engineer scalable devices utilizing quantum defects to fabricate efficient single chip devices (Ellis et al., 2011; Mizuochi et al., 2012a; Stevenson et al., 2012).

Diamond planar or membrane devices are the next steppingstones toward optically integrated circuits. While polycrystalline diamond membranes have been available technology since the 1990s (Windischmann and Epps, 1992), these were synthesized via MPCVD deposition into silicon wafers and could reach diameters of up to 75 mm. However, the ability to fabricate single crystal diamond membranes of sufficient quality for optical resonators wasn't realized until 2008 (Fairchild et al., 2008). The fabrication of free-standing single crystal diamond membranes has led to high-quality waveguides, ring resonators and nano electromechanical systems (NEMS). The free-standing membranes led to the collection efficiency from single photon emitters embedded in cavity structures (i.e., whispering gallery modes and ring resonators) to be enhanced through the Purcell effect (Purcell et al., 1946; Su et al., 2008). Free-standing single crystal diamond membranes have been fabricated to be as thin as 200 nm (Bormashov et al., 2017; Fairchild et al., 2008; Magyar et al., 2014) and numerous devices have been fabricated, including photonic crystals (Lee et al., 2013), cavities (Hausmann et al., 2013a), nanomechanical resonators (MacQuarrie et al., 2016; Tao et al., 2014), diodes (Berhane et al., 2015) and other novel structures. For quantum applications it is favorable to fabricate structures and devices from high-quality single crystal diamond.

Furthermore, to exploit the optical properties of diamond color centers, it is imperative to understand their internal energy level structure, in particular SiV color centers. It was first shown by Sittas et al. (1996) that the electron excitation energy for the SiV defect is 1.68 eV. This was determined by photoluminescence excitation (PLE) from 490 to 690 nm using a 250 W halogen lamp. PLE measurements provide information about dephasing mechanisms and optical transitions of emitters at cryogenic temperatures (<10 K).

Another benefit of using diamond is its biocompatibility and surface tuneability. This feature of diamonds can be exploited for a range of therapeutic and diagnostic applications from targeting various processes, monitoring pathways and even drug delivery. There has been an abundance of research utilizing diamonds containing NV color centers for bio-imaging and drug delivery. However, the NV defects possess a wide fluorescent band emission in the visible range that contributes to tissue absorption and autofluorescence (Alhaddad et al., 2011; Balasubramanian et al., 2014; Fu et al., 2007; Li et al., 2010; McGuinness et al., 2011; Merson et al., 2013). Therefore, using diamond containing the near-infrared SiV color center is ideal.

2. Diamond devices

Diamond materials with hosted color centers are promising options for myriad of integrated nanophotonics, optoelectronics and quantum information processing applications (Aharonovich and Neu, 2014; Childress and Hanson, 2013; Hausmann et al., 2012; Riedrich-Möller et al., 2011). As mentioned previously, diamond possesses numerous unique and interesting properties, including a wide bandgap (5.5 eV), high breakdown voltage ($\sim 10^7$ V cm^{-1}) (Isberg et al., 2002) and a high carrier mobility (~ 0.2 m^{2}V^{-1} s^{-1}), high thermal conductivities (>2000 Wm^{-1}K^{-1}) and a breakdown field strength of 10 MVcm^{-1} (Wort and Balmer, 2008). These properties make diamond ideal for high power, high energy, and high frequency optoelectronic devices. Therefore, diamond holds many advantages over its competitors, namely silicon carbide (SiC) and gallium nitride (GaN) (Table 1). The main advantage of using diamond for all the mentioned applications is that it allows operation under extreme conditions due to its materials hardness, chemical inertness, wear resistance and high thermal conductivity.

To produce electroluminescence (EL) from a diamond defect, the p-i-n diode device requires fabrication of a high-quality n-type (electron) and p-type (holes) conducting material and an intrinsic (i) layer. Diamonds are generally doped with high concentrations of boron and phosphorous to create the p-type and n-type layer, respectively (Koizumi et al., 2001; Makino et al., 2011). The doped diamond layers are typically created using

Table 1 Material properties for silicon (Si), 4H-silicon carbide (4H-SiC), gallium nitride (GaN), natural diamond and CVD grown diamond (Isberg et al., 2002; Wort and Balmer, 2008).

Material	Si	4H-SiC	GaN	Natural diamond	CVD diamond
Bandgap (eV)	1.1	3.2	3.44	5.47	5.47
Breakdown field ($MVcm^{-1}$)	0.3	3	5	10	10
Electron saturation velocity ($\times 10^7 cm s^{-1}$)	0.86	3	2.5	2	2
Hole saturation velocity ($\times 10^7 cm s^{-1}$)	n/a	n/a	n/a	0.8	0.8
Electron mobility ($cm^2 V^{-1} s^{-1}$)	1450	900	450	200–2800	4500
Hole mobility ($cm^2 V^{-1} s^{-1}$)	480	120	200	1800–1200	3800
Thermal conductivity ($Wcm^{-1} K^{-1}$)	1.5	5	1.3	22	24

microwave plasma-enhanced CVD (MPCVD) growth techniques (Achard et al., 2011; Kato et al., 2007a,b; Iwasaki et al., 2013). There has been a lot of research and great progress in diamond electronics, where EL has been observed in NV color centers (Berhane et al., 2015; Lohrmann et al., 2011; Mizuochi et al., 2012a) and more recently SiV color centers (Bray et al., 2018a; Mizuochi et al., 2012b). However, the fluorescent intensity of the electrically pumped devices is $\leq 10^4$ counts s^{-1}, which is significantly lower than if the emitters were optically excited (Aharonovich et al., 2011a; Pezzagna et al., 2010). LEDs fabricated from diamond can be advantageous for biosensors, on-chip applications, and photodetectors. Therefore, it is desirable to possess a narrow EL bandwidth or to increase the EL signal. Electrical excitation of quantum emitters is needed to allow the design, scalability and fabrication of efficient on-chip designs (Ellis et al., 2011; Mizuochi et al., 2012a; Stevenson et al., 2012). Also, electrical excitation allows the study of the charge dynamics and injection pathways between various charge states in a particular defect (Doi et al., 2014; Kato et al., 2013; Lohrmann et al., 2011; Mizuochi et al., 2012b).

Furthermore, the literature shows a variety of material bases that have been optimized for optoelectronic devices by integrating different single photon emitters in classical LED structures. For instance, by embedding Indium Gallium Nitride (InGaN) QDs into the intrinsic layer of GaN p-i-n diode or, similarly, Indium Arsenide (InAs) QDs in the intrinsic layer of Gallium Arsenide (GaAs), single photon EL has been demonstrated with onset voltages of ~1.5–3.5 V and a current density up to ~30 A/cm^2 (Deshpande et al., 2014; Yuan et al., 2002). Single photon EL has also been observed in isolated carbon nanotubes (Ross et al., 2014), Silicon Carbide (SiC) (Ikeda et al., 1979; Khramtsov et al., 2018), single molecules (Zhang et al., 2017), monolayer Tungsten di-Selenide (WSe_2) with a layer of boron nitride (Ross et al., 2014) and monolayer Molybdenum di-Sulfide (MoS_2) (Sundaram et al., 2013; Ye et al., 2014). Nevertheless, EL of single photons from these devices require cryogenic temperatures due to the necessity to confine carriers. Later in this chapter we show the realization of a stable, room temperature, electrically triggered SiV color center in a nanoscale diamond membrane diode structure.

3. Diamond membrane fabrication

The fabrication of diamond membranes is imperative for a variety of advanced nanophotonics, nanoelectronics and optomechanics applications. Exploiting the coveted properties of diamond for nanophotonic and nanoelectronics applications requires high-quality, thin, single crystal diamond membranes containing color centers (Faraon et al., 2011, 2012). The thickness of single crystal diamond membranes is very important for the confinement of light and should be on the order of nanometers. The ideal thickness is determined by the emitter wavelength divided by the refractive index of the material (Tomljenovic-Hanic et al., 2006). For a SiV center, with a ZPL of 737 nm, and refractive index of diamond as 2.4, the required thickness is ~300 nm. There remain challenges when scalable fabrication of high-quality single crystal diamond devices for use in a wide variety of quantum applications.

Fabrication of single crystal diamond plates have been a difficult and elusive process, numerous groups have attempted to fabricated diamond planar structures, waveguides and cavities through the use of femtosecond lasers and ion irradiation (Eaton et al., 2019). Basset et al. (Parks et al., 2018) was able to reliably and precisely fabricate (111) faced single crystal diamond plates from commercially available, CVD grown, electronic grade type IIa single crystal diamond substrates purchased from element 6 with (100) faces. Their method uses a 532 nm nanosecond-pulsed Yb-doped fiber laser to nucleate and propagate a mechanical cleave in a chosen (111) crystal plane creating faces up to 3 × 0.3 mm large. Basset et al. began the laser-nucleated cleaving (LNC) by creating nucleation sites on the bottom surface of the diamond plate, which acts as a seed for the initial cleave.

LNC starts when the laser beam is aligned with the nucleation site and subsequently scanned in one direction (x) at a speed of $8\,\text{cm}^{-1}$ by a predetermined distance, d.

$$d = \frac{h}{\tan 55^{o}} \tag{1}$$

where h is the starting single crystal diamond plate thickness, and 55° is the angle between the (111) and (100) lattice.

To extend the (111) crystal face, this lateral scanning process is repeated but in the y direction near the edge of the freshly generated (111) face, this continues until the cleave has propagated across the width of the single crystal diamond plate, leading to an exposed (111) face.

Our work uses ion implantation and selective etching techniques to fabricate diamond membranes from bulk, single crystal diamond. we make use of a flip and thin etching step to remove damaged material created through the implantation process. This technique results in 300 nm thick single crystal diamond membranes that are up to $1000\,\mu m \times 2000\,\mu m$. Through MPCVD, the diamond membranes are able to host a variety of color centers including NV (Fairchild et al., 2008; Magyar et al., 2011), SiV (Aharonovich et al., 2012; Bray et al., 2018a) and GeV (Iwasaki et al., 2015). With the availability of various color centers in the visible to NIR along with a high yield and simple growth technique, it allows for the high aspect ratio diamond membranes to be utilized for a variety of nanophotonics, nanoelectronics and quantum information science processes.

3.1 Fabrication of diamond membranes

Single crystal diamond membranes were created from a bulk type IIa CVD diamond from Element 6™, with a nitrogen concentration <1 ppm. The crystal was implanted with 1 MeV He^{+} ions to a dose of 5×10^{16} ions/cm^2. Stopping range of ions in matter (SRIM) simulations predicted a stopping range of 1.71 μm. The implanted diamond was subsequently annealed at 900 °C in vacuum to create a thin amorphous carbon layer 1.7 μm below the surface, as shown in Fig. 1. The amorphous layer enables lift-off of the 1.7 μm thick diamond membrane by electrochemical etching. To selectively remove the graphitic residual, the sample was then immersed in deionized water and electrochemical etching was carried out using a constant forward bias of 60 V. During etching, a positively biased tip was

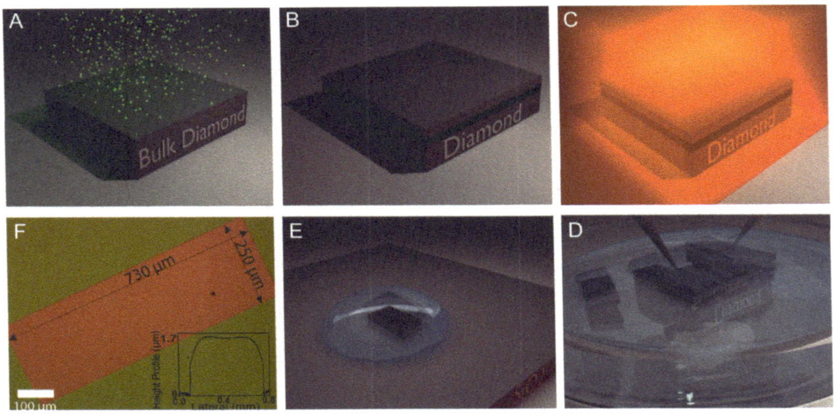

Fig. 1 Schematic of the process used to engineer single crystal diamond membranes. (A) Single crystal, undoped CVD diamond is implanted with He^+ ions (1 MeV, 5×10^{16} ions/cm^2) to create a damaged layer ~1.7 μm below the surface (B). (C) The sample is annealed at 900 °C and (D) electrochemically etched to lift off the diamond membranes. (E) The membranes are transferred onto a Ti coated silica substrate using a liquid transfer process. (F) A top-down optical image (false color) of a membrane that has been lifted off. Inset: height profile that shows the membrane is ~1.7 μm thick.

contacted onto the top surface of the diamond while the negatively biased tip was positioned slightly above the substrate. The positive biased contact was coated in a thin layer of PMMA, except for the tip. This allows the injected current to be localized into the bulk diamond sample to etch away the graphitic layer faster. The dissociated diamond membranes were cleaned using a 3:1 (sulfuric acid - hydrogen peroxide) piranha solution and transferred using a liquid droplet onto a Ti coated silica substrate. The diamond membranes range in size from [~50–1000 μm] × [~100–2000 μm] in width and length, respectively (Fig. 1). Fig. 1F shows an optical image of a single crystal diamond membrane with an area of ~0.2 mm^2 and a thickness of 1.7 μm.

The lifted off diamond membranes are generally not flat, especially as they increase in size. The inherent strain in the material causes the edges of the membranes to curl at different angles as shown in Fig. 2, which leads to the extremities of the sample to be suspended in air. The curling of the diamond membranes generates numerous issues for fabrication of devices due to uneven growth and etching. This leads to non-uniform or roughening of the diamond surfaces. Therefore, the diamond membranes need to be flatten down using a layer of PMMA or titanium. The flattening can be achieved by transferring the membranes with tweezers or using a clean piece of silicon or sapphire to press the membrane down gently. Handling the

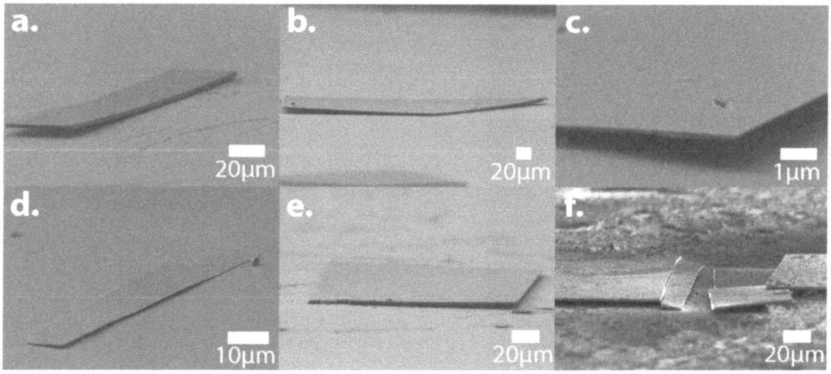

Fig. 2 SEM images of the lifted diamond membranes showing (A–E) the various degrees of induced strain of the diamond membranes to cause curling of the sample. (F) SEM image of a diamond membrane that has shattered due to induced stress from handling the sample.

membranes is also a very delicate process. The diamond membranes are fragile and prone to shattering when transferring the membranes between growth and etching procedures (Fig. 2F).

3.2 Diamond overgrowth

To incorporate a variety of color centers, we utilized a MPCVD chamber. The chamber was pumped down to 1×10^{-6} Torr, to remove all atmospheric gasses and cleaned using a hydrogen plasma at 900 W with a pressure of 30 Torr for several hours to remove all unwanted contaminants. The diamond membranes are overgrown using a MPCVD chamber (Fig. 3) with a hydrogen/methane ratio of 100:1 at 60 Torr, a microwave power of 900 W for 8 min to fabricate a ~100 nm intrinsic single crystal diamond layer. Silicon doping occurs naturally with the silicon source from the substrate and is incorporated into the membrane creating SiV color centers, shown by the ZPL at 737 nm (Fig. 3D). Fig. 5E shows a Scanning Electron Microscopy (SEM) image of an overgrown diamond membrane with a homogeneous single crystal diamond layer.

3.3 Fabrication of p-type diamond membranes

Similarly, to the Section 3.1, boron doped membranes were created from a bulk boron doped ($\sim 3 \times 10^{20}$ atoms/cm^3) diamond crystal grown by MPCVD. Boron has been reliably used to achieve p-type doping in diamond (Achard et al., 2011; Mizuochi et al., 2012a). The crystal was

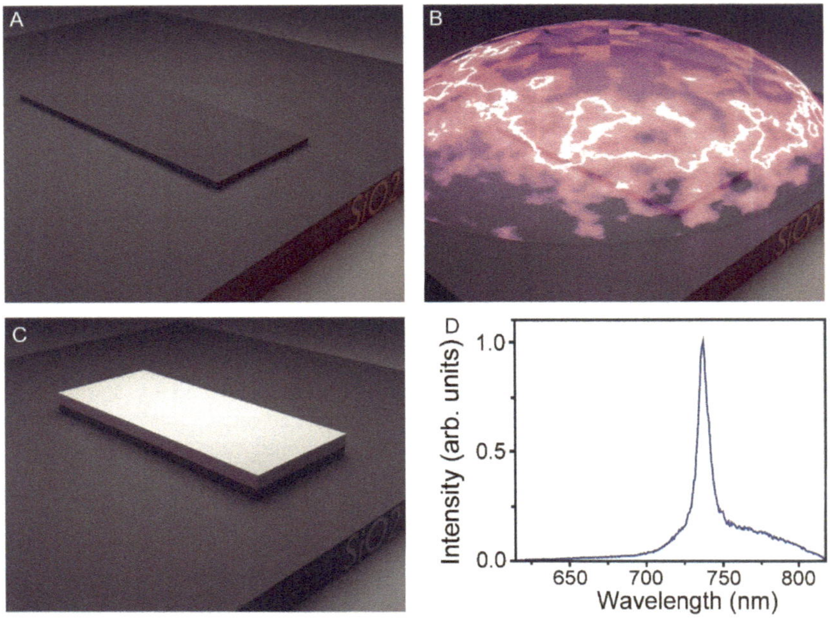

Fig. 3 (A) The transferred diamond membrane is placed on a SiO_2 substrate to act as a source for SiV color centers. (B) The sample is epitaxially overgrown using an MPCVD to create (C) a layer of single crystal diamond. (D) The overgrown layer contains SiV color centers, as shown by the peak at 737 nm.

implanted with 1 MeV He^+ ions and subsequently annealed at 900 °C in a vacuum to create a thin amorphous carbon layer. The membranes were selectively removed through electrochemical etching in deionized water with a constant forward bias of 60 V. This optimized implantation and electrochemical etch method leads to the lift off of membranes with high surface area. The diamond membranes were cleaned using piranha solution and transferred to a silicon substrate coated with ~150 nm of titanium which acts as a sticking layer and a bottom contact for electrical measurements, detailed below. The membranes similarly range in size from [~100 to 600 μm] × [~120 to 800 μm].

3.4 Flip and thin

It is ideal to fabricate high-quality single crystal diamond membrane for engineering of devices. Therefore, the original damaged diamond sample needs to be removed as it may be host to residual sp2 carbon, a rough surface and a damaged lattice structure from ion implantation. To remove the

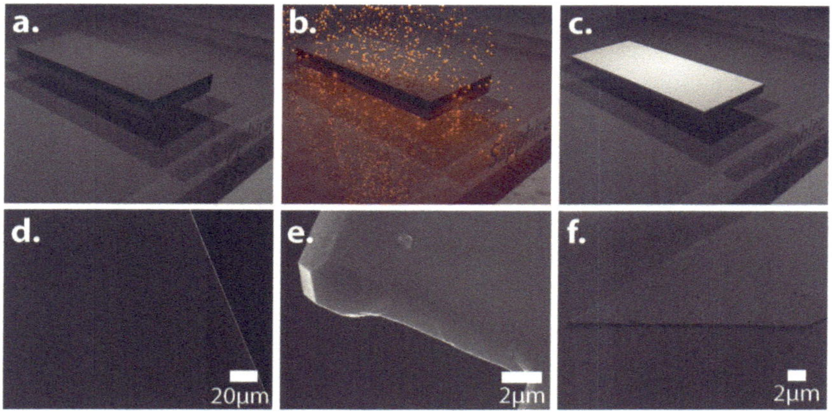

Fig. 4 (A) The membrane is flipped and placed onto a sapphire substrate. (B) The damaged layer is removed by ICP-RIE in the presence of CF_4 and O_2, leaving (C) only the epitaxially grown single crystal diamond layer. (D) SEM image of the lifted diamond membrane after piranha cleaning. (E) SEM image of the diamond membrane after CVD grown and (F) shows the membrane after ICP-RIE thinning.

original damaged diamond, the membranes were flipped and placed onto a sapphire substrate to reduce sputtering and masking in the chamber. The membranes were thinned by Inductively Coupled Plasma—Reactive Ion Etching (ICP-RIE) in the presence of tetrafluoromethane (CF_4) and oxygen (O_2) in a ratio of 1:3 at a pressure of 20 Pa, with a forward power of 200 W for 7 min as shown in Fig. 4B. Using ICP-RIE, thin membranes were reliably produced to achieve thicknesses smaller than 300 nm. Thinning the membranes using this method ensures that only the high-quality overgrown single crystal layer remains. A SEM image of the 300 nm thick diamond membrane is shown in Fig. 4F, the surface is still relatively clean and can be used as a base for other applications discussed below.

The diamond membranes were probed using Raman spectroscopy with a 633 nm HeNe laser. A peak at $\sim$1332 cm^{-1} corresponding to a single crystal diamond is shown in Fig. 5A with a FWHM of 8.3 cm^{-1} (Faggio et al., 2012). The surface roughness of the membrane before and after etching was determined by an atomic force microscope (AFM) to be below 3 nm, suitable for photonic and electronic devices (Fig. 5B). To unambiguously show that diamond overgrowth is indeed a single crystal diamond, the overgrown membranes were transferred to a TEM grid. The typical membrane morphology after etching and epitaxial overgrowth is shown in the TEM micrographs in Fig. 5C. Select-area electron diffraction confirms the membrane

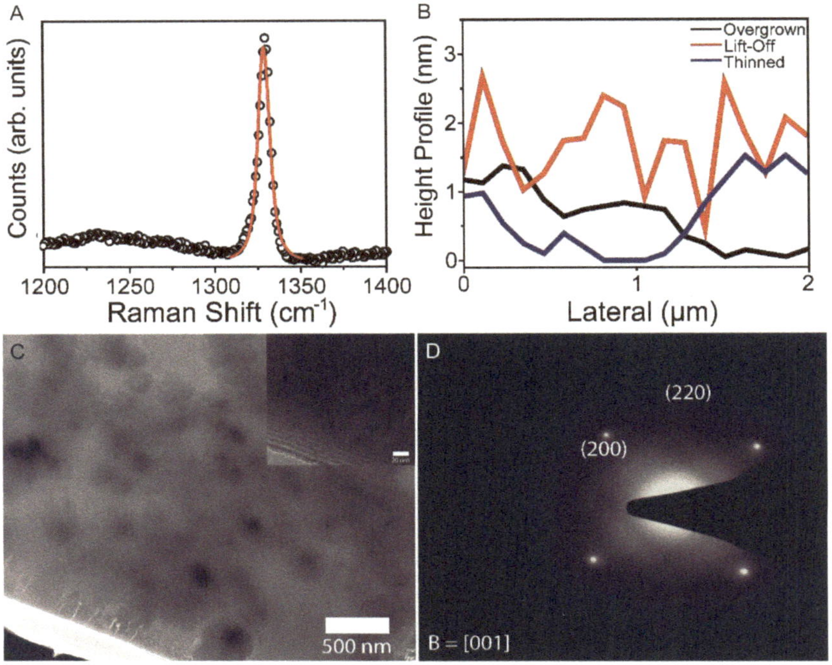

Fig. 5 Structural characterization. (A) Raman spectroscopy of the diamond membrane using a 633 HeNe laser shows the characteristic single crystal diamond peak at 1332 cm^{-1}. The curve was fit with a Voight function to show a FWHM of 8.3 cm^{-1}. (B) Lateral AFM scan show the surface roughness of the membrane after lift-off, after thinning and after an overgrowth to be ~2.5, ~1.5 and ~1.5 nm, respectively. (C) TEM image of an overgrown diamond membrane (~250 nm in thickness) transferred to a copper TEM grid. Inset: higher magnification image of edge region. Scale bar 20 nm. (D) Diffraction pattern from the diamond membrane in (C) indexed to the [001] zone axis.

microstructure is a single crystal (Fig. 5D). The diffraction intensity pattern is indexed to the [001] zone axis of a diamond cubic structure. The measured plane spacing of (200) and (220) planes were 1.275 and 0.903 Å, respectively, in good agreement with bulk diamond. TEM and SAED characterization were conducted on a FEI Tecnai T20 microscope at 200 keV with a LaB 6 filament.

3.5 Optical characterization of diamond membranes

The fabrication of thin, optically active diamonds membranes is important for advancing diamond device fabrication. The scalability of this process can drive the use of diamond from a laboratory setting into a large variety of devices spanning high power electronics, sensing and photonics such as resonators and waveguides. Membrane fabrication is scalable because the size of the initial bulk diamond can be increased and implanted with He^+ ions. The single crystal bulk diamond can then be overgrown with an intrinsic layer (containing SiV or GeV color centers) and a phosphorus doped layer using CVD techniques. The overgrown bulk diamond can then be patterned with HSQ, EBIE and RIE techniques where is can be subsequently lifted off through electrochemical etching, allowing a defined p-i-n single crystal diamond membrane to be transferred, flipped, and thinned in preparation for electrical stimulation.

If a device set-up or processes could be developed to allow the transfer and accurate positioning of multiple diamond membranes, then this will also allow a significantly increased throughput of p-i-n diamond devices for characterization. Currently, the diamond membranes can be lifted off with a 100% success rate. However, the rate of etching varies between ~2 and ~8 weeks. This variation in etching rate may be due to a poor electrical etching set-up, i.e., poor contacts, lack of ion channels in the deionized water or inhomogeneity of the initial He^+ ion implantation and subsequent annealing. The transferring, growing, flipping, and thinning of the membranes has a success rate of ~80%. The ~20% include the loss or destruction of the membranes through transferring, CVD growth, RIE etching, optical and electrical characterization. The main cause of diamond membrane loss occurs at the transferring between different substrates and flipping of the diamond membranes. We will now discuss why it is that high-aspect ratio diamond membranes are necessary for photonic, optoelectronic and quantum information processes.

3.6 Fabrication of diamond color centers

The integration of color centers with photonic resonators is a primary goal of diamond fabrication for several reasons. Firstly, it enables the enhancement of photon emission flux from the single optically active defects, as shown by Neu's work on diamond pillars (Neu et al., 2014). Secondly, it provides the means to interconnect potential quantum nodes in large networks, by integrating emitters into defined cavities connected in a wave guide, allowing the construction of large photonic networks (Wang et al., 2018; Wehner et al., 2018). Finally, if the resonator is designed carefully, with coupled direction emission, the highest quality cooperativity between emission and structure can be achieved, realizing advanced quantum phenomena such as a single photon switch in a solid state system (Sun et al., 2018).

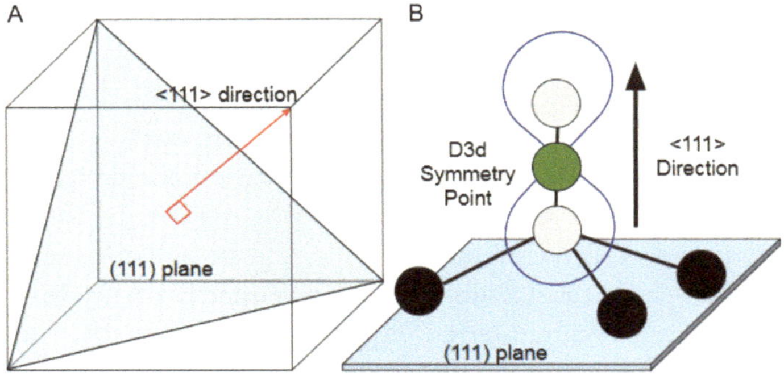

Fig. 6 Simplified Single crystal diamond Crystal structure and orientation. (A) Depicting the 111 plane in a standard cubic cell, red arrow noting ⟨111⟩ emission direction. (B) Substitution of group IV atom (green circle) and vacancies (gray) in carbon (black) unit cell of diamond. Blue field represents dipole orientation.

The simplified approach for incorporation of diamond color centers into diamond membranes is to introduce specific contaminations such as Silicon, Germanium or Tin to produce emitters. During CVD overgrowth, dissociated ions from the source material will incorporate into the diamond during overgrowth, reliably producing color centers. It is necessary to incorporate the desired emitters homogenously in the sample for ease of fabricating structures on a larger scale. This has been demonstrated in several diamond membrane devices by using a germanium source through MPCVD growth.

Diamond's lattice structure is comprised of face centered cubic (FCC) unit cells formed from sp3 bonded carbon atoms (Fig. 6). This crystal structure is responsible for diamonds extraordinary physical properties, boasting the highest known natural material hardness. However, the limited potential orientations and the physical differences between crystal vectors limits the availability of locations for the atomic substitutions necessary in the formation of diamond color centers. In polarization and photoluminescent emission studies of Group IV color centers, the defect structures are shown to align with the <111> crystal vectors, having their strongest dipole moment along this symmetry axis, which aligns with the known vacancy D3d symmetry site.

4. Fabrication of optical resonators from diamond membranes

Diamond membranes can serve as a platform for the fabrication of optical resonators. Utilizing their size as transferrable parent material for modular integrated diamond photonics without the cumbersome and economically unviable use of bulk diamond. Having an understanding of the emission directionality allows the design of cavities in diamond. However, despite the remarkable progress in nanofabrication of diamond cavities (Hausmann et al., 2013b; Khanaliloo et al., 2015b), all photonic resonators to date were fabricated from (100)-oriented diamond, primarily because this is the most common orientation provided by commercial suppliers and the difficulty in engineering and polishing (111) oriented crystals (Parks et al., 2018). This limits the potential coupling strength of many color centers to cavities since most centers studied to date have dipoles oriented along the <111> direction, and the dipole overlap with the cavity field is therefore not optimal in cavities fabricated from (100)-oriented diamond.

To provide more insight into the coupling efficiency between the emitters in diamond membrane and the cavity modes, a finite-difference time domain (FDTD) simulation was performed to inform the effect of emission orientation and Purcell enhancement. We focus here on the group IV defects, and predominantly the SiV defect. Its structure is a split vacancy with a silicon in an interstitial location along the <111> symmetry axis, resulting in a D_3d point group symmetry. Fig. 7A shows the schematic drawing of the four (i, ii, iii and iv) potential orientation of SiV defect in a (111) oriented diamond (Lesik et al., 2015; Rogers et al., 2014). These orientations are further simplified with red arrows indicating dipole orientations and α indicating angles between the dipole and the <111> vector.

One SiV dipole orientation makes a 0 degree while the rest of the dipoles make equal angle of 70.5 degree with the top diamond surface. The same SiV dipole in (100) membrane would form an angle of 54.7 degrees.

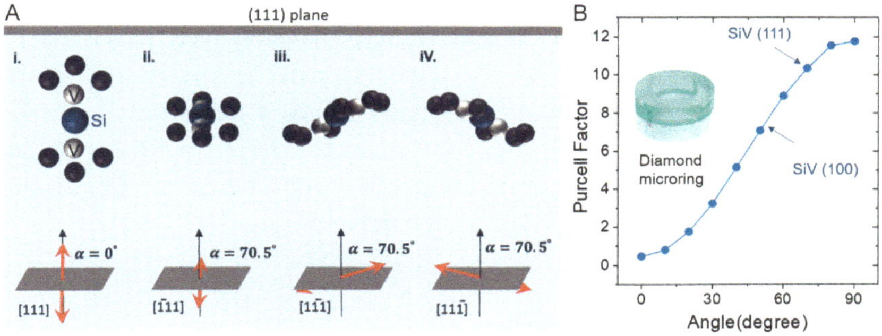

Fig. 7 Dipole orientation in (111)-oriented diamond and Purcell factors. (A) Simplified schematics of the four possible orientations of SiV centers in (111)-oriented diamond and the corresponding dipole orientations (red arrows). The angle α between the <111> direction and the dipole is shown in each case. (B) Purcell factor calculated as a function of dipole angle.

Most of the photonic cavities are designed for transverse electric (TE)-like modes, meaning dipoles with larger angle α would have a better electric field overlap. Consequently, Purcell factor with respect to the dipole angle α is calculated as shown in Fig. 7B for a diamond micro-ring. The simulation result shows that more in-plain components dipole has the higher Purcell factor is expected. Note that we conducted simulation for SiV center in this case, but this would be true also for other group IV defects in diamond Due to similar structures.

Given the advantages of the (111) membranes, ring resonators have been engineered from this material. Electron beam lithography was employed to pattern a hydrogen silsesquioxane (HSQ) layer spin coated onto the diamond membrane, forming an SiO_2 hard mask array of micro-ring resonators with external diameter of $\sim$400 nm and a radius of 5 μm. The pattern is transferred into the diamond membrane through a second ICP-RIE step with Oxygen at 10 mTorr. Finally, the mask is removed with another RIE step of SF_6 at 40 mTorr to directly etch the SiO_2 mask and undercut into the SiO_2 substrate. Fig. 8A shows the SEM of the patterned micro-rings, while Fig. 8B shows a high resolution image of an individual micro-ring.

The diamond cavities were then probed using a green laser in a standard confocal microscope. Fig. 8C, shows the whispering gallery modes (WGMs) from the resonator (manifested by the periodic peaks in the spectrum). With the designated micro-ring size, the free spectral range should be $\sim$14 nm. The spectrum of the ring resonators is augmented by the modes of the optical cavity surrounding the SiV peak at $\sim$738 nm. Fig. 8C further shows an emission enhancement form the fabricated micro-ring cavities as compared to the membrane. An enhancement of $\sim$25 times is observed after measuring a number of devices. A high resolution spectrum centered at the SiV region is shown in Fig. 8D. Quality factors as high as $\sim$3000 were observed from these devices (measured as $\lambda/\lambda\Delta$). The losses occur due to an increased roughness from the imperfect HSQ mask removal and from the leakage to the underlying SiO2 substrate. The enhancement of brightness and high quality factor are desirable and advantageous properties for the application of color centers to practical nanophotonic devices.

4.1 Photonic crystal cavities

Photonic crystal cavities are structures that have generated great interest, as it possesses a periodic structure and the ability of confining and manipulating photons via alternating refractive indexes (Hepp et al., 2014; Lesik et al., 2015). Diamond is considered to be a great candidate for

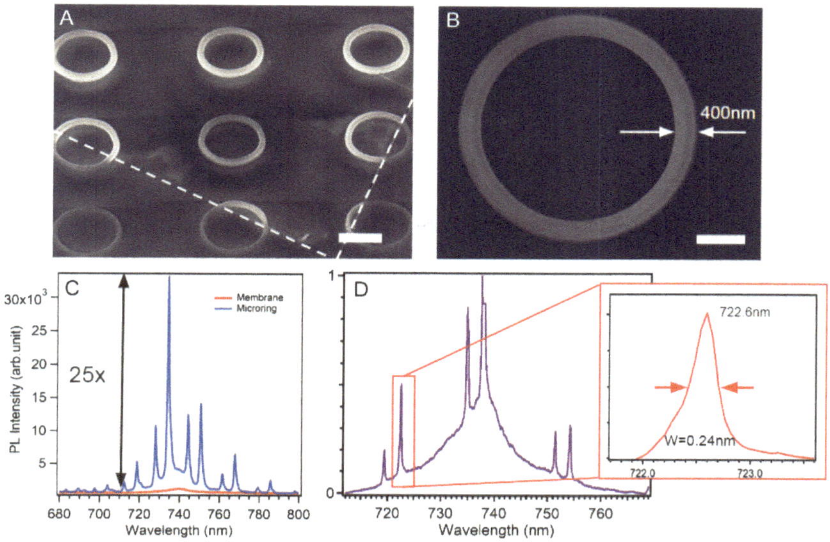

Fig. 8 Diamond ring resonators fabricated from (111) diamond membrane containing SiV color centers. (A) shows a wide area show at 45 degree inclination depicting the array of micro-rings. Inset scale bar corresponds to 3 μm (B) Shows a top view of an individual micro-ring. Inset scale bar correlates to 1 μm. (C) PL spectrum recorded from the diamond micro-ring cavity under 532 nm excitation at room temperature. WGMs are clearly visible. The ZPL of the SiV center is observed at 738 nm. Comparison of recorded emission from SiV color center recorded from the membrane and from a micro-ring cavity, showing an enhancement of ~25 times. (D) fine grating Spectral analysis of whispering gallery resonators, individual marked mode with FWHM of ~0.24 nm, Q factor ~3000.

photonic crystal cavities due to their optically narrow emission linewidths and long spin coherence times. Optimal photonic crystal cavity hybrid devices have been fabricated where, for example, NDs or bulk diamonds have been coupled to GaP cavities (Bochmann et al., 2013; Hausmann et al., 2012). The coupling showed enhancement in the NV^- ZPL, but there were limitations such as scattering and losses from the cavity material, spatial mismatch as well as line-broadening mechanisms (Sipahigil et al., 2016; Watanabe et al., 2009). As a natural extension of this interest, investigation into the use of diamond membranes as a platform for photonic crystal cavities, due to their potential for modular diamond components, has led to the development of numerous fabrication techniques to form photonic crystal cavities from isolated diamond platforms.

Furthermore, there has been recent exploration into tunable photonic crystals. By manipulating the structural arrangement or material refractive index, it allows the properties of the bandgap to be tuned. Tunable structures are necessary for optoelectronic integrated circuits and optical interconnects (Choi et al., 2015; Deshpande et al., 2013; Michler et al., 2000). Photonic crystal cavities have been fabricated in silicon and coupled to a graphene layer and triggered using an external bias to reduce the linewidth of the emitter by 1.5 nm and shift the cavity resonance by 1–2 nm (Sandorfy et al., 2002) and increase the quality factor by threefold (Sze, 1995). Shifts of >70 nm have been observed in silicon photonic crystals when a bias is applied (Cristea, 2011; Fedyanin and Agio, 2016a; Shields, 2007).

While the primary desired outcome of membrane fabrication would be to form free-standing transferrable cavities, there have been multiple approaches done to achieve the desired outcome, ranging from direct write techniques such as focused ion beam milling, to RIE mask transferring.

Utilizing a similar fabrication method as shown in the <111> fabrication section to make diamond micro-rings/disks, Photonic crystal cavities can be fabricated using direct EBL patterning followed by RIE etching. However due to the limitations of reproducibility and difficulty in fabrication of diamond membranes, other avenues have simultaneously been explored. Fig. 9 is a demonstration of electron beam lithography mask deposition showing that high-quality structures are still viable using this more common method.

Fig. 10 shows an approach employing FIB milling with a metallic hard mask, which generates a high-quality structure and minimizes the damage to the crystal structure through a sophisticated post process cleaning procedure. However even with an appropriate cleaning procedure, FIB fabrication procedures damage the crystal structure, causing a reduction in the quality of fabricated devices. Despite this, FIB milling is still an avenue worthy of further research, as a direct write technique can be used on complex surfaces will little regard for surrounding material, making modular components more easily fabricated before and after transferring.

The alternative technique of RIE etching with a defined hard mask has been demonstrated previously. However the ability to fabricate PCCs into a diamond membrane required a reassessment of direct write techniques as the low conductivity of diamond causes deflection of the electron beam during EBL, limiting the pattern resolution. An example technique to mitigate this issue is a transferable hard mask that may be separately patterned and then transferred to a diamond membrane. Fig. 11 demonstrates such an approach, further, without the damage commonly caused by FIB milling the overall device quality is greater.

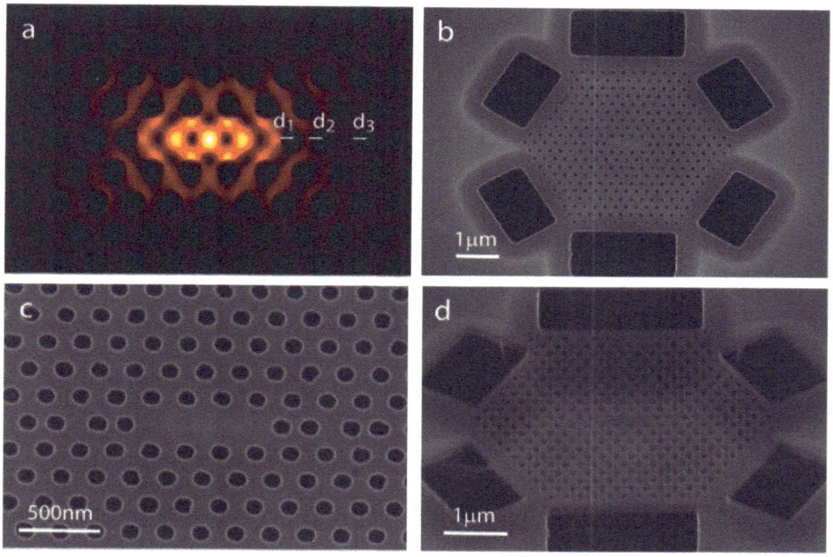

Fig. 9 (A) Color plot showing the electric field energy density in the fundamental mode of a linear three-hole defect cavity. The lateral three holes are shifted laterally by d1, d2, and d3 to increase the quality factor. (B)–(D) Scanning electron microscope images of a photonic crystal cavity fabricated in monocrystalline diamond. The black rectangles in (B) are openings etched in the membrane to facilitate etching underneath the membrane. The darker gray region around the openings indicates the extent of the etched Si under the diamond. *Images courtesy of Faraon, A., Santori, C., Huang, Z. ,Acosta, V.M., Beausoleil, R.G., 2012. Coupling of nitrogen-vacancy centers to photonic crystal cavities in monocrystalline diamond. Phys. Rev. Lett. 109 (3), 2–6.*

While this technique does produce high-quality structures, it is limited by the finesse required to transfer a thin film mask material to the thin film diamond. Further, the structure quality would decrease over repeated use of the transferable hard mask due to physical etching, meaning that the difficult process of fabricating the transferable mask would be an enduring bottleneck.

5. Electroluminescence of diamond

EL is a valued technique of optically active materials. Electrically driven excitation of emitters requires photons to incorporate into the active region of a device junction where the emitters will trap the efficient transport of excess carriers resulting in radiative recombination (Berhane et al., 2015; Choi et al., 2015; Deshpande et al., 2013; Michler et al., 2000;

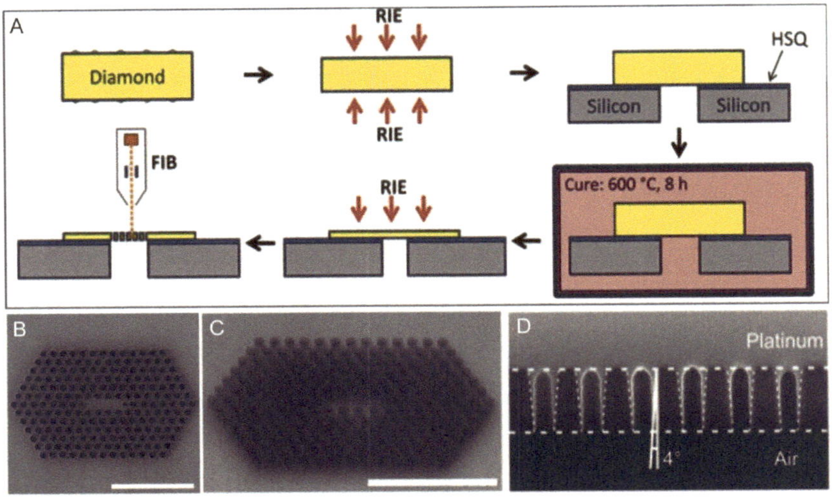

Fig. 10 (A) Fabrication methodology of two-dimensional air-suspended PhC-cavities. Details at source (B, C) SEM images post processing of FIB milled PCCs from diamond membranes. The picture quality is lower due to the reduced surface conductivity after removing the chromium layer. To minimize sample charging, the electron beam accelerating voltage was set to 20 kV, causing a majority of beam electrons to transmit the membrane without depositing extra charge. However, this induced a significant "see through" effect in the tilted view (C). All scale bars for SEM images correspond to 2 mm. The structures show reasonable defined edges and accurate structure sizes. (D) cross sectional analysis of PCC showing inclination artifact from fib based technique. *Images courtesy of Edmonds, A.M., Jung, T., Kreiner, L., Pauly, C., Frank, M., 2016. Reproducible fabrication and characterization of diamond membranes for photonic crystal cavities. 3264 (12), 3254–3264.*

Mizuochi et al., 2012a; Sandorfy et al., 2002). This technique allows electrically controlled single photon generation and is integral for the realization of large-scale integrated quantum devices. Additionally, using an external voltage supply to inject current into a single photon emitting system provides insight into fundamental properties, for example, charge state switching in active emitters.

For electrically driven emission of photons, it is imperative that the recombination of excess charge carriers happens between the energy levels of the emitter. Therefore, incorporation of emitters into devices is unavoidable for excess carriers to be introduced into the neighborhood of the energy levels. A simplified review of the basics of common semiconductor junctions is covered below to explain the charge flow and driving forces behind excess carrier injection.

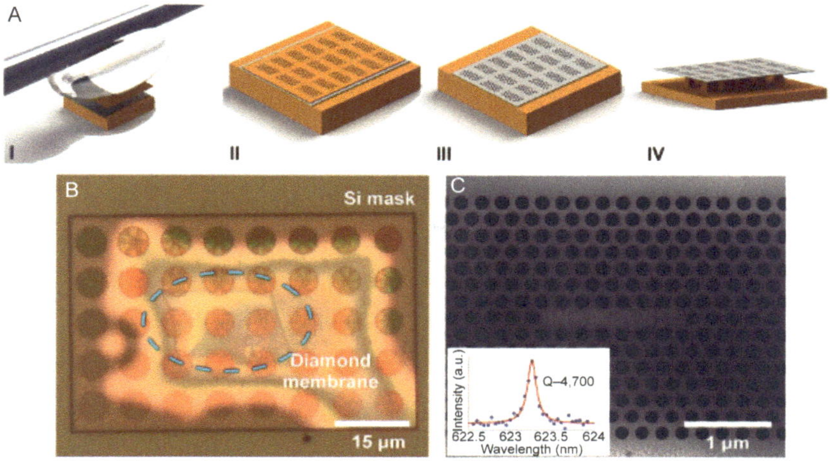

Fig. 11 Patterning of a diamond membrane using a silicon membrane as a contact etch mask. (A) Illustration of fabrication method: (I) A patterned silicon mask was transferred onto a diamond membrane using a macro-PDMS adhesive. (II) The silicon membrane on top of the diamond membrane served as an etch mask for oxygen plasma etching. (III) The diamond membrane was patterned with nanostructures during oxygen etching after subsequent mask removal. (IV) A SF6 isotropic dry etching removed the silicon underneath and suspended the diamond membrane at devices' locations. (B) Optical image of a silicon mask covering a diamond membrane that is circled by the blue dotted line. (C) SEM of a suspended diamond L7 PC cavity. Inset: Measured cavity resonance (blue dots) are 623.3 nm with a Lorentzian fit, yielding a quality factor of 4700 (red line). *Figure courtesy of Li, L., et al., 2015. Nanofabrication on unconventional substrates using transferred hard masks. Sci. Rep. 5, 7802.*

Two similar semiconductors that are either n-doped or p-doped, are brought into close contact (Fig. 12A). The dotted line in the above band diagrams illustrates the Fermi energy levels of the n-doped and p-doped semiconductors which are the majority carrier electrons (donors) and majority carrier holes (acceptors), respectively. The acceptor density is generally denoted as N_A while the donor density is N_D. When the differently doped regions are brought into contact, holes can diffuse to the n-doped region, across the junction leaving behind acceptor impurities in the p-doped region and becoming a minority carrier as shown in Fig. 12B (Choi et al., 2015; Sandorfy et al., 2002). Furthermore, current will flow readily in one direction (forward biased) but not in the other (reverse biased), creating a basic diode. This non-reversing behavior arises from the nature of the charge transport process. This process lowers the Fermi energy level of the n-doped region and raises it on the p-doped region.

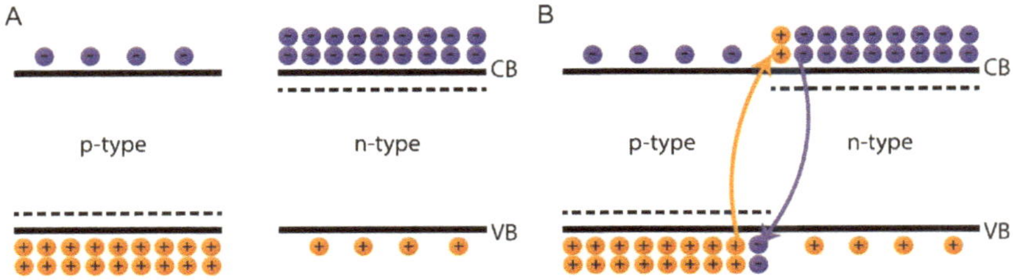

Fig. 12 Energy band schematics of two materials that are p-doped and n-doped. (A) The majority of holes are located in the valance band (VB) of the p-doped material while the majority of electrons reside in the conduction band (CB) of the n-doped material. (B) When the materials come into contact, the majority carrier in the n-doped and p-doped material diffuse to the opposite side, to become minority carrier before they recombine.

Similarly, the electron-minority flow moving toward the p-doped region leaves behind donor-impurities in the n-doped region. This can be expected since holes are the absence of electrons. This diffusion process near the junction will result in recombination of majority carrier to minority carriers resulting in conservation of charge through photon emission as well as momentum.

Due to the electric field build up across the p-n junction between the fixed donor and acceptor impurities, the diffusion of carriers cannot continue indefinitely (Cristea, 2011; Sze, 1995). This intrinsic field will inspect charge flow, pushing electrons to the n-doped region and holes to p-doped region. This region is referred to as the depletion region or the space charge region (SCR) and it is empty of carriers. The minority carrier flow must overcome a potential barrier, V_b, due to the built-in electric field, caused by the work function difference between the n-type and p-type semiconductors. This leads to band bending in the band energy schematic (Fig. 13). At thermal equilibrium, the Fermi level of the system is uniform, as indicated by the broken black line in the center.

The built-in potential can be quantitatively described, at thermal equilibrium, using the difference in the energy between the Fermi level, E_F, and the conduction band, E_C, and valance band, E_V, by (Cristea, 2011; Sze, 1995):

$$E_c - E_F = kTln\left(\frac{N_c}{N_D}\right) \tag{2}$$

and

$$E_F - E_v = kTln\left(\frac{N_v}{N_A}\right) \tag{3}$$

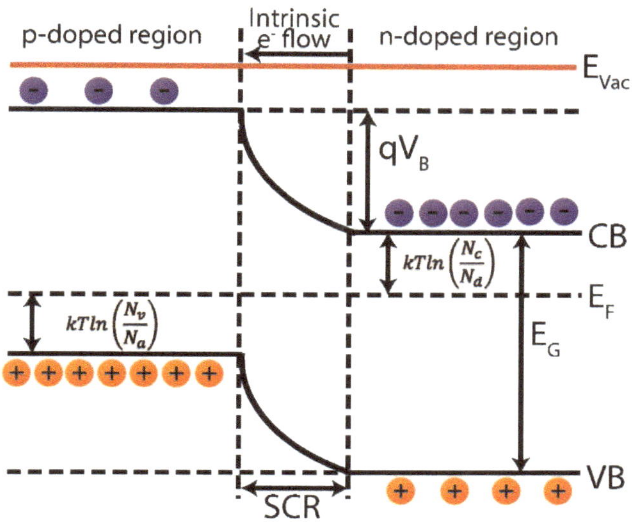

Fig. 13 A band energy schematic showing band bending when an n-doped and p-doped material are in contact and at equilibrium. The space charge region (SCR) is empty of carriers. E_G is the band gap of the material.

Where kT is the Boltzmann constant ($1.38 \times 10^{-23} m^2 kgs^{-2} K^{-1}$) multiplied by room temperature to obtain 25 meV. N_c and N_v are the intrinsic carrier concentration and are dependent on the material properties. N_D is the electron carrier, or donor, density while N_A is the hole carrier, or acceptor, densities. N_D and N_A are reliant on the doping concentrations of the n-type and p-type semiconductors.

Fig. 13 shows us that:

$$\begin{aligned} qV_b &= E_G - (E_c - E_F) - (E_F - E_v) \\ &= E_G - kTln\left(\frac{N_c}{N_D}\right) - kTln\left(\frac{N_v}{N_A}\right) \\ &= E_g - kTln\left(\frac{N_c N_v}{N_D N_A}\right) \end{aligned} \tag{4}$$

where q is the charge of the minority carriers. This can be used to determine the depletion region, ω, of a p-n junction. The electric field is related to charge density, by using Gauss's law, as:

$$E(x) = \frac{1}{\varepsilon} \int \rho(x) dx \tag{5}$$

where ε is the dielectric constant. Considering a Dirac-delta function assumption for the p-doped region, the integral can be represented as:

$$E(x) = \frac{1}{\varepsilon}\int_{0}^{\omega} qN_D dx \tag{6}$$

Therefore, the maximum field in the depletion region can be represented as:

$$E_{max} = \frac{qN_D\omega}{\varepsilon} \tag{7}$$

If we assume that the maximum electric field is reached in the center of the depletion region at $\omega/2$, the potential drop due to band bending is:

$$V_b = \frac{E_{max}\omega}{2} = \frac{qN_D\omega^2}{2\varepsilon} \tag{8}$$

This can be simplified to:

$$\omega = \sqrt{\frac{2\varepsilon V_b}{qN_D}} \tag{9}$$

Similarly, for a p-n junction:

$$\omega = \sqrt{\frac{2\varepsilon V_b}{q^2}\left(\frac{1}{N_A} + \frac{1}{N_D}\right)}$$
$$\omega = \sqrt{\frac{2\varepsilon V_b}{q^2}\left(\frac{N_D + N_A}{N_A N_D}\right)} \tag{10}$$

As mentioned previously, p-n junctions can be used to give insight into semiconductor device properties, including current flow and band bending. Numerous other device structures can be manufactured by changing the desired semiconductor material, fabrication technique and application of the device.

It is imperative to ensure that the injected excess carriers will lead to the photon emitting sources avoiding unwanted recombination channels. Therefore, when fabricating junctions, it is important to use a p-i-n device structure where the intrinsic layer contains the photon emitting defects (Kato et al., 2013; Mizuochi et al., 2012b; Sandorfy et al., 2002; Shields, 2007).

Fig. 14 shows the emission of an electrically triggered photon. A photon may have a defect state in a semiconductor with a single electron in its

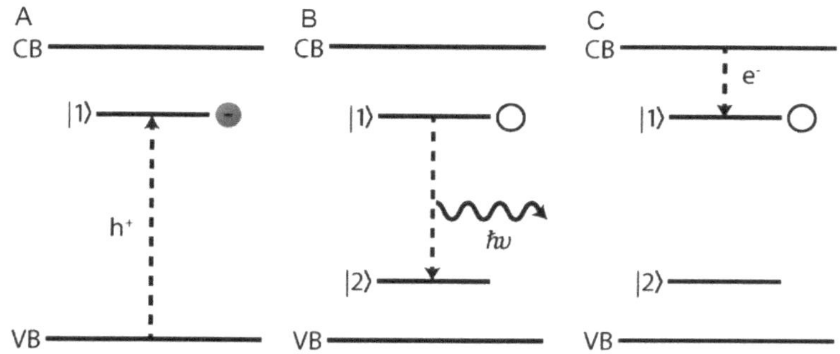

Fig. 14 Schematic illustration of a photon being emitted at a charged defect site. (A) A hole is trapped by a defect state that has an electron in its ground state. (B) An immediate decay from the excited state to the ground state, of the trapped state, emits a photon. This turns the charge state neutral. (C) An electron travels to the decay state, changing it from a neutral to a negatively charged state to repeat the cycle.

ground state (for example NV^-). This state would cause an excess of minority holes to be attracted to this state and combine as shown in Fig. 14A. Subsequently, this will cause the state to relax by sending the trapped hole to the ground state. This will lead to EL of the emitter (Fig. 14B). Radiative recombination causes the electrically triggered photon to lose its electron to the ground state, changing the charge state neutral (Fedyanin and Agio, 2016b). This causes the electron to move from the conduction band to the photon defect state, trapping it and turning the state negatively charged (Fig. 14C).

To summarize optical pumping of an emitter through an external laser source uses sub-bandgap direct excitation that drives electrons to transition into an isolated photon emitter energy state. This is juxtaposed to electrical pumping through excess carrier injection by an external voltage supply, where it involves the valance and conduction bands of the semiconducting material where subsequent trapping of photons can occur.

6. Single crystal diamond membranes for nanoelectronics

The availability of conductive diamond membranes for applications in nanoelectronics so far has remained as an unreachable goal. In this summary, we present a complete nanofabrication methodology for engineering high aspect ratio, electrically active single crystal diamond membranes. The membranes have large lateral directions, exceeding $\sim 500 \times 500\,\mu m^2$ and are only several hundreds of nanometers thick. We further realize vertical single crystal p-n junctions, made from the diamond membranes that exhibit onset voltages of $\sim$10 V and a current of several mA. Moreover, we deterministically introduce optically active color centers into the membranes and demonstrate for the first time a single crystal nanoscale diamond LED.

The majority of modern photonic and optoelectronic devices including photodetectors, LEDs, lasers, MEMS and sensors rely on efficient doping (p-type and n-type) for electrical triggering or readout. Many devices also require robust nanofabrication protocols that enable engineering of nanoscale (tens to hundreds of nanometers) suspended membranes. These membranes are fundamental building blocks of advanced nanophotonic components, such as waveguides or photonic crystal cavities, and also crucial to enable mechanical motion in MEMs (Bochmann et al., 2013; Hausmann et al., 2013a; MacQuarrie et al., 2016; Sipahigil et al., 2016; Tao et al., 2014).

The exceptional properties of diamond, including excellent thermal conductivity, high young modulus and wide optical transparency (Aharonovich et al., 2011b; Koizumi et al., 2001; Simon et al., 2014; Watanabe et al., 2009; Wort and Balmer, 2008), make it an ideal platform for a vast majority of these applications. Diamond is also poised to be the leading candidate for modern nanoelectronic devices due to its ability to sustain high temperatures and high electric fields before breakdown (Matsumoto et al., 2016; Schneider et al., 2005; Shikata, 2016). However, broad adoption of diamond-based devices has so far been hindered by a lack of large area conductive diamond membranes and efficient nanoscale p-n junctions. While progress has been made to demonstrate proof of concept experiments and develop nanofabrication protocols to sculpt diamond (Aharonovich et al., 2012; Buja et al., 2014; Drijkoningen et al., 2016; Hausmann et al., 2013a; Khanaliloo et al., 2015a; Krauss et al., 2001; Macpherson, 2015; Piracha et al., 2016b; Rath et al., 2013, 2015), large-scale conductive diamond membranes that are suitable for efficient p-n or p-i-n junction engineering are currently beyond reach. This is due, in part, by the challenges associated with epitaxial growth onto non-diamond sacrificial substrates that can be subsequently chemically removed—a process that is well established for silicon and gallium arsenide.

In this section, we demonstrate a robust method to fabricate p-type diamond membranes and engineer vertical p-n and p-i-n junctions that are suitable for further development of on-chip nanoelectronic devices. The nanoscale diamond membranes produced in the work have large lateral dimensions, over $500 \times 500\,\mu m^2$, excellent electronic properties, and are fabricated using a highly robust and technologically mature process, making them attractive candidates for broad adoption and scalable device fabrication.

6.1 Characterization of p-type diamond membranes

To test the electrical properties of the p-type diamond membranes, we measured the current—voltage (I—V) curves for a thick (1.7 μm) and a thinned (~200 nm) membrane. An optical image of the latter is shown in Fig. 15A, while the electrical measurements are presented in Fig. 15B, C. As expected, semiconducting behavior (Schottky junction) is observed from both membranes. The thicker membrane exhibits a forward threshold voltage of ~10 V and a breakdown voltage of −5 V, while the thin membrane exhibits a slightly reduced threshold of 8 V. In both cases, currents on the order of ~ mA (~1.5–2 A/cm^2) are recorded. These values are comparable to bulk

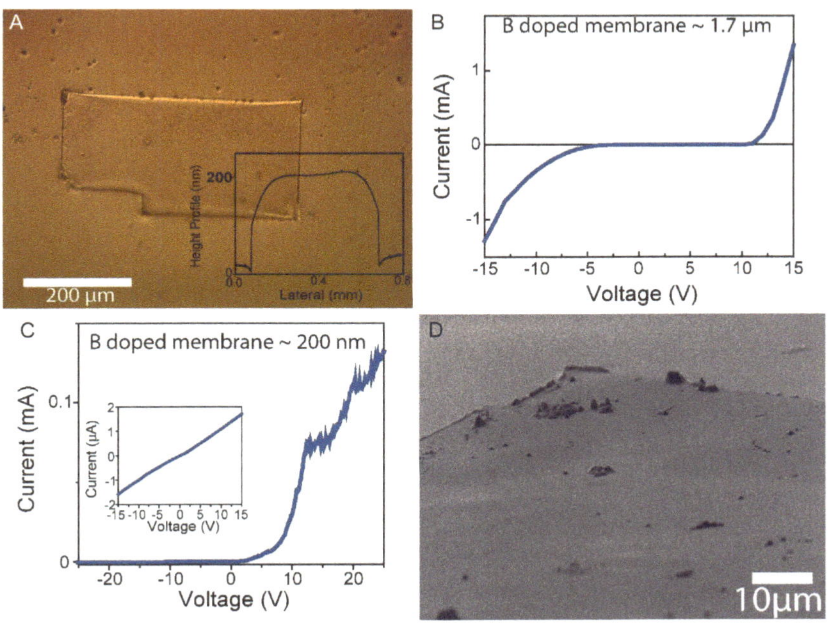

Fig. 15 Boron-doped diamond membranes. (A) Optical image of a thin boron doped diamond membrane. Inset: height profile that confirms the membrane is ~200 nm thick. (B) An I—V curve of the lifted off boron doped membrane (~1.7 μm) exhibits a threshold of ~10 V. (C) An I—V curve of a 200 nm thick, boron doped membrane exhibits an onset threshold of ~8 V. Both I—V curves show semiconducting behavior. Inset: IV curve of an undoped diamond membrane showing insulating behavior with negligible μA currents. (D) SEM image of the diamond membrane.

and polycrystalline diamond devices (Liang et al., 2007; Mizuochi et al., 2012a). For reference, the inset of Fig. 15C shows the same measurement performed on an undoped diamond membrane with the same thickness (200 nm), that results in a small leakage current of several ~ μA (~2 mA/cm^2) flowing through the junction. These results confirm that the boron doped diamond membranes maintain their electrical properties even when they are thinned down to several hundred nanometers.

For the electrical measurements, the membranes were positioned onto a titanium (~150 nm) coated silicon dioxide wafer. The top contact was titanium/gold metal that was sputtered at a base pressure of 1×10^{-6} Torr and a 100 W forward bias, through a lithographic mask (thickness ~150 nm and 20 μm diameter). To contact the devices, two xyz nanomanipulators were used, one positioned on top of the membrane and the second one positioned in a close proximity to it. The voltage was generated and controlled using the Keithley 617 electrometer and the current passing through the device was measured using the same tool. The data was recorded using custom-built software and all the measurements were done at room temperature.

6.2 Fabrication of vertical nanoscale p-n junctions

The availability of p-type membranes enables the exploration of p-n junction engineering using the single crystal diamond membranes. To realize the p-n junctions, we employed a modified fabrication protocol in which a 200 nm n-type layer was grown epitaxially using MPCVD on top of He^+ ion implanted boron doped single crystal bulk diamond, before the electrochemical etching. For the n-type layer, Phosphine (PH_3) diluted with H_2 was used for phosphorus doping at a ratio (PH_3/CH_4) of 5% and the final doping is $\sim 8 \times 10^{18}$ cm^{-3} of phosphorus atoms. The sample then underwent electrochemical etching to dissociate the p-n diamond membrane that is ~1.9 μm thick. The p-layer was thinned by ICP-RIE to achieve a total membrane thickness of ~300 nm, as described previously.

6.3 Characterization of vertical nanoscale p-n junctions

The overgrown n-type layer was verified using a secondary ion mass spectrometry (SIMS) measurement, to quantify the phosphorus concentration (Fig. 16). The measurement was done on an epitaxial layer that was grown under identical conditions for a longer time. The doping concentration was determined to be $\sim 1 \times 10^{18}$ atoms/cm^2 (blue curve). The red curve represents the detection limit of the boron atoms.

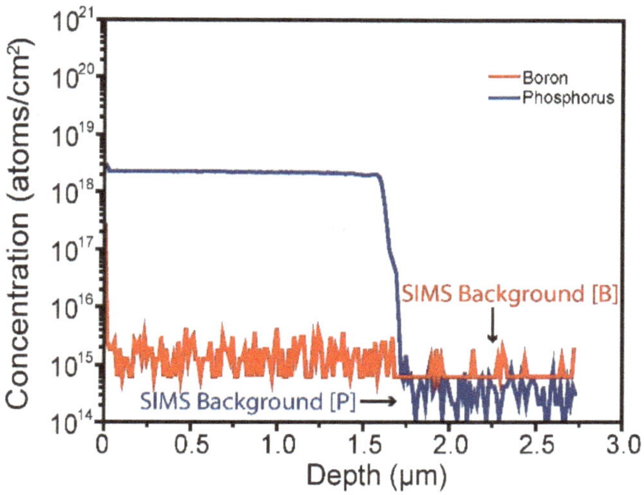

Fig. 16 SIMS analysis from a bulk element 6 doped diamond. Secondary ion mass spectrometry measurement of CVD grown phosphorus doped diamond. Hydrogen, nitrogen, and boron exhibit background noise while Phosphor shows a doping concentration of $\sim 10^{18}$ atoms/cm^2 (blue curve) while the boron shows background noise (red curve).

The experimental conditions allow a slow growth that achieves uniform n-type doping across the sample. The overgrown diamond was then processed as described previously to produce a single crystal membrane with a total thickness of $\sim 1.9\,\mu m$ that is a vertical p-n junction (see Fig. 17A, B). An optical image of the membrane is shown in the inset of Fig. 17B. It exhibits diode-like behavior with a forward threshold voltage of ~ 10 V with $\sim$mA (~ 3A/cm^2) current, and a negligible current ($\sim 1 \times 10^{-7}$ A/cm^2) under a reverse bias of >40V (Fig. 17c). This leads to an extremely high rectification ratio of $\sim 10^4$ which is ideal for diode fabrication (Butler et al., 2003; Makino et al., 2009). This value is comparable to bulk diamond devices (Mizuochi et al., 2012a).

Upon subsequent thinning of the membrane to ~ 300 nm by RIE (yielding a p-type layer of ~ 100 nm and an n-type layer of ~ 200 nm), p-n membrane still behaves as a diode with a forward voltage of ~ 8 V with $\sim$mA (~ 2 A/cm^2) current, and negligible current ($\sim 2 \times 10^{-7}$ A/cm^2) under reverse bias (Fig. 17D). The rectification ratio of the thin membrane is lower, ~ 30, and may be caused by roughening during the RIE process or by heterogeneous boron doping near the p-n interface. The lifted off p-n diamond membranes show rectifications ratios $\sim 10^4$. Unfortunately, upon thinning

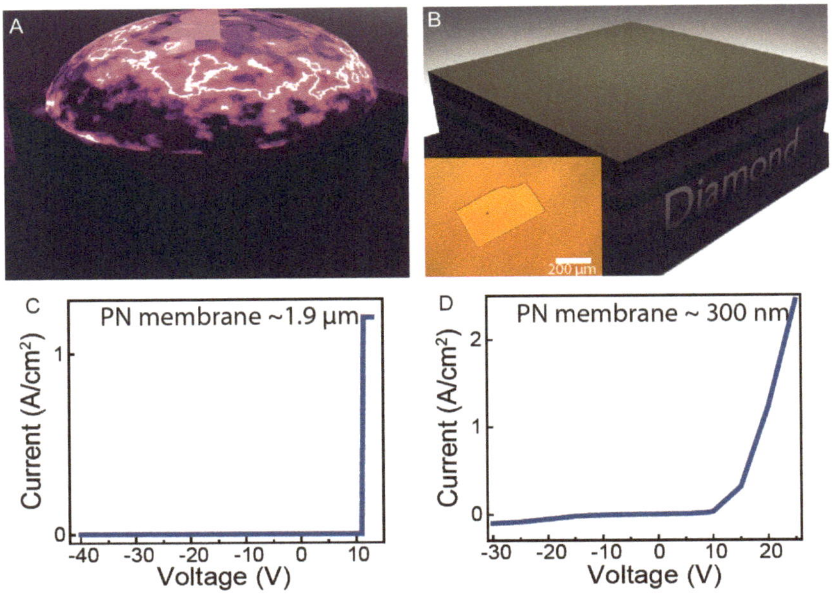

Fig. 17 Engineering of vertical p-n junctions. (A–B) Schematic of the process used to produce single crystal p-n diamond membranes. (A, B) Bulk boron doped diamond that was implanted with He ions to create a graphitic channel is overgrown with diamond to create a phosphorus doped n-type layer represented by the dark green color in (B). Inset, the lifted off entire p-n diamond membranes. (C) Electrical measurements of the lifted off, 1.9 μm thick PN membrane showing diode behavior at ~10 V and (D) an I—V curve of the thinned (~300 nm) p-n membrane, showing a similar onset voltage.

the p-n membranes show a similar trend as the previously mentioned same, with rectification ratios of ~30 and 20.

To determine the thickness of the depletion region, ω, of the p-n diamond junction, the following equation was employed,

$$\omega = \sqrt{\frac{2\varepsilon_s}{q}\left(\frac{1}{N_A} + \frac{1}{N_D}\right)(\phi_i - V_a)} \quad (11)$$

where ε_s is the permittivity, q is the electron charge, N_A and N_B are the acceptor ([B] $\sim 3 \times 10^{20}/\text{cm}^3$) and donor ([P] $\sim 8 \times 10^{18}/\text{cm}^3$) concentrations, respectively, ϕ_i is the built-in potential (~5.5 V) of the device and V_a is the applied voltage (calculation is done for $V_a = 0$ at equilibrium). Under the experimental conditions, a depletion layer of ~21 nm exists. The short depletion layer indicates that only a minimal voltage is needed to overcome this barrier potential and allow for recombination of electrons and holes flowing from the n- and p+ layers, respectively, and is important for practical devices. Furthermore, the p-n single crystal diamond membrane devices show a maximum electric field,

$$\mathcal{E} = \frac{2(\phi_a - V_i)}{w} \tag{12}$$

of ~500 MV/m. While this field is not as high as millimeter scale bulk diamond, it is still an order of magnitude greater than silicon (300 kV/m) which is ordinarily used as the base material for nanoelectronic devices. A higher electric field allows the diamond membranes to withstand higher voltages (up to ~100 V for a 500 nm vertical device) and stresses before the device electrically breaks down and the material becomes conductive. Overall, the short depletion layer and the high breakdown field make these nanoscale single crystal diamond membranes extremely attractive for real nanoelectronic devices.

6.4 Characterization of vertical p-i-n junctions

There exists great potential for the use of diamond membranes within optoelectronic and nanophotonic applications. To realize a working diode device, we fabricated a p-i-n structure into a single crystal diamond membrane containing a uniform layer of SiV color centers (Fig. 19). The SiVs color centers were chosen intentionally, as they are emerging fluorescent defects in diamond that hold great potential for quantum photonic applications (Neu et al., 2011a; Sipahigil et al., 2016). To realize this structure, a He^+ ion implanted boron doped single crystal diamond was overgrown in the presence of a silicon source to introduce a 100 nm layer of SiV color centers (into the intrinsic layer). This growth was performed using a MPCVD chamber with a hydrogen/methane ratio of 100:1 at 60 Torr, a microwave power of 900 W for 8 min. A 200 nm n-type diamond ($\sim 8 \times 10^{18}\,cm^{-3}$ phosphorous) was subsequently epitaxially overgrown and the sample then underwent electrochemical etching to dissociate the p-i-n diamond membrane as described previously. It should be noted that neither the intrinsic layer nor the n-layer is exposed to the ion damage.

An optical image of the standalone p-i-n single crystal membrane device is shown in Fig. 18A and the schematic illustration of the structure and the device are shown in Fig. 18B. A schematic energy band diagram is shown in the Fig. 20. The formed single crystal diamond membrane device displays excellent electrical characteristics, as shown in Fig. 18C.

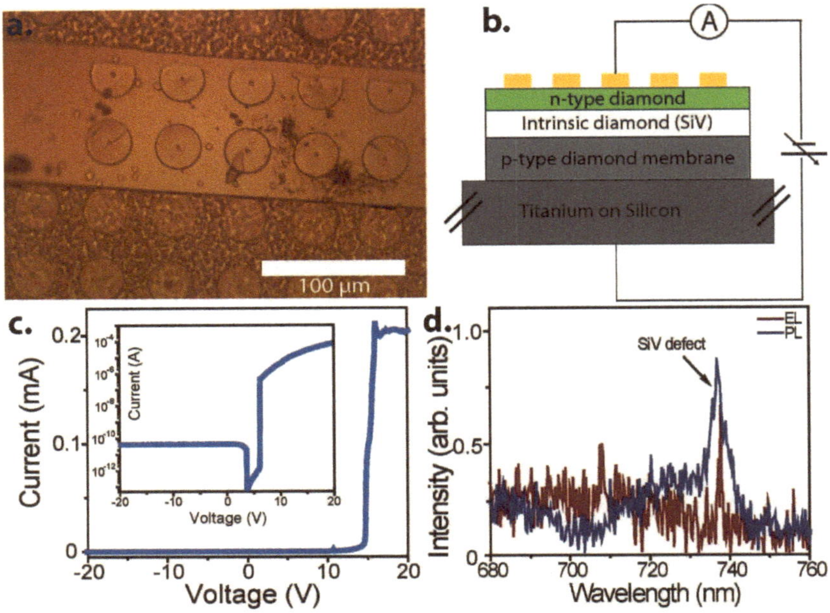

Fig. 18 Vertical standalone single crystal diamond p-i-n device. (A) An optical image of the p-i-n membrane containing a layer of SiV emitters in the intrinsic region. The circles are the metal contacts on the top n-type surface. (B) Schematic diagram of the cross section of the p-i-n diamond device. (C) The p-i-n membrane shows diode-like behavior with an onset voltage of ~11 V. Inset: Log (I) vs (V) rectification curve of the p-i-n diamond membrane showing a rectification ratio of $\sim 10^6$.

The diamond membranes exhibit diode-like behavior with a threshold voltage at ~11 V and negligible current under a reverse bias of >20V. This was confirmed by the semi-log I—V curve with a high rectification ratio of $\sim 10^6$ (Fig. 18c). The low threshold voltage is advantageous for efficient device operation and is comparable to many standard nano-optoelectronic devices made of GaN, SiC and ZnO (Choi et al., 2010; Deshpande et al., 2013; Lohrmann et al., 2015). It is possible that the phosphorous doping of the overgrown n-type diamond layer, may be causing issues with the efficient recombination of species and as an extension won't show electroluminescence of our SiV emitters in the intrinsic layer of the diamond membrane diode. The CVD recipe for uniform n-type growth may be tweaked by manipulating several growth parameters which may lead to a more homogenous doping and/or a higher doping of n-type species to allow for recombination to occur. By modifying the gas rate ratios and gas flow of the phosphorous gas. This will modify the incorporation rate of n-doped species and therefore the doping density of the n-doped layer which may lead to an enhancement of the diamonds electrical properties. It will be a very time-consuming process and can't currently be conducted in-house.

A high-quality single crystal diamond layer is necessary to prevent the loss of electrons through grain boundaries and other dislocations in the diamond lattice. However, it was observed that a few of the diamond membrane samples possessed a rough overgrown layer after MPCVD epitaxial overgrowth. The dislocations in the diamond surface are possibly due to MPCVD growth on the damaged side of the diamond membrane, i.e., where the graphitic layer formed after He^{+} ion implantation and the presence of sp2 carbon. Also, the dislocations may be due to the presence of residual organic compounds after piranha cleaning or after transferring the membranes between substrates prior to growth. The non-smooth surface caused an array of squares of varying heights to be produced (Fig. 19). This pattern is possibly due to the previously mentioned micro masking of organics, other contaminants, or a rough surface causing the diamond planes to grow at different rates. This non-single crystalline surface will cause the electrons to preferentially flow through the deformities, grain boundaries and dislocations on the diamond surface. The grain boundary lattice sites will have more available sp2 sites for recombination to occur away from a SiV defect. To improve the quality of the diamond surface, the diamond

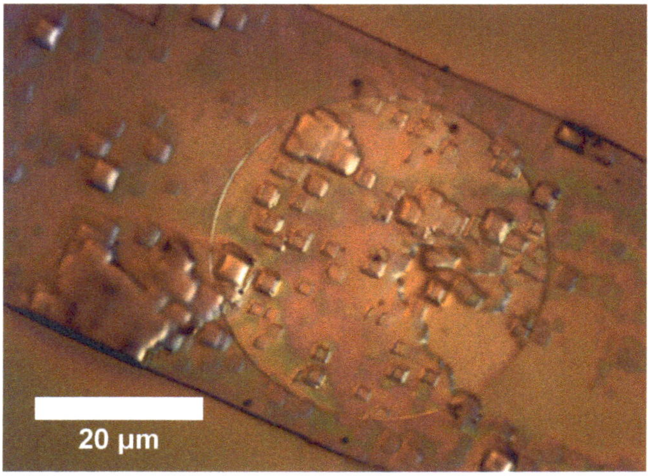

Fig. 19 Rough surface. Optical microscopy image of a membrane that underwent MPCVD overgrowth to produce an array of squares of varying dimensions.

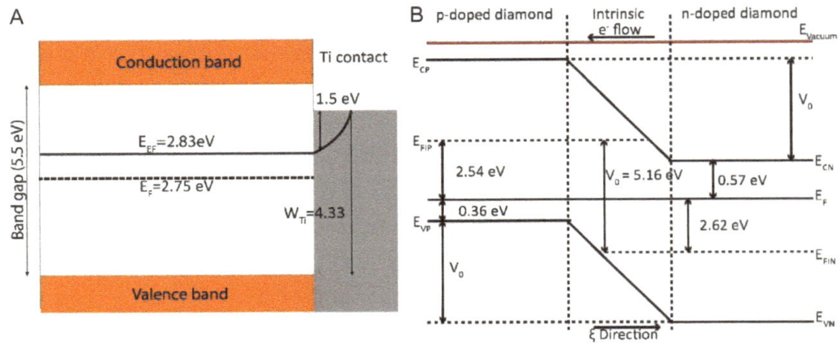

Fig. 20 Band diagram of the p-i-n diamond device. (A) Diagram illustrating when the Titanium comes into contact with the n-layer of the diamond membrane. E_{EF} is the effective fermi level, E_F is the fermi level and W_{Ti} is the work function of titanium. (B) Schematic energy band diagram of the p-i-n diamond membrane at equilibrium. The boron and phosphorous energy levels were taken as 0.36 and 0.57 eV, respectively.

samples and equipment need to be cleaned of all unwanted contaminants through both wet and dry-cleaning techniques, which include piranha solution as well as cleaning the sample in a hydrogen or a weak oxygen plasma at high pressure to burn away organics.

A band diagram of a typical all diamond p-i-n junctions is presented below (Fig. 20). W_{Ti} corresponds to the work function of titanium, E_F and E_{EF} are the Fermi and effective Fermi level after doping, respectively. V_O is the step-in energy between the bands of the p and n doped diamond regions. The boron and phosphorous energy levels were taken as 0.36 and 0.57 eV, respectively. E_{FIP} and E_{FIN} are the effective Fermi energies for the p-doped and n-doped diamond layers.

Low rectification curves may be caused by the charge mobility across the p-i-n junction due to the low conductivity of phosphorus doped diamond. Phosphorous doped diamond films only have a carrier mobility of $\sim 660\,cm^2\,V^{-1}\,s^{-1}$ which is approximately three time slower than boron doped diamond films at $\sim 1800\,cm^2\,V^{-1}\,s^{-1}$ (Katagiri et al., 2004; Kato et al., 2007b).

The fabricated p-i-n diamond membranes possess an average threshold voltage of ~10 V with a forward bias current of ~4 mA. The current under reverse bias is negligible under a voltage >40 V. The devices boast a high rectification ratio of 1×10^6. These properties make them ideal for electrical applications and are comparable to other semiconductor p-i-n diode structures in the literature. For instance, p-i-n ZnO structure has been studied for decades boasting a threshold voltage of ~3.2 V with a forward bias current of up to 80 mA. The device exhibits high breakdown voltages of up to 30 V with negligible current on the order of 1×10^{-7} A, which leads to high rectification ratios of 8×10^5 (Choi et al., 2010). Furthermore, by embedding InGaN QDs into the intrinsic layer of GaN p-i-n diode, diode-like behavior has been demonstrated with onset voltages of ~3.5 V and a current density up to ~100 mA with breakdown voltages and leakage current of between 4 and 20 V and 1×10^{-12} A, respectively, leading to a rectification ratio of ~2000 (Deshpande et al., 2013, 2014). SiC is another readily studied wide band gap material and exhibits a threshold voltage of ~3.1 V with currents in the range of mA. The diode structures show a very small reverse leakage current of $\sim 1 \times 10^{-10}$ A, which leads to very high rectification ratios of 140×10^6 (Ikeda et al., 1979; Khramtsov et al., 2018; Lohrmann et al., 2015). The fabricated diamond membranes boast higher rectification ratios and similar leakage current, making them an ideal substitute to ZnO and GaN diode devices. The forward bias current of the diamond membranes is lower than the ZnO and GaN devices. The lower current is due to power transport of carries between the n-doped and p-doped region. To improve the transport of carriers through the device the more sophisticated fabrication protocols need to be developed to create a higher quality n-doped single crystal layer and more conductive metal contacts to allow high current densities to flow into the device.

Furthermore, diodes have been engineered out of layered 2D chalcogenide materials such as MoS_2 that possesses a lower forward voltage of ~0.5 V with 0. 35–1 μA current under a positive bias. They also have an unfavorably low breakdown voltage ~1.5–10 V with a leakage current of $\sim 5 \times 10^{-8} - 5 \times 10^{-9}$ A (Sundaram et al., 2013; Ye et al., 2014). MoS_2 is known to have poor optical quality than other group-VIB transition-metal dichalcogenides (TMDs), possibly due to impurities. In contrast, it has been shown previously that monolayer WSe_2 has better optical properties compared to other group-VIB TMDs (Jones et al., 2013; Zhao et al., 2013). WSe_2 has been shown to possess a threshold voltage of 6.5 V with a current of ~7 nA. The breakdown voltage for this device was not explicitly shown but is >2V (Ross et al., 2014).

Diode fabrication is not limited to bulk and monolayer materials, it has also been demonstrated for single molecules. For example, single tris (1-phenylisoquinoline)iridium ($Ir(piq)_3$) molecules (Hedley et al., 2009) were used as emitters inside wide bandgap PMMA and poly(9-vinylcarbazole):2-(4-*tert*-butylphenyl)-5-(4-biphenylyl)-1,3,4-oxadiazole (PVK:PBD) polymers.

Iridium-based emitters are known for their short triplet-state lifetimes compared with other organometallic complexes (Nothaft et al., 2012; Yersin, 2008). These embedded single molecules exhibit a lower forward threshold voltage of ~5 V and a high current of ~100 mA. These types of devices do not possess the same variability as the fabricated diamond membranes, in the sense that they cannot be further processed to create resonators or waveguides as well as they cannot be easily transferred to different substrates or other desired structures.

The fabricated diamond membranes are also comparable to other diamond diode devices that have been fabricated out of polycrystalline and single crystalline diamond. For instance, polycrystalline diamond was grown onto a lightly phosphorous doped silicon substrate using MPCVD then a layer of ITO was magneton sputtered. The device displays a low threshold voltage of ~1.5 V with a forward bias current of ~60 mA. The diode device exhibits a small breakdown voltage and large current of ~2.5 V and ~8 mA, respectively. This leads to a low rectification ratio of ~10 (Liang et al., 2007). Alternatively, bulk single crystal diamond p-i-n structures have been engineered through MPCVD growth of a bulk boron doped diamond crystal with an independently grown intrinsic layer containing either NV or SiV color centers and a phosphorus doped n-type layer. The bulk diamond diodes show a forward threshold voltage of ~30–40 V with 3–12 mA current and a breakdown voltage and leakage current of ~43 V and 1×10^{-7} A, respectively. This leads to very high rectification ratios of up to 1×10^{9} (Berhane et al., 2015; Mizuochi et al., 2012a). The data is summarized in Table 2 Diamond membranes are the next step in the development of diamond nanoelectronics since they possess similar electrical characteristics to their bulk counterparts and have a high-quality diamond surface. This will allow the realization of membranes for quantum communication, quantum information processes and on-chip applications.

To summarize, we show the first reliable and robust engineering of nanoscale conductive diamond membranes. Moreover, we engineered the first-of-its-kind vertical p-n and p-i-n nanoscale devices using entirely single crystal diamond. The membranes and the devices have high aspect ratio of lateral to vertical dimensions, are standalone and can be positioned onto a substrate of choice. The diamond membrane devices are therefore suitable for a variety of nanophotonic, nanoelectronics and quantum information applications. Their suitability will be shown and discussed in the following section, demonstrating EL from nanoscale diamond membrane diodes.

Table 2 A summary of the electrical properties of varying diode structures in the literature, compared to the fabricated diamond membranes.

Material	Threshold voltage (V)	Current (mA)	Rectification ratio	Breakdown voltage (V)	References
ZnO	~3.2	80	800,000	30	Choi et al. (2010)
SiC	~3.1	~14	140×10^6	*greater than* 0	Lohrmann et al. (2015)
GaN	~3.5	~100	~2000	*greater than* 4 − ~20	Deshpande et al. (2013, 2014)
MoS_2	~0.5	$\sim 0.35 \times 10^{-3} - \sim 1 \times 10^{-3}$	7–1000	~1.5–10	Sundaram et al. (2013) and Ye et al. (2014)
Single molecule	~5	~100	n/a	*greater than* 0	Nothaft et al. (2012)
WSe_2	6.5	$\sim 7 \times 10^{-6}$	n/a	*greater than* 2	Ross et al. (2014)
Polycrystalline diamond	~1.5	~60	~10	~2.5	Liang et al. (2007)
Bulk diamond	~30 − 40	~3 − 12	$1 \times 10^4 - 1 \times 10^9$	~43	Berhane et al. (2015) and Mizuochi et al. (2012a)
Diamond membranes	~10 V	~4 mA	1×10^6	*greater than* 40	Bray et al. (2018a)

7. Diamond electroluminescence

To demonstrate a working diamond LED, we fabricated a p-i-n structure into a single crystal diamond membrane containing SiV color centers (Section 6.4); that display both PL and EL (Fig. 18). The inclusion of the SiVs into the intrinsic layer between the p- and the n-doped regions allows the recombination process to occur at the SiV centers, thus generating EL signal.

Color centers in diamond have become the platform of choice to study quantum nanophotonics and light matter interactions at the nanoscale. Amongst many different color centers the SiV color center has garnered a significant attention due to its photostability, ease of fabrication and high Debye Waller factor (Becker et al., 2018; Bradac et al., 2019; Evans et al., 2018; MacHielse et al., 2019; Meesala et al., 2018; Neu et al., 2011b; Pingault et al., 2017; Zhou et al., 2017). Recently, the SiV was used to demonstrate basic quantum photonic components including quantum spin registers and photon-photon interactions in a photonic crystal cavity (Nguyen et al., 2019).

One of the outstanding challenges in the field, is detailed studies of electrical triggering of diamond color centers, and in particular the SiV (Berhane et al., 2015; Mizuochi et al., 2012a). Electrical excitation is vital for scaling up nanophotonic components and integrate diamond photonics with mature CMOS technology (Lohrmann et al., 2011, 2015; Widmann et al., 2019). However, growing diamond *p-n* or *p-i-n* junction devices is challenging due to lack of efficient *n-type* doping (Fedyanin and Agio, 2016b; Khramtsov and Fedyanin, 2019). To this extent, availability of single crystal ultrathin diamond membranes that host optically active color centers is important (Bray et al., 2018a, b; Magyar et al., 2011). These membranes can be integrated with other materials to realize a hybrid architecture.

We have demonstrated such a hybrid device, consisting of a thin diamond membrane sandwiched between an *n-type* indium tin oxide (ITO) and *p-type* silicon substrate. We show efficient electroluminescence from the SiV and study in details its excitation mechanisms.

The literature of diamond membranes is summarized in Table 3 and provides the key characteristics of size and optical activity. Notably, the membranes reported in this work are amongst the largest in terms of surface area, while still being relatively thin—on the order of ~300 nm thick. Furthermore, the membranes have the advantage of being free-standing single crystal diamond that can be transferred and accurately positioned

Table 3 A table summarizing the performance of diamond membranes.

Length × width (μm)	Thickness (nm)	PL	EL	Free standing	References
15 × 3	200	No	No	No	Babinec et al. (2010)
2000 × 2000	3000	No	No	No	Pomorski et al. (2013)
3000 × 3000	2000	No	No	No	Suk et al. (2017)
400 × 400	1000	Yes, (NV)	No	No[a]	Piracha et al. (2016a,b)
1100 (diameter)	1200	Yes, (NV)	No	No	Ali Momenzadeh et al. (2016)
300 × 300	200	No	No	Yes	Fairchild et al. (2008)
4500 × 4500	15,000	No	No	Yes	Bormashov et al. (2017)
10 × 10	200	Yes, (NV)	No	Yes	Ali Momenzadeh et al. (2016)
220 × 220	200	Yes (NV, SiV)	No	Yes	Aharonovich et al. (2012) and Magyar et al. (2014)
1000 × 500	300	Yes, (SiV)	Yes, (SiV)	Yes	Bray et al. (2018a)

[a]The membrane was attached to a large polycrystalline diamond frame.
The main parameters of size, optical activity (PL), EL and whether the membrane is free standing or attached are provided.

on a substrate of choice. The membranes engineered in the current work also host optically active defects that can be triggered both optically and electrically. In combination, all these attributes are unique and ideal for future fabrication of integrated nanophotonic, optoelectronic and optomechanical circuits using single crystal diamond membranes. Most of the works referenced in Table 3 claim to be using membranes. The work most similar to my work are references (Piracha et al., 2016a,b) since they make use of a similar fabrication protocol. The "no" in the EL column for references (Aharonovich et al., 2012; Magyar et al., 2014) means that the presented referenced work did not demonstrate EL upon the characterization of their materials.

References

Achard, J., et al., 2011. Thick boron doped diamond single crystals for high power electronics. Diamond Relat. Mater. 20 (2), 145–152.

Aharonovich, I., 2010. Novel Single Photon Emitters Based on Color Centers in Diamond. November.

Aharonovich, I., Neu, E., 2014. Diamond nanophotonics. Adv. Opt. Mater. 2, 1–18.

Aharonovich, I., Castelletto, S., Simpson, D.A., Su, C.H., Greentree, A.D., Prawer, S., 2011a. Diamond-based single-photon emitters. Reports Prog. Phys. 74 (7).
Aharonovich, I., Greentree, A.D., Prawer, S., 2011b. Diamond photonics. Nat. Photonics 5 (7), 397–405.
Aharonovich, I., et al., 2012. Homoepitaxial growth of single crystal diamond membranes for quantum information processing. Adv. Mater. 24 (10), OP54–OP59.
Aharonovich, I., Lee, J.C., Magyar, A.P., Bracher, D.O., Hu, E.L., 2013. Bottom-up engineering of diamond micro- and nano-structures. Laser Photonics Rev. 7 (5), 61–65.
Alhaddad, A., et al., 2011. Nanodiamond as a vector for siRNA delivery to Ewing sarcoma cells. Small 7 (21), 3087–3095.
Ali Momenzadeh, S., et al., 2016. Thin circular diamond membrane with embedded nitrogen-vacancy centers for hybrid spin-mechanical quantum systems. Phys. Rev. Appl. 6 (2), 024026.
Alléaume, R., et al., 2004. Experimental open-air quantum key distribution with a single-photon source. New J. Phys. 6, 1–14.
Arute, F., et al., 2019. Quantum supremacy using a programmable superconducting processor. Nature 574 (7779), 505–510.
Awschalom, D.D., Epstein, R., Hanson, R., 2007. The diamond age of spintronics. Sci. Am. 297 (4), 84–91.
Babinec, T.M., Choy, J.T., Smith, K.J.M., Khan, M., Lončar, M., 2010. Design and focused ion beam fabrication of single crystal diamond nanobeam cavities. J. Vac. Sci. Technol. B 29 (010601), 1–6.
Balasubramanian, G., Lazariev, A., Arumugam, S.R., Duan, D.W., 2014. Nitrogen-vacancy color center in diamond-emerging nanoscale applications in bioimaging and biosensing. Curr. Opin. Chem. Biol. 20 (1), 69–77.
Becker, J.N., et al., 2018. All-optical control of the silicon-vacancy spin in diamond at Millikelvin temperatures. Phys. Rev. Lett. 120 (5), 1–6.
Berhane, A.M., et al., 2015. Electrical excitation of silicon-vacancy centers in single crystal diamond. Appl. Phys. Lett. 106 (17), 171102.
Bochmann, J., Vainsencher, A., Awschalom, D.D., Cleland, A.N., 2013. Nanomechanical coupling between microwave and optical photons. Nat. Phys. 9 (11), 712–716.
Bormashov, V.S., et al., 2017. Thin large area vertical Schottky barrier diamond diodes with low on-resistance made by ion-beam assisted lift-off technique. Diamond Relat. Mater. 75, 78–84.
Bradac, C., Gao, W., Forneris, J., Trusheim, M., Aharonovich, I., 2019. Quantum Nanophotonics with Group IV defects in Diamond Carlo. https://arxiv.org/abs/1906.10992.
Bray, K., et al., 2018a. Single crystal diamond membranes for nanoelectronics. Nanoscale 10 (8), 4028–4035.
Bray, K., et al., 2018b. Single crystal diamond membranes and photonic resonators containing germanium vacancy color centers. ACS Photonics 5 (12), 4817–4822.
Buckley, S., Rivoire, K., Vučković, J., 2012. Engineered quantum dot single-photon sources. Reports Prog. Phys. 75 (12), 126503.
Buja, F., Sumant, A.V., Kokorian, J., van Spengen, W.M., 2014. Electrically conducting ultrananocrystalline diamond for the development of a next generation of micro-actuators. Sensors Actuators A Phys. 214, 259–266.
Butler, J.E., et al., 2003. Exceptionally high voltage Schottky diamond diodes and low boron doping. Semicond. Sci. Technol. 18 (3), S67–S71.
Childress, L., Hanson, R., 2013. Diamond NV centers for quantum computing and quantum networks. MRS Bull. 38 (2), 134–138.
Choi, Y.-S., Kang, J.-W., Hwang, D.-K., Park, S.-J., 2010. Recent advances in ZnO-based light-emitting diodes. IEEE Trans. Electron Devices 57 (1), 26–41.

Choi, S., Berhane, A.M., Gentle, A., Ton-that, C., Phillips, M.R., Aharonovich, I., 2015. Electroluminescence from localized defects in zinc oxide: toward electrically driven single photon sources at room temperature. ACS Appl. Mater. Interfaces 7 (10), 5619–5623.

Cristea, M.J., 2011. Unified model for p-n junction current-voltage characteristics. Cent. Eur. J. Eng. 1 (1), 113–116.

Deshpande, S., Heo, J., Das, A., Bhattacharya, P., 2013. Electrically driven polarized single-photon emission from an InGaN quantum dot in a GaN nanowire. Nat. Commun. 4, 1675.

Deshpande, S., Frost, T., Hazari, A., Bhattacharya, P., 2014. Electrically pumped single-photon emission at room temperature from a single InGaN/GaN quantum dot. Appl. Phys. Lett. 105 (14), 141109.

Doi, Y., et al., 2014. Deterministic electrical charge-state initialization of single nitrogen-vacancy center in diamond. 011057, 1–11.

Drijkoningen, S., Janssens, S.D., Pobedinskas, P., Koizumi, S., Van Bael, M.K., Haenen, K., 2016. The pressure sensitivity of wrinkled B-doped nanocrystalline diamond membranes. Sci. Rep. 6 (1), 35667.

Eaton, S.M., et al., 2019. Quantum micro–nano devices fabricated in diamond by femtosecond laser and ion irradiation. Adv. Quantum Technol. 2 (5–6), 1900006.

Ellis, B., et al., 2011. Ultralow-threshold electrically pumped quantum-dot photonic-crystal nanocavity laser. Nature 5, 297–300.

Evans, R.E., et al., 2018. Photon-mediated interactions between quantum emitters in a diamond nanocavity. Science 362 (6415), 662–665.

Faggio, G., Messina, G., Santangelo, S., Prestopino, G., Ciancaglioni, I., Marinelli, M., 2012. Raman scattering in boron-doped single-crystal diamond used to fabricate Schottky diode detectors. J. Quant. Spectrosc. Radiat. Transf. 113 (18), 2476–2481.

Fairchild, B.A., et al., 2008. Fabrication of ultrathin single-crystal diamond membranes. Adv. Mater. 20 (24), 4793–4798.

Faraon, A., Barclay, P.E., Santori, C., Fu, K.M.C., Beausoleil, R.G., 2011. Resonant enhancement of the zero-phonon emission from a colour centre in a diamond cavity. Nat. Photonics 5 (5), 301–305.

Faraon, A., Santori, C., Huang, Z., Acosta, V.M., Beausoleil, R.G., 2012. Coupling of nitrogen-vacancy centers to photonic crystal cavities in monocrystalline diamond. Phys. Rev. Lett. 109 (3), 2–6.

Fedyanin, D.Y., Agio, M., 2016a. Ultrabright single-photon source on diamond with electrical pumping at room and high temperatures. New J. Phys. 18 (7).

Fedyanin, D.Y., Agio, M., 2016b. Ultrabright single-photon source on diamond with electrical pumping at room and high temperatures. New J. Phys. 18 (7), 073012.

Fu, C.-C., et al., 2007. Characterization and application of single fluorescent nanodiamonds as cellular biomarkers. Proc. Natl. Acad. Sci. U. S. A. 104 (3), 727–732.

Gisin, N., Ribordy, G., Tittel, W., Zbinden, H., 2002. Quantum cryptography. Rev. Mod. Phys. 74 (1), 145–195.

Hauf, M.V., Simon, P., Seifert, M., Holleitner, A.W., Stutzmann, M., Garrido, J.A., 2014. Low dimensionality of the surface conductivity of diamond. Phys. Rev. B 89 (11), 1–5.

Hausmann, B.J.M., et al., 2012. Integrated diamond networks for quantum nanophotonics. Nano Lett. 12 (3), 1578–1582.

Hausmann, B.J.M., et al., 2013a. Integrated high-quality factor optical resonators in diamond. Nano Lett. 13 (5), 1898–1902.

Hausmann, B.J.M., et al., 2013b. Coupling of NV centers to photonic crystal nanobeams in diamond. Nano Lett. 13 (12), 5791–5796.

Hedley, G.J., Ruseckas, A., Samuel, I.D.W., 2009. Ultrafast intersystem crossing in a red phosphorescent indium complex. J. Phys. Chem. A 113 (1), 2–4.

Hepp, C., et al., 2014. Electronic structure of the silicon vacancy color center in diamond. Phys. Rev. Lett. 112 (3), 036405.
Ikeda, M., Hayakawa, T., Yamagiwa, S., Matsunami, H., Tanaka, T., 1979. Fabrication of 6H-SiC light-emitting diodes by a rotation dipping technique: electroluminescence mechanisms. J. Appl. Phys. 50 (12), 8215–8225.
Isberg, J., et al., 2002. High carrier mobility in single-crystal plasma-deposited diamond. Science 297 (5587), 1670–1672.
Iwasaki, T., Hoshino, Y., Tsuzuki, K., Kato, H., Makino, T., Ogura, M., 2013. High-temperature operation of diamond junction field-effect transistors with lateral p-n junctions. 34 (9), 1175–1177.
Iwasaki, T., et al., 2015. Germanium-vacancy single color centers in diamond. Sci. Rep. 5 (1), 12882.
Jelezko, F., Wrachtrup, J., 2004. Read-out of single spins by optical spectroscopy. J. Phys. Condens. Matter 16 (30).
Jones, A.M., et al., 2013. Optical generation of excitonic valley coherence in monolayer WSe2. Nat. Nanotechnol. 8 (9), 634–638.
Katagiri, M., Isoya, J., Koizumi, S., Kanda, H., 2004. Lightly phosphorus-doped homoepitaxial diamond films grown by chemical vapor deposition. Appl. Phys. Lett. 85 (26), 6365–6367.
Kato, H., Makino, T., Yamasaki, S., Okushi, H., 2007a. N-type diamond growth by phosphorus doping on (001)-oriented surface. J. Phys. D Appl. Phys. 40 (20), 6189–6200.
Kato, H., Makino, T., Yamasaki, S., Okushi, H., 2007b. N-type diamond growth by phosphorus doping on (0 0 1)-oriented surface. J. Phys. D Appl. Phys. 40 (20), 6189–6200.
Kato, H., et al., 2013. Tunable light emission from nitrogen-vacancy centers in single crystal diamond PIN diodes. Appl. Phys. Lett. 102 (15), 1–5.
Khanaliloo, B., Jayakumar, H., Hryciw, A.C., Lake, D.P., Kaviani, H., Barclay, P.E., 2015a. Single-crystal diamond nanobeam waveguide optomechanics. Phys. Rev. X 5 (4), 041051-1–041051-21.
Khanaliloo, B., Mitchell, M., Hryciw, A.C., Barclay, P.E., 2015. High-Q/V monolithic diamond microdisks fabricated with quasi-isotropic etching. Nano Lett. 15, 5131–5136.
Khramtsov, I.A., Fedyanin, D.Y., 2019. Superinjection in diamond homojunction P-I-N diodes. Semicond. Sci. Technol. 34 (3).
Khramtsov, I.A., Vyshnevyy, A.A., Fedyanin, D.Y., 2018. Enhancing the brightness of electrically driven single-photon sources using color centers in silicon carbide. NPJ Quantum Inf. 4 (1), 15.
Kieling, K., Rudolph, T., Eisert, J., 2007. Percolation, renormalization, and quantum computing with nondeterministic gates. Phys. Rev. Lett. 99 (13), 2–5.
Knill, E., 2010. Quantum computing. Nature 463 (January), 441–443.
Koizumi, S., Watanabe, K., Hasegawa, M., Kanda, H., 2001. Ultraviolet emission from a diamond pn junction. Science 292 (June), 1899–1901.
Krauss, A.R., et al., 2001. Ultrananocrystalline diamond thin films for MEMS and moving mechanical assembly devices. Diamond Relat. Mater. 10 (11), 1952–1961.
Lee, J.C., Magyar, A.P., Bracher, D.O., Aharonovich, I., Hu, E.L., 2013. Fabrication of thin diamond membranes for photonic applications. Diamond Relat. Mater. 33, 45–48.
Lesik, M., et al., 2015. Preferential orientation of NV defects in CVD diamond films grown on (113)-oriented substrates. Diamond Relat. Mater. 56, 47–53.
Li, J., Zhu, Y., Li, W., Zhang, X., Peng, Y., Huang, Q., 2010. Nanodiamonds as intracellular transporters of chemotherapeutic drug. Biomaterials 31 (32), 8410–8418.
Liang, X., et al., 2007. Enhanced red electroluminescence from a polycrystalline diamond film/Si heterojunction structure. Appl. Phys. Lett. 90 (16), 16–19.
Lodahl, P., Mahmoodian, S., Stobbe, S., 2015. Interfacing single photons and single quantum dots with photonic nanostructures. Rev. Mod. Phys. 87 (2), 347–400.

Lohrmann, A., et al., 2011. Diamond based light-emitting diode for visible single-photon emission at room temperature. Appl. Phys. Lett. 99 (25), 251106.
Lohrmann, A., et al., 2015. Single-photon emitting diode in silicon carbide. Nat. Commun. 6, 7783.
Lounis, B., Orrit, M., 2005. Single-photon sources. Reports Prog. Phys. 68 (5), 1129–1179.
Lu, W., Lieber, C.M., 2007. Nanoelectronics from the bottom up. Nat. Mater. 6 (11), 841–850.
MacHielse, B., et al., 2019. Quantum interference of electromechanically stabilized emitters in nanophotonic devices. Phys. Rev. X 9 (3), 1–11.
Macpherson, J.V., 2015. A practical guide to using boron doped diamond in electrochemical research. Phys. Chem. Chem. Phys. 17 (5), 2935–2949.
MacQuarrie, E.R., Otten, M., Gray, S.K., Fuchs, G.D., 2016. Cooling a mechanical resonator with a nitrogen-vacancy center ensemble using a room temperature excited state spin-strain interaction. Nat. Publ. Gr. 8, 1–10.
Magyar, A.P., et al., 2011. Fabrication of thin, luminescent, single-crystal diamond membranes. Appl. Phys. Lett. 99 (8), 081913.
Magyar, A.P., Bracher, D., Lee, J.C., Aharonovich, I., Hu, E.L., 2014. High quality SiC microdisk resonators fabricated from monolithic epilayer wafers. Appl. Phys. Lett. 104 (5), 051109.
Makino, T., et al., 2009. Diamond Schottky p-n diode with high forward current density and fast switching operation. Appl. Phys. Lett. 94, 1–3.
Makino, T., et al., 2011. Enhancement in emission efficiency of diamond deep-ultraviolet light emitting diode. Appl. Phys. Lett. 99 (6), 25–28.
Matsumoto, T., et al., 2016. Inversion channel diamond metal-oxide-semiconductor field-effect transistor with normally off characteristics. Sci. Rep. 6 (1), 31585.
McGuinness, L.P., et al., 2011. Quantum measurement and orientation tracking of fluorescent nanodiamonds inside living cells. Nat. Nanotechnol. 6 (6), 358–363.
Meesala, S., et al., 2018. Strain engineering of the silicon-vacancy center in diamond. Phys. Rev. B 97 (20), 1–13.
Merson, T.D., Castelletto, S., Aharonovich, I., Turbic, A., Kilpatrick, T.J., Turnley, A.M., 2013. Nanodiamonds with silicon vacancy defects for nontoxic photostable fluorescent labeling of neural precursor cells. Opt. Lett. 38 (20), 4170–4173.
Michler, P., Kiraz, A., Becher, C., Schoenfeld, W.V., 2000. A quantum dot single-photon turnstile device. 290 (December), 2282–2286.
Mizuochi, N., et al., 2012a. Electrically driven single-photon source at room temperature in diamond. Nat. Photonics 6 (5), 299–303.
Mizuochi, N., et al., 2012b. Electrically driven single-photon source at room temperature in diamond. 6 (April), 299–303.
Mouradian, S.L., et al., 2015. Scalable integration of long-lived quantum memories into a photonic circuit. Phys. Rev. X 5 (3), 031009.
Neu, E., et al., 2011a. Single photon emission from silicon-vacancy colour centres in chemical vapour deposition nano-diamonds on iridium. New J. Phys. 13 (2), 025012.
Neu, E., Agio, M., Becher, C., 2012. Photophysics of single silicon vacancy centers in diamond: implications for single photon emission. Opt. Express 20 (18), 19956–19971.
Neu, E., et al., 2011. Single photon emission from silicon-vacancy colour centres in chemical vapour deposition nano-diamonds on iridium. New J. Phys. 13, 025012.
Neu, E., et al., 2014. Photonic nano-structures on (111)-oriented diamond. Appl. Phys. Lett. 104 (15).
Nguyen, C.T., et al., 2019. An integrated nanophotonic quantum register based on silicon-vacancy spins in diamond. Phys. Rev. B 165428, 1–19.
Northup, T.E., Blatt, R., 2014. Quantum information transfer using photons. Nat. Photonics 8 (5), 356–363.

Nothaft, M., Höhla, S., Jelezko, F., Frühauf, N., Pflaum, J., Wrachtrup, J., 2012. Electrically driven photon antibunching from a single molecule at room temperature. Nat. Commun. 3, 628.
O'Brien, J.L., Furusawa, A., Vuckovic, J., 2009. Photonic quantum technologies. Nat. Photonics 3, 687–695.
Parks, S.M., Grote, R.R., Hopper, D.A., Bassett, L.C., 2018. Fabrication of (111)-faced single-crystal diamond plates by laser nucleated cleaving. Diam. Relat. Mater. 84 (December 2017), 20–25.
Pezzagna, S., Naydenov, B., Jelezko, F., Wrachtrup, J., Meijer, J., 2010. Creation efficiency of nitrogen-vacancy centres in diamond. New J. Phys. 12.
Pingault, B., et al., 2017. Coherent control of the silicon-vacancy spin in diamond. Nat. Commun. 8 (May), 1–7.
Piracha, A.H., Rath, P., Ganesan, K., Kühn, S., Pernice, W.H.P., Prawer, S., 2016a. Scalable fabrication of integrated Nanophotonic circuits on arrays of thin single crystal diamond membrane windows. Nano Lett. 16 (5), 3341–3347.
Piracha, A.H., et al., 2016b. Scalable fabrication of high-quality, ultra-thin single crystal diamond membrane windows. Nanoscale 8, 6860–6865.
Pomorski, M., Caylar, B., Bergonzo, P., 2013. Super-thin single crystal diamond membrane radiation detectors. Appl. Phys. Lett. 103 (11), 112106.
Purcell, E.M., Torrey, H.C., Pound, R.V., 1946. Resonance absorption by nuclear magnetic moments in a solid. Phys. Rev. 69 (37), 37–38.
Rath, P., Khasminskaya, S., Nebel, C., Wild, C., Pernice, W.H.P., 2013. Diamond-integrated optomechanical circuits. Nat. Commun. 4, 1690.
Rath, P., Ummethala, S., Nebel, C., Pernice, W.H.P., 2015. Diamond as a material for monolithically integrated optical and optomechanical devices. Phys. Status Solidi 15, 2385–2399.
Riedrich-Möller, J., et al., 2011. One- and two-dimensional photonic crystal microcavities in single crystal diamond. Nat. Nanotechnol. 7, 69.
Rogers, L.J., et al., 2014. Electronic structure of the negatively charged silicon-vacancy center in diamond. Phys. Rev. B 89 (23), 1–8.
Ross, J.S., et al., 2014. Electrically tunable excitonic light-emitting diodes based on monolayer WSe2 p–n junctions. Nat. Nanotechnol. 9 (4), 268–272.
Sandorfy, C., et al., 2002. Electrically driven single-photon source. Science 295 (January).
Schneider, H., Sanchez, J.L., Achard, J., 2005. 2005 European Conference on Power Electronics and Applications.
Shields, A.J., 2007. Semiconductor quantum light sources. Nat. Photonics 1 (4), 215–223.
Shikata, S., 2016. Single crystal diamond wafers for high power electronics. Diamond Relat. Mater. 65, 168–175.
Sipahigil, A., et al., 2016. An integrated diamond nanophotonics platform for quantum-optical networks. Science 354 (6314), 847–850.
Sittas, G., Kanda, H., Kiflawi, I., Spear, P.M., 1996. Growth and characterization of Si-doped diamond single crystals grown by the HTHP method. Diam. Relat. Mater. 5 (6–8), 866–869.
Stevenson, R.M., et al., 2012. Indistinguishable entangled photons generated by a light-emitting diode. Phys. Rev. Lett. 108 (4), 1–5.
Su, C., Greentree, A.D., Hollenberg, L.C.L., 2008. Towards a picosecond transform-limited nitrogen-vacancy based single photon source. Opt. Express 16 (9), 2722–2725.
Suk, J., et al., 2017. Fabrication of thin diamond membranes by using hot implantation and ion-cut methods. Appl. Phys. Lett. 110 (10).
Sun, S., Kim, H., Luo, Z., Solomon, G.S., Waks, E., 2018. A single-photon switch and transistor enabled by a solid-state quantum memory. Science 45 (July), 57–60.

Sundaram, R.S., et al., 2013. Electroluminescence in single layer MoS2. Nano Lett. 13 (4), 1416–1421.
Sze, 1995. Physics of Semiconductor Devices Physics of Semiconductor Devices.
Tao, Y., Boss, J.M., Moores, B.A., Degen, C.L., 2014. Single-crystal diamond nanomechanical resonators with quality factors exceeding one million. Nat. Commun. 5, 1–8.
Tomljenovic-Hanic, S., Steel, M.J., de Sterke, C.M., Salzman, J., 2006. Diamond based photonic crystal microcavities. Opt. Express 14 (8), 3556.
Wang, C., Kurtsiefer, C., Weinfurter, H., Burchard, B., 2006. Single photon emission from SiV centres in diamond produced by ion implantation. J. Phys. B At. Mol. Opt. Phys. 39 (1), 37–41.
Wang, J., et al., 2018. Multidimensional quantum entanglement with large-scale integrated optics. Science 360 (6386), 285–291.
Watanabe, H., Nebel, C.E., Shikata, S., 2009. Isotopic homojunction band engineering from diamond. Science 342 (June), 1425–1428.
Wehner, S., Elkouss, D., Hanson, R., 2018. Quantum internet: A vision for the road ahead. Science 362 (6412).
Widmann, M., et al., 2019. Electrical charge state manipulation of single silicon vacancies in a silicon carbide quantum optoelectronic device. Nano Lett. 19 (10), 7173–7180.
Windischmann, H., Epps, G.F., 1992. Free-standing diamond membranes: optical, morphological and mechanical properties. Diamond Relat. Mater. 1 (5–6), 656–664.
Wort, C.J.H., Balmer, R.S., 2008. Diamond as an electronic material. Mater. Today 11 (1–2), 22–28, Elsevier Ltd.
Ye, Y., et al., 2014. Exciton-dominant electroluminescence from a diode of monolayer MoS2. Appl. Phys. Lett. 104 (19), 193508.
Yersin, H., 2008. Highly Efficient OLEDs with Phosphorescent Materials. Highly Effic. OLEDs with Phosphorescent Mater., no. 3, pp. 1–438.
Yuan, Z., et al., 2002. Electrically driven single-photon source. Science 295 (5552), 102–105.
Zhang, L., et al., 2017. Electrically driven single-photon emission from an isolated single molecule. Nat. Commun. 8 (1), 580.
Zhao, W., et al., 2013. Evolution of electronic structure in atomically thin sheets of WS 2 and WSe 2. ACS Nano 7 (1), 791–797.
Zhou, Y., Rasmita, A., Li, K., Xiong, Q., Aharonovich, I., Gao, W.B., 2017. Coherent control of a strongly driven silicon vacancy optical transition in diamond. Nat. Commun. 8, 1–7.

CHAPTER SIX

Diamond quantum nanophotonics and optomechanics

Cleaven Chia[a,*], Bartholomeus Machielse[b], Benjamin Pingault[a], Michelle Chalupnik[b], Graham Joe[a], Eliza Cornell[a], Sophie Weiyi Ding[a], Stefan Bogdanović[a,c], Kazuhiro Kuruma[a], Afaq Habib Piracha[a], Smarak Maity[a], Thomas M. Babinec[a], Srujan Meesala[d], and Marko Lončar[a]

[a]John A. Paulson School of Engineering and Applied Sciences, Harvard University, Cambridge, MA, United States
[b]Department of Physics, Harvard University, Cambridge, MA, United States
[c]X, The Moonshot Factory, Mountain View, CA, United States
[d]Thomas J. Watson, Sr., Laboratory of Applied Physics, California Institute of Technology, Pasadena, CA, United States
*Corresponding author: e-mail address: loncar@seas.harvard.edu

Contents

1. Introduction 220
2. Photonic devices 222
   2.1 Surface-etched devices for photon extraction into free space 223
   2.2 Freestanding photonic devices for enhanced spin-photon interactions and efficient photon collection 225
   2.3 Summary 230
3. Mechanical devices 230
   3.1 Frequency tuning and stabilization of emitted photons 231
   3.2 Classical qubit control using mechanical vibrations 233
   3.3 Toward phonons as on-chip quantum information carriers 235
   3.4 Summary 238
4. Optomechanical devices 239
   4.1 Optomechanical crystals 239
   4.2 Microdisks 241
   4.3 Waveguides 242
   4.4 Summary 242
5. Conclusions and outlook 243
Acknowledgments 244
References 244

Semiconductors and Semimetals, Volume 104
ISSN 0080-8784
https://doi.org/10.1016/bs.semsem.2020.08.005

1. Introduction

Quantum information technology promises to revolutionize computation, communication, and sensing, but realizing this promise requires control over the way quantum information is transferred onto, between, and away from sets of qubits (Kimble, 2008; Reiserer and Rempe, 2015; Wehner et al., 2018). Pioneering experiments in the field relied on carefully controlled interactions between isolated collections of atoms and their environment, allowing for the implementation of fundamental quantum technology protocols. These demonstrations spurred a search for quantum platforms that could be scaled up to large quantum circuits capable of handling many qubits and quantum operations. Optically addressable spins in solid state that can be interconnected using quantum photonic links emerged as a leading candidate for this role (Aharonovich et al., 2016; Atatüre et al., 2018; Awschalom et al., 2018).

Among solid state systems, diamond has long been considered as a promising material for realization of scalable quantum technology: it hosts a wide range of color centers (Bradac et al., 2019; Goss et al., 1996; Iwasaki et al., 2017; Rogers et al., 2014a; Siyushev et al., 2017; Tchernij et al., 2017). within its wide bandgap, many of which have long spin coherence times (Abobeih et al., 2018; Bar-Gill et al., 2013) and bright coherent optical transitions. These color centers can be introduced into the diamond during the crystal growth process (Lesik et al., 2014; Michl et al., 2014; Teraji et al., 2015), including delta (Ohno et al., 2012; Osterkamp et al., 2015) or spot (Chen et al., 2017, 2019; Ohno et al., 2014) doping, or generated in already grown diamond through ion implantation (Schröder et al., 2017; Ziegler et al., 2010), vacancy generation (Capelli et al., 2019; Chen et al., 2017; Ohno et al., 2014; Pezzagna et al., 2010), or annealing (Naydenov et al., 2010; Yamamoto et al., 2013). In general, these techniques allow for spatially targeted creation of color centers with accuracy down to tens of nanometers.

Historically, the most prominent of diamond's color centers has been the nitrogen vacancy (NV) center, which benefits from long coherence times even at room temperature (Abobeih et al., 2018; Balasubramanian et al., 2009) and spin-selective optical transitions that can be used to entangle the spin with photons (Bernien et al., 2013; Hensen et al., 2015; Pfaff et al., 2014; Togan et al., 2010). Unfortunately, the NV center was found to suffer from spectral diffusion and charge instability when inside diamond nanostructures, which limits its usefulness for quantum information applications (Chu et al., 2014; Cui et al., 2015; Faraon et al., 2012) Fixing the NV stability problem remains an active and important area of research (Bluvstein et al., 2019; Radtke et al., 2019; Ruf et al., 2019; Sangtawesin, 2019; van Dam et al., 2019). In the meantime, group IV color centers, such as the negatively-charged silicon vacancy (SiV) center, have emerged as a promising alternative because they are more stable in diamond nanostructures (Evans et al., 2016). Furthermore, most of these color centers have microwave transitions which can be used to coherently control their spin states, allowing for straightforward implementation of single qubit gates (Childress et al., 2006; Jelezko et al., 2004; Pingault et al., 2017).

The speed and fidelity of information transfer between different color center memories dramatically depend on the efficiency with which photons can interact with, and be distributed between, the quantum memories (both on chip and between different chips). This has spurred research activities aimed at realization of low-loss waveguides and optical resonators in diamond. All of these devices require optical isolation provided by refractive index contrast from the surrounding bulk material. One way to achieve this is through free-standing structures that rely on diamond as a device layer and air (or vacuum) as low index cladding. In traditional nanophotonic material platforms (Dietrich et al., 2016; Lipson, 2005) this is accomplished through heteroepitaxial growth or bonding of the device layer on top of a sacrificial layer, which is then etched away to suspend nanophotonic devices. However, the lack of techniques for heteroepitaxial growth of pure diamond thin films (Schreck et al., 2017) on foreign substrates meant that nanofabrication techniques for bulk diamond had to be developed. Focused ion beam milling was initially explored, but this approach was limited by device yield and reproducibility (Babinec et al., 2011; Bayn et al., 2011; Riedrich-Möller et al., 2011). To overcome this, a thin film diamond platform was developed and used in initial diamond quantum photonic experiments (Hausmann et al., 2012, 2013). However, this approach turned out to be incompatible with the post-fabrication processing (annealing, acid cleaning, etc.) needed to recover properties of color centers embedded within, particularly their optical coherence and stability. As a result, bulk diamond nanostructuring based on crystallographic (Khanaliloo et al., 2015a; Xie et al., 2018) or angled etching has been developed (Atikian et al., 2017; Burek et al., 2012). Suspended waveguides and photonic crystal cavities that were fabricated using these methods enabled the first demonstration of memory-enhanced quantum communication (Bhaskar et al., 2020)—an important benchmark for creation of long distance quantum networks.

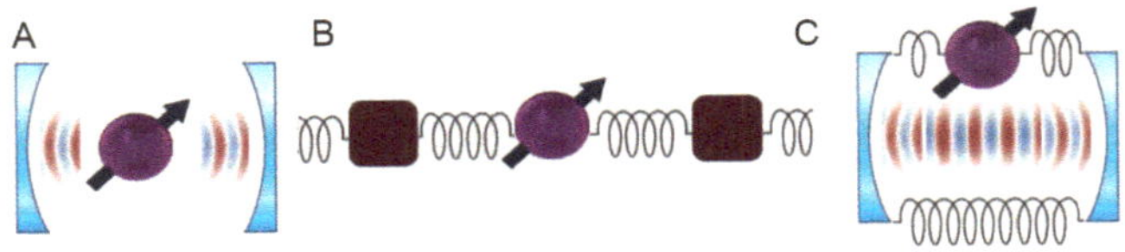

Fig. 1 Three types of diamond spin-qubit interactions that are covered in this chapter: (A) spin-photon interaction mediated by a photonic cavity fabricated in diamond; (B) spin-phonon interaction mediated by a diamond nanomechanical resonator; (C) photon-phonon interaction in optomechanical cavity that can also be combined with spin-mechanical coupling.

Importantly, the suspended devices that provide optical isolation also provide mechanical isolation and allow for acoustic energy to be concentrated. Thus, the fabrication techniques used to create suspended photonic devices in diamond are ideally suited for realizing low-loss diamond nanomechanical devices as well. This has stimulated interest in, and exploration of, mechanical degrees of freedom for controlling and interfacing diamond spin qubits. Of particular interest are group IV spin qubits that feature large strain susceptibilities (Maity et al., 2018; Meesala et al., 2018), while cantilevers and phononic crystals have emerged as promising platforms to control and enhance spin-strain coupling (Barfuss et al., 2015; Hong et al., 2012; Lee et al., 2017; Meesala et al., 2018; Sohn et al., 2018). Development of electromechanically deformable waveguides (Machielse et al., 2019), surface acoustic wave resonators (Maity et al., 2020), and optomechanical cavities (Burek et al., 2016) promises to enable coherent spin-phonon interactions and coherent control of single- and multi-qubit interactions mediated by phonons.

In this chapter, we discuss the progression of integrated diamond devices used to couple color centers to single photons (Section 2, Fig. 1A) and single phonons (Section 3, Fig. 1B), as well as optomechanical devices (Section 4, Fig. 1C) that can enable coherent interactions between optical and mechanical degrees of freedom.

2. Photonic devices

The rapid development of fabrication technology for coupling individual diamond color centers to itinerant photons has allowed the diamond photonics platform to emerge as a leading candidate for the implementation of quantum networking protocols. Pioneering experiments with the NV center (Hensen et al., 2015), together with the recent demonstration of memory-enhanced quantum communication based on the SiV center (Bhaskar et al., 2020), have laid the groundwork for implementation of the full quantum repeater protocol (Duan et al., 2001)—a technology which would make the creation of long-distance quantum networks possible. Here we trace the development of integrated diamond nanophotonic devices that have enabled the creation of the spin-photon interfaces that form the backbone of these technologies.

2.1 Surface-etched devices for photon extraction into free space

The first structures that were developed to overcome total internal reflection at the diamond-air interface, and thus increase the collection efficiency for photons emitted by color centers, were diamond nanopillars (Fig. 2A). In these devices, light from color centers is coupled to the optical mode of a nanopillar, emitted into free space at the pillar's facet, and collected by a lens.

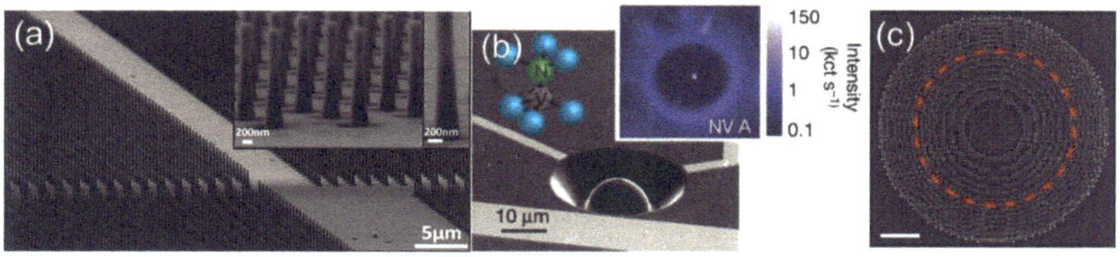

Fig. 2 Surface-etched devices in diamond for photon extraction into free space. (A) Scanning electron microscope (SEM) image showing an array of ~2 μm tall and ~200 nm wide diamond nanowires containing NV centers. (B) SEM image of solid immersion lens (SILs). Focused ion beam is used to etch the SILs around a specific NV-center. Inset scanning confocal microscope image of NV A (logarithmic color scale). kct, 1000 counts. (C) SEM image of the immersion metalens consisting of subwavelength pillars extending from the surface of a single-crystal diamond substrate designed to create a high-numerical-aperture immersion lens for coupling NV center photoluminescence to a collimated beam in air The metalens is fabricated using electron-beam lithography and O_2-based dry etching. *Panel A: reprinted from Hausmann, B.J.M. et al. 2011. Single-color centers implanted in diamond nanostructures. New J. Phys. 13, 045004, © IOP Publishing and Deutsche Physikalische Gesellschaft. Reproduced by permission of IOP Publishing. CC BY-NC-SA. Panel B: reprinted by permission from Springer Nature Customer Service Centre GmbH: Robledo, L. et al. 2011. High-fidelity projective read-out of a solid-state spin quantum register. Nature 477, 574–578, © 2011. Panel C: reprinted from Huang, T.-Y. et al. 2019. A monolithic immersion metalens for imaging solid-state quantum emitters. Nat. Commun. 10, 2392 licensed under the terms of the Creative Commons Attribution 4.0 license.*

Nanopillars could be fabricated on the surface of diamond using reactive ion etching (RIE) (Babinec et al., 2010; Hausmann et al., 2011), implanted with ions and subsequently annealed to create color centers inside them. Alternatively, they could be fabricated in substrates with sufficient natural abundance of color centers such that each nanopillar probabilistically contained individual emitters. Diamond nanopillars allowed for a 10-fold increase in photon collection efficiency when compared to bulk diamond and have found many applications in quantum sensing and scanning magnetometry in particular (Batzer et al., 2020; Maletinsky et al., 2012). To enhance the light-matter interaction (e.g., via the Purcell effect) needed for quantum information applications, diamond pillars could be surrounded by metal (Choy et al., 2011, 2013). This, however, came at the expense of increased optical losses introduced by the metal layer.

These early device experiments uncovered an important shortcoming of NV centers: although they are stable emitters in bulk diamond, their optical properties degrade substantially when they are close to etched diamond surfaces. To address this problem while still overcoming the total internal reflection of emitted photons, solid immersion lenses (SILs) have been fabricated around individually targeted color centers (Fig. 2B). This could be done either by patterning and then reflowing resist masks prior to etching (Lee et al., 2007) or by using focused ion beam (FIB) milling to pattern desired structures (Marseglia et al., 2011). Importantly, NV centers in SILs are separated from the etched surfaces by several microns and thus are less susceptible to fabrication-induced damage. As a result, SILs preserve the stability of optical transitions of NVs while enabling dramatic improvements in photon extraction efficiency. These structures were workhorses in many of the seminal NV center quantum optics experiments (Hensen et al., 2015; Humphreys et al., 2018; Kalb et al., 2017; Robledo et al., 2011).

Metasurfaces have also recently emerged as a promising platform for achieving high collection efficiency (Fig. 2C), combining the scalability of the planar nanofabrication technology used for diamond nanopillars with the optical stability characteristic of color centers inside SILs. By carefully tuning the shape and spacing of structures fabricated with top-down etching technology, the emission of a single color center buried microns below the surface of the diamond can be efficiently routed into collection optics (Huang et al., 2019).

Research on surface-etched diamond nanophotonic structures has allowed the development of a variety of techniques for collecting light from subsurface emitters without compromising their optical properties. As a result, these approaches are primarily used in concert with color centers which are sensitive to electric fields and other forms of etch-induced damage, such as the NV center. However, these techniques are limited in the overall fraction of emitted photons they can collect by the unavoidable photon emission into the substrate and thus are not sufficient for the realization of efficient spin-photon interfaces.

2.2 Freestanding photonic devices for enhanced spin-photon interactions and efficient photon collection

One established route toward an interface between nanophotonic devices and color centers consists of a single-mode waveguide implanted with individual color centers. When properly designed, waveguides enhance the photonic density of states near the emission wavelength of color centers. In addition to diamond nanopillars discussed above, this approach has been pursued in the case of SiV and germanium vacancy (GeV) color centers embedded inside angle-etched diamond waveguides (Bhaskar et al., 2017; Burek et al., 2017).

In order to extract emitted photons from the waveguide into an optical fiber, an interface between these two modes must be constructed. This is commonly accomplished through either physical fiber-to-waveguide contact or through vertical couplers designed to efficiently scatter light into a collection objective. Examples of vertical couplers include notches (Burek et al., 2014) (Fig. 3A), photonic crystal mirrors (Hausmann et al., 2012) (Fig. 3B), and grating couplers (Faraon et al., 2013) (Fig. 3C). While easy to implement, these approaches suffer from low collection efficiencies. However, recent progress with inverse design techniques (Dory et al., 2019) (Fig. 3D) suggests that further improvements are possible.

Direct fiber collection has so far proved more efficient than vertical coupling, as tapered diamond-to-fiber interfaces have demonstrated the ability to collect more than 95% of the waveguide-coupled photons (Burek et al., 2017) (Fig. 3E). This is achieved by tapering the ends of both the optical fiber and the diamond waveguide, creating an adiabatic effective mode transition from the fiber's mode into the waveguide's mode. However, this technique can currently only be used to couple to individual diamond devices, as it requires the use of bulky nanopositioning stages. Permanent bonding of fiber tips to diamond devices—fiber pigtailing—would make it possible to address multiple photonic devices simultaneously.

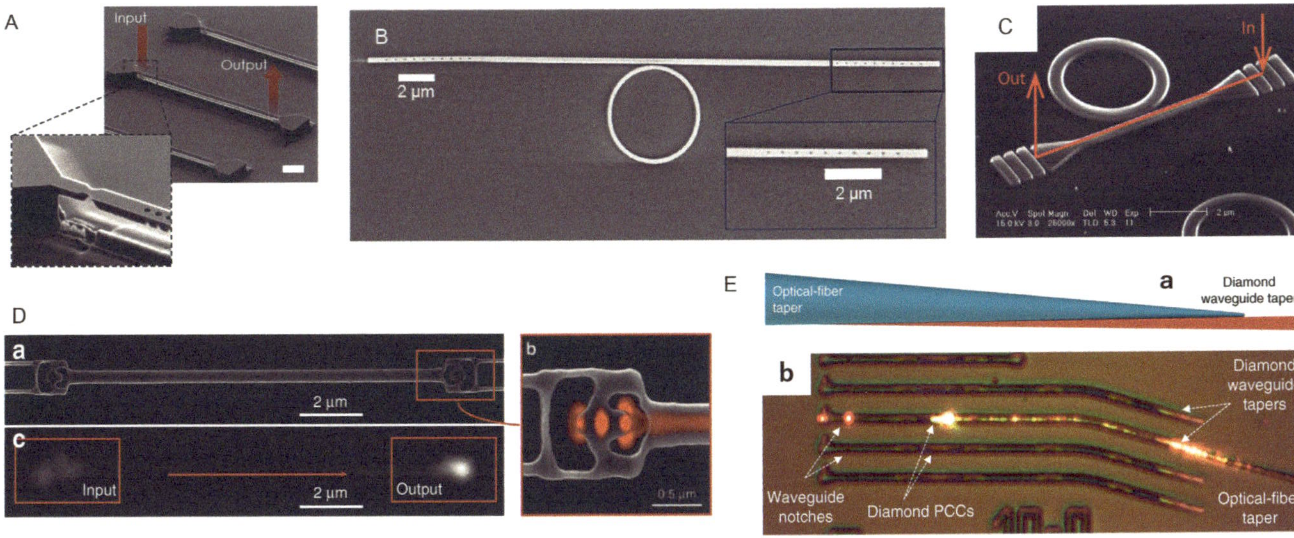

Fig. 3 Different photon extraction strategies for integrated diamond quantum photonics. (A) Scanning electron microscope (SEM) image of suspended 1-dimensional photonic crystal nanobeam cavity with notches. Inset shows a magnified view of the notch used to extract light. (B) SEM image of a device composed of a diamond microring resonator and a waveguide with photonic crystal mirrors. Inset shows close-up view of photonic crystal mirror. (C) SEM image of a device composed of a diamond microring resonator and a waveguide with grating couplers. (D) a. SEM image of a suspended diamond waveguide that is terminated with inverse-designed vertical couplers; b. SEM image of closer view of the coupler; c. Optical microscope image of the device. (E) a. Schematic of the adiabatic coupling between a single-ended conical optical-fiber taper (blue) and a diamond waveguide taper (red) b. Optical microscope image taken when an optical-fiber taper in contact with a waveguide taper. *Panel A: reprinted by permission from Springer Nature Customer Service Centre GmbH: Burek, M.J. et al. 2014. High quality-factor optical nanocavities in bulk single-crystal diamond. Nat. Commun. 5, 5718, © 2014. Panel B: reprinted with permission from Hausmann, B.J.M. et al. 2012. Integrated diamond networks for quantum nanophotonics. Nano Lett. 12, 1578–1582, Copyright 2012 American Chemical Society. Panel C: reprinted from Faraon, A. et al. 2013. Quantum photonic devices in single-crystal diamond. New J. Phys. 15, 025010 licensed under the terms of the Creative Commons Attribution 3.0 license. Panel D: reprinted from Dory, C. et al. 2019. Inverse-designed diamond photonics. Nat. Commun. 10, 3309 licensed under the terms of the Creative Commons Attribution 4.0 license. Panel E: reprinted figures with permission from Burek, M.J. et al. 2017. Fiber-coupled diamond quantum nanophotonic interface. Phys. Rev. Appl. 8, 024026, Copyright 2017 by the American Physical Society.*

A final technique for extracting photons from diamond waveguides involves using a pick-and-place process to couple them to nanophotonic devices patterned in other materials (Najafi et al., 2015). In this technique, diamond nanophotonic devices are patterned in such a way that they can be lifted individually or in groups as large as 128 devices (Wan et al., 2019) from the diamond substrate and attached to waveguide devices patterned in more traditional materials. In exchange for added fabrication complexity, this allows for on-chip integration of diamond color centers with materials with nonlinear optical or piezoelectric properties. As a result, this technique shows promise for offering new device functionalities, including complex photon routing and multiplexing, while maintaining the efficient photon extraction required for quantum optics experiments.

Realization of near unity photon collection efficiency as well as high-cooperativity spin-photon interfaces requires the introduction of resonant optical structures to enhance and direct photon emission. Although initial demonstrations using ring resonators and whispering gallery mode structures achieved measurable emitter-cavity coupling (Lee et al., 2012; Santori et al., 2010), photonic crystal cavities have emerged as the dominant technology for resonant enhancement due to their high optical quality factors and small mode volumes. Both one- and two-dimensional photonic crystal cavities (Fig. 4A–F) have been demonstrated (Evans et al., 2018; Hausmann et al., 2013; Mouradian et al., 2017; Riedrich-Möller et al., 2011; Wan et al., 2018), but the former has been more extensively implemented.

One-dimensional photonic crystal cavities created using an angled etch approach (Fig. 4E and F) were used to demonstrate spin-photon interfaces with cooperativities $C_a > 100$ (Bhaskar et al., 2020). Here, $C_a = g_a^2/2\kappa_o\gamma_s$ is defined as the ratio of the spin-photon coupling rate g_a (roughly 10 GHz for state of the art devices) to the optical loss rate κ_o and spin decay rate γ_s (Reiserer and Rempe, 2015). For the demonstrated devices, more than 99% of emitted resonant photons are emitted into the cavity. These photons can be efficiently coupled into the waveguide and, using the aforementioned tapered fiber interface, extracted into the fiber. Microwave coherent control of the color center spin state, another requirement for many quantum optics experiments, can be achieved with on-chip coplanar microwave waveguides (Nguyen et al., 2019).

Combining these technologies has allowed for the demonstration of high-fidelity interfaces between photons and long-lived SiV color center spins (Nguyen et al., 2019). These interfaces can be used to entangle photons and quantum memories with high efficiency—a key step to implementing a

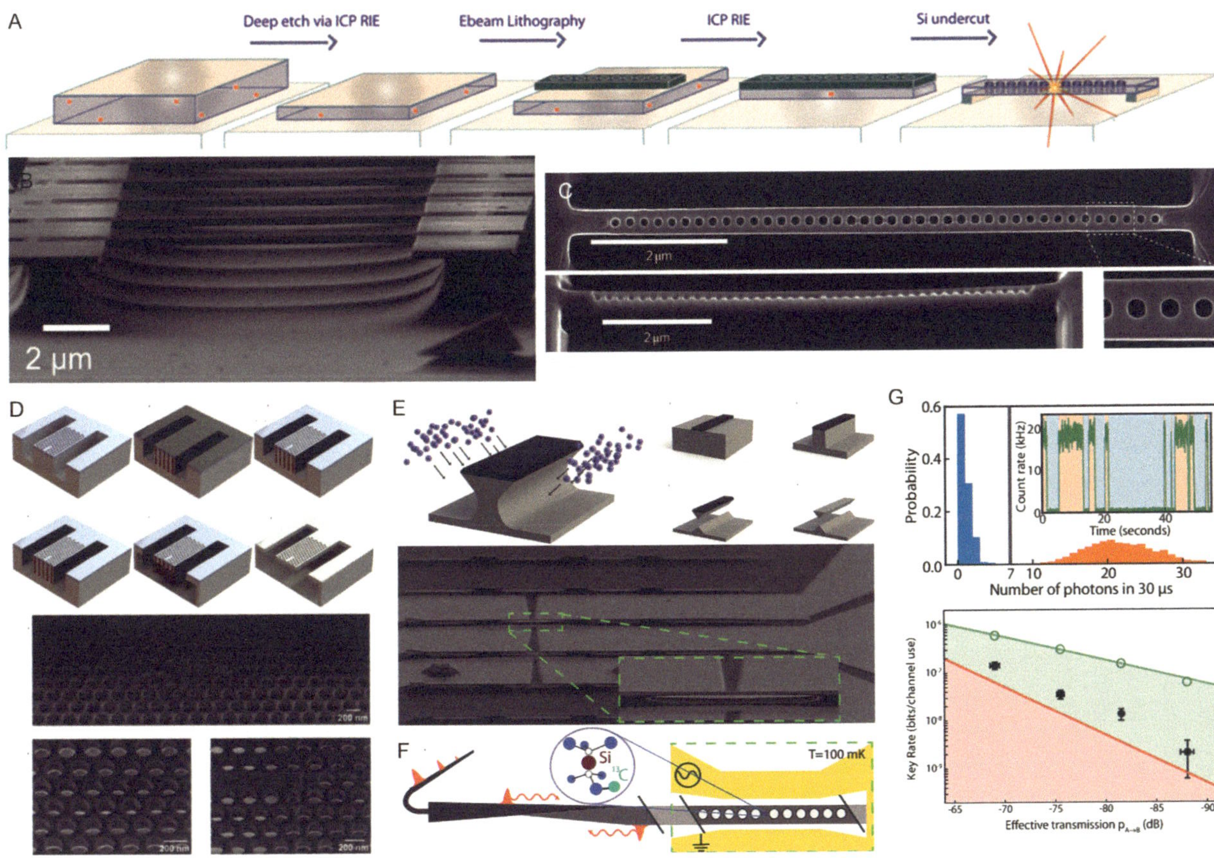

Fig. 4 See figure legend on opposite page.

Fig. 4 Progression of diamond nanophotonic crystal devices. (A) Fabrication schematic for thin film bonding diamond techniques. A thin film of diamond is bonded (or grown) onto a sacrificial substrate. The diamond film is then thinned and patterned, before being chemically undercut. (B) Scanning electron micrograph (SEM) of suspended diamond nanophotonic crystals made using the thin film bonding technique. (C) SEMs of nanophotonic crystal cavities focussed ion beam (FIB) patterned into thin film diamond grown on iridium/yttria-stabilized zirconia (Ir/YSZ) carrier wafer. (D) Fabrication process and SEMs for crystallographic undercut etched photonic crystals carved out of bulk diamond. Here the shape of the photonic crystal is patterned onto the surface of bulk diamond before being crystallographically undercut using a low-bias oxygen ICP-RIE etch. This technique offers exceptional flexibility and can be used to realize 2-dimensional photonic crystals. (E) Angled undercut fabrication technique. Here photonic crystal designs are patterned into the surface of bulk diamond before being undercut using angled oxygen ion beams. This technique results in triangular cross-section 1D nanophotonic crystals that have thus far demonstrated the highest performance. Subsequent deposition of microwave coplanar waveguides (CPWs) allows for coherent control of SiV spin state. (F) Schematic of experimental setup used in Bhaskar et al. (2020) and Nguyen et al. (2019). SiV centers are embedded within triangular cross-section photonic crystals, coupled to waveguide modes which can transfer photons into optical fibers. CPWs are used to control SiV spin while the entire device is operated inside a dilution refrigerator at 100 mK. (G) Experimental results from Bhaskar et al. (2020). Top: High-fidelity spin-photon interface leads to large contrast in cavity reflection coefficient based on SiV spin state. In the spin down (up) spin state, shown in blue (orange), the SiV reflects few (most) of the photons routed towards the cavity, allowing for single-shot, high-fidelity readout of the SiV spin state (inset). Bottom: Demonstration of memory-enhanced communication beyond the repeaterless bound. Red line represents the theoretical limit on the communication rate set by photon loss through the communication channel. Green line is the raw communication rate enabled by the diamond SiV-photon interface, while the black dots represent the communication rate after correction for errors. *Panels A and B: reprinted with permission from Hausmann, B.J.M. et al. 2013. Coupling of NV centers to photonic crystal nanobeams in diamond. Nano Lett. 13, 5791–5796, Copyright 2013 American Chemical Society. Panel C: reprinted by permission from Springer Nature Customer Service Centre GmbH: Riedrich-Möller, J. et al. 2011. One- and two-dimensional photonic crystal microcavities in single crystal diamond. Nat. Nanotechnol. 7, 69–74, © 2014. Panel D: reprinted from Wan, N.H., Mouradian, S., Englund, D. 2018. Two-dimensional photonic crystal slab nanocavities on bulk single-crystal diamond. Appl. Phys. Lett. 112, 141102 with the permission of AIP Publishing. Panel E: reprinted by permission from Springer Nature Customer Service Centre GmbH: Burek, M.J. et al. 2014. High quality-factor optical nanocavities in bulk single-crystal diamond. Nat. Commun. 5, 5718, © 2014. Panel F: reprinted figures with permission from Nguyen, C.T. et al. 2019. Quantum network nodes based on diamond qubits with an efficient nanophotonic interface. Phys. Rev. Lett. 123, 183602, Copyright 2019 by the American Physical Society. Panel G: reprinted by permission from Springer Nature Customer Service Centre GmbH: Bhaskar, M.K. et al. 2020. Experimental demonstration of memory-enhanced quantum communication. Nature 580, 60–64, © 2020.*

wide variety of quantum information technologies. For example, a recent demonstration (Nguyen et al., 2019) of memory-enhanced quantum communication using this SiV-nanophotonic cavity system is a key step towards the implementation of a full quantum repeater protocol and the creation of large quantum networks (Bhaskar et al., 2020; Wehner et al., 2018) (Fig. 4G). In addition, this technology shows great promise for generating large cluster states (Nielsen, 2004; Raussendorf et al., 2003) with nontrivial entanglement topologies. Pioneering work with the nitrogen vacancy center (Abobeih et al., 2019; Bradley et al., 2019) and the recent preliminary efforts with group IV color centers (Nguyen et al., 2019) have demonstrated that diamond color centers can be efficiently linked with nearby nuclear spins. This result in combination with high fidelity spin-photon interfaces could make it possible to form cluster states that can be used for quantum computation or long distance communication (Gimeno-Segovia et al., 2019).

2.3 Summary

Great progress in diamond photonics over the past decade has enabled new benchmark quantum optics experiments and made reproducible production of individual diamond photonic devices possible. Broad application of these technologies is in large part prevented by the difficulty in scaling these techniques to fabricate and address larger collections of devices in an economical manner. Fortunately several newly developed techniques—including thin film (Piracha et al., 2016) and isotropic etching (Khanaliloo et al., 2015a; Wan et al., 2018)—provide promising routes towards realization of these goals. Open challenges in the field include routing photons on-chip, improving compatibility with non-inversion symmetric color centers such as the NV center, tuning photonic crystal cavities individually, and enabling higher temperature operation of optical devices.

3. Mechanical devices

Diamond displays remarkable mechanical properties such as the highest Young's modulus among all materials and low thermoelastic damping (Tao et al., 2014). As a result, the MEMS community has been historically interested in diamond for high frequency mechanical resonators up to several GHz (Burek et al., 2016; Dow et al., 2013; Gaidarzhy et al., 2007; Kirsch et al., 2006; Rodriguez-Madrid et al., 2012). From a quantum technology standpoint, low dissipation mechanical devices are at the heart of proposals for hybrid quantum systems (Schuetz et al., 2015; Wallquist et al., 2009) that utilize phonons to coherently interface disparate qubit platforms. Since color centers, being embedded in a solid-state matrix, are sensitive to lattice deformations (Davies et al., 1976; Sternschulte et al., 1994), there has been growing interest in exploring their interactions with acoustic degrees of freedom. In this section, we review the recent development of diamond mechanical devices for tuning and classical control of color centers, and their potential for phononic quantum information processing.

3.1 Frequency tuning and stabilization of emitted photons

To take advantage of optically active color centers as spin-photon interfaces, the photons emitted by multiple centers must be indistinguishable (Bernien et al., 2012; Hensen et al., 2015; Sipahigil et al., 2012, 2014). However, color centers generally display an inhomogeneous distribution of their emission wavelengths due to local variations of strain caused by lattice defects. This considerably hampers efforts to collectively interface groups of color centers unless the emitters can be tuned with respect to each other. While electric field tuning using the Stark effect is the most common technique (Bassett et al., 2011; Tamarat et al., 2006), not all color centers can be tuned this way. In particular, centro-symmetric group IV color centers, such as the SiV, GeV and tin vacancy (SnV) centers, are insensitive to electric fields to first order and cannot be tuned by Stark shift. Their large susceptibility to strain (Maity et al., 2018; Meesala et al., 2018) instead makes strain tuning an ideal technique. The neighboring nuclei and electronic orbitals are modified by strain in the crystal (Hughes and Runciman, 1967; Lee et al., 2016; Maze et al., 2011; Meesala et al., 2018), which shifts the energy levels of the orbitals and consequently tunes the transition wavelengths.

A diamond cantilever structure with embedded NV centers and dynamically driven by a piezoelectric transducer was used to achieve stroboscopic frequency tuning of the NV fluorescence, and demonstrate overlap of the optical transitions of two different NVs (Lee et al., 2016). Similarly, static tuning was demonstrated on color centers implanted into a diamond cantilever (Fig. 5A). Electrodes were patterned on and below the cantilever to apply a voltage-dependent force. The controllable bending of the cantilever applied a static strain to the color centers (Maity et al., 2018; Meesala et al., 2018; Sohn et al., 2018), and tuned their optical transitions. The applied static strain can also improve the spin coherence properties of SiV centers (Sohn et al., 2018), and likely other group IV centers as well (Siyushev et al., 2017; Trusheim et al., 2020), by reducing their coupling to thermal phonons.

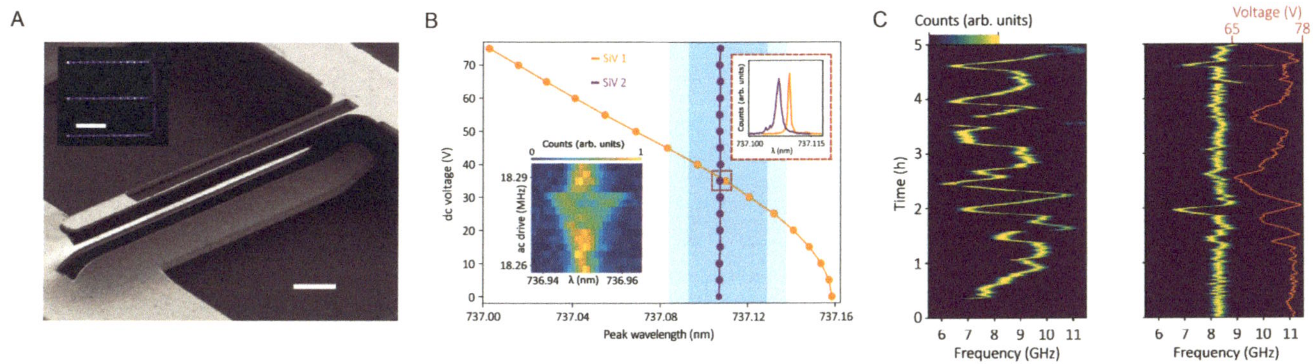

Fig. 5 (A) Scanning electron microscopy image of a diamond cantilever and surrounding electrodes (light gray). The scale bar corresponds to 2 μm. The inset is a confocal photoluminescence image of three adjacent cantilevers, with the bright spots along the cantilevers corresponding to emission from implanted SiV centers. The inset scale bar corresponds to 10 μm. (B) Using strain, emission wavelength of SiV implanted close to the clamp (the position of the highest strain) can be tuned and overlapped with another SiV positioned further away from the clamp. *Panel A: reprinted from Sohn, Y.I. et al. 2018. Controlling the coherence of a diamond spin qubit through its strain environment. Nat. Commun. 9, 1–6, licensed under the terms of the Creative Commons Attribution 4.0 International License. Panels B and C: reprinted from Machielse, B. et al. 2019. Quantum interference of electromechanically stabilized emitters in nanophotonic devices. Phys. Rev. X 9, 031022 licensed under the terms of the Creative Commons Attribution 4.0 International License.*

Building on this achievement, strain has been used to tune two SiV centers into resonance with each other, and likewise with two GeV centers (Fig. 5B) (Machielse et al., 2019; Maity et al., 2018). Superradiant emission has been demonstrated with two SiV centers in resonance, a signature that the emitted photons are indeed indistinguishable (Machielse et al., 2019). Furthermore, emitters in nanostructures can display some amount of spectral wandering due to their close proximity to surfaces, where charge traps and dangling bonds generate electric field noise. By combining strain tuning with a feedback mechanism, the emission from color centers can be stabilized spectrally, and thus reduce slow spectral wandering at frequencies below 5 Hz (Fig. 5C) (Machielse et al., 2019).

3.2 Classical qubit control using mechanical vibrations

Mechanical coherent control of color center qubits can be used as an alternative to microwave (Childress et al., 2006; Jelezko et al., 2004; Pingault et al., 2017) and all-optical control (Becker et al., 2018; Pingault et al., 2014; Rogers et al., 2014b; Yale et al., 2013). Fast microwave control of spin qubits typically relies on the small magnetic dipole moment and hence requires high power, which can lead to undesirable heating in cryogenic conditions. Optical control relies on a lambda scheme, which is difficult to combine with single-shot readout that relies on cycling transitions (Neumann et al., 2010; Robledo et al., 2011; Sukachev et al., 2017). These are not limitations for mechanical control.

The first realizations of coupling between the NV spin and a mechanical resonator was mediated with an external magnetic field gradient (Arcizet et al., 2011; Kolkowitz et al., 2012; Rabl et al., 2009, 2010; Rugar et al., 2004) (Fig. 6A). The motion of the NV spin in such magnetic field leads to a coupling directly proportional to the gradient strength and to the amplitude of the mechanical oscillations, and reaches 10–70 Hz per phonon (Arcizet et al., 2011; Kolkowitz et al., 2012; Rugar et al., 2004). A microwave field resonant with the spin transition when the system is at rest drives the spin depending on the position of the NV in the gradient field when the mechanical system is oscillating. This can synchronize the spin and the oscillation and result in an effective coupling between the mechanical resonance and the spin (Hong et al., 2012; Pigeau et al., 2015).

The NV ground state corresponds to an orbital singlet, and because the spin transition between the $m_s = -1$ and $m_s = +1$ states is magnetic

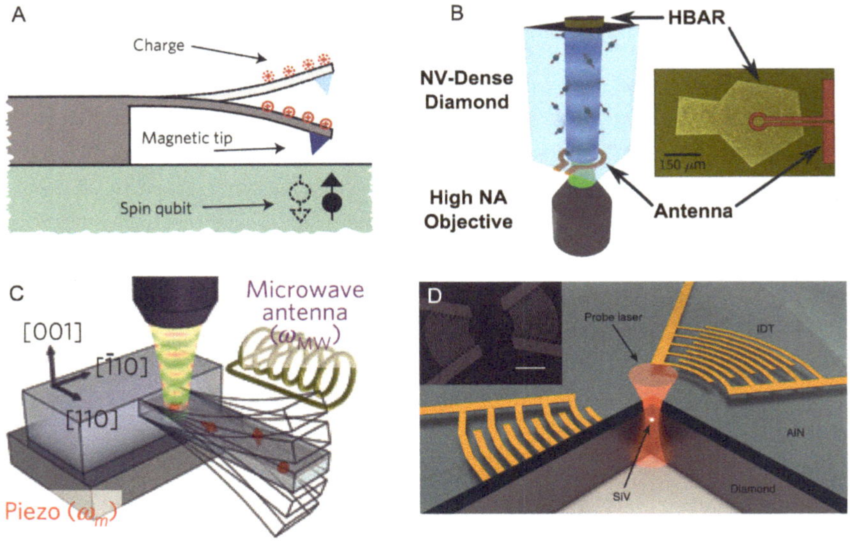

Fig. 6 (A) Illustration of the working principle of magnetically-mediated coupling between a single NV spin and a mechanical resonator. (B) Schematic and an optical micrograph of a diamond slab with a high-overtone bulk acoustic resonator (HBAR) with a microwave loop antenna in red. (C) Schematic of NV centers (in red) in a diamond cantilever driven mechanically by a piezoelectric transducer. (D) Schematic of a surface acoustic wave (SAW) device on diamond. *Panel A: reprinted by permission from Springer Nature Customer Service Centre GmbH: Rabl, P. et al. 2010. A quantum spin transducer based on nanoelectromechanical resonator arrays. Nat. Phys. 6, 602–608, © 2010. Panel B: reprinted with permission from MacQuarrie, E.R. et al. 2015. Coherent control of a nitrogen-vacancy center spin ensemble with a diamond mechanical resonator. Optica 2, 233, © The Optical Society. Panel C: reprinted by permission from Springer Nature Customer Service Centre GmbH: Barfuss, A., Teissier, J., Neu, E., Nunnenkamp, A., Maletinsky, P. Strong mechanical driving of a single electron spin. Nat. Phys. 11, 820–824, © 2015. Panel D: reprinted from Maity, S. et al. 2020. Coherent acoustic control of a single silicon vacancy spin in diamond. Nat. Commun. 11, 193 licensed under the terms of the Creative Commons Attribution 4.0 International License.*

dipole-forbidden, it cannot be easily driven with a microwave field (MacQuarrie et al., 2013). It can, however, be driven by strain (MacQuarrie et al., 2013, 2015a; Meesala et al., 2016; Teissier et al., 2014). This direct spin-strain coupling relies on a strain-induced variation of the zero-field splitting constant which corresponds to the dipolar interaction between the two electron spins forming the NV $S=1$ ground state (Doherty et al., 2012; Maze et al., 2011). This yields a orbital strain susceptibility of about 10 GHz/strain (Barson et al., 2017), leading to a single phonon coupling of up to 2 Hz in the best devices (Meesala et al., 2016). In contrast, the NV excited state possesses a much larger orbital strain susceptibility of 1 PHz/strain (Davies et al., 1976) which enables strain coupling to electronic orbitals (Lee et al., 2016), but it lacks the coherence properties of the ground state spin.

Microwave-driven dynamical decoupling sequences of an NV spin can be synchronized with the mechanical resonance of the cantilever in which it is embedded, and this was first used to directly couple the mechanical motion to the NV spin (Ovartchaiyapong et al., 2014). Direct coherent control of the NV spin was then demonstrated using a stress wave generated by a piezoelectric transducer on bulk diamond (Fig. 6B) (MacQuarrie et al., 2015a) and by driving the oscillation of a diamond cantilever (Fig. 6C) (Barfuss et al., 2015). The mechanical dressing of the spin states also leads to a decoupling from the magnetic environment and thus to an extension of the coherence time (Barfuss et al., 2015; MacQuarrie et al., 2015b).

Surface acoustic waves (SAWs) have also been used to control diamond spins. SAWs are acoustic vibrations that are confined to the surface of the substrate, and thus offer a stronger concentration of acoustic energy than bulk modes. They are usually generated by interdigital transducers (IDTs) (Fig. 6D). SAWs can be used as a classical coherent phonon source to control the state of color centers, as demonstrated in recent works (Golter et al., 2016; Maity et al., 2020). This coupling between SAWs and color center spins can be enhanced by using acoustic cavities (Schuetz et al., 2015). Compared to the more common microwave control, acoustic control of spin qubits allows the source of the driving field to be spatially well-separated from the spin qubit (tens of microns, as opposed to a few microns for microwave). This ensures that the heat generated by the source can be more efficiently dissipated and does not result in thermally-induced decoherence of the color center spin, which is an even more important consideration when scaling up the number of devices on chip (Maity et al., 2020).

3.3 Toward phonons as on-chip quantum information carriers

Up to this point we have primarily considered the interactions of color centers in diamond with classical strain fields. However, much like electromagnetic systems, mechanical systems can operate at the individual phonon level. In fact, using phonons as information carriers in hybrid quantum systems can be advantageous. Phonons can couple to various quantum systems including superconducting qubits, color centers, trapped ions, and quantum dots (Lee et al., 2017). Importantly, phonons can be better confined in resonators than photons which may scatter into the continuum of vacuum modes.

To achieve coherent quantum state transfer between phonons and color center spins, the spin-phonon cooperativity $C_b = 4g_b^2/(n_{th}+1)\gamma_s\gamma_m$ should be greater than 1, where g_b is the single-phonon coupling rate, n_{th} is the thermal phonon population in the cavity, γ_s is the qubit decoherence rate, and γ_m is the phonon loss rate. Several systems capable of reaching this coherent regime have been proposed in diamond. Shallow NV centers placed underneath a cantilever with a magnetic tip (Kolkowitz et al., 2012; Rabl et al., 2009, 2010) could in principle achieve spin-phonon cooperativities above 1, assuming a sufficient magnetic field gradient, a large zero-point motion of the tip, and the use of dynamical decoupling on the NV spin. A similar scheme has also been proposed for the SiV center (Kepesidis et al., 2016). A direct spin-strain coupling approach for the group IV (SiV, GeV, SnV, and lead vacancy (PbV)) color centers takes advantage of their large spin-strain susceptibility on the order of 100 THz/strain arising from strong spin-orbit interaction (Maity et al., 2018; Meesala et al., 2018). The SiV displays long spin coherence times reaching 10 ms at milli-Kelvin temperatures though, like the NV, it requires dynamical decoupling (Sukachev et al., 2017). Static strain can also be used to increase the spin coherence of the SiV, though this reduces its strain susceptibility (Meesala et al., 2018).

A promising approach for strong spin-phonon coupling is to embed an SiV in a phononic crystal cavity (Burek et al., 2016; Lemonde et al., 2019; Li et al., 2019; Meesala et al., 2018; Neuman et al., 2020). Such a structure supports localized phononic modes in the bandgap of a one-dimensional phononic crystal (Fig. 7A). Fig. 7B–E plot the simulated displacement profiles of two of its localized mechanical modes with frequencies in the phononic crystal bandgap. These cavities are designed such that the strain induced by a single phonon is maximized at the location of the color center as illustrated in Fig. 7C and E, or equivalently such that the mechanical mode volume is minimized (Li et al., 2019). The spin transition of the color center is brought into resonance with one of the localized mechanical modes through Zeeman tuning, creating resonant spin-phonon coupling. Current fabrication techniques allow the design of phononic crystals with acoustic wavelength-scale mode volumes, giving spin-phonon coupling rates of a few MHz (Meesala et al., 2018). Spin-phonon cooperativity above 1 can be reached with experimentally measured mechanical quality factors of

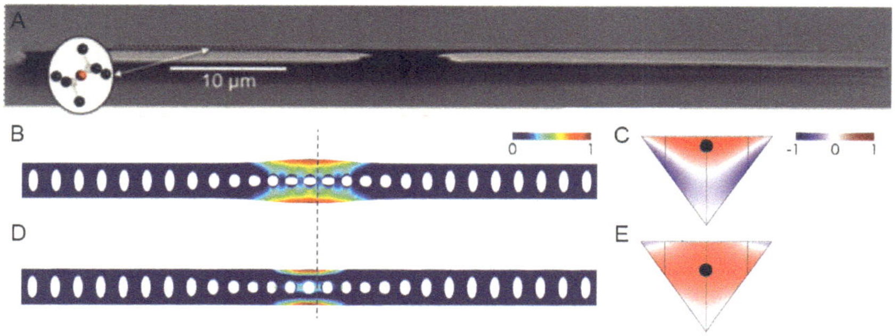

Fig. 7 (A) Scanning electron microscope (SEM) image of a diamond optomechanical (photonic and phononic) cavity (as detailed in Section 4.1), with a taper on the right for photon extraction (as described in Section 2). An SiV would be positioned at the center of the phononic cavitiy, as illustrated. (B) Mechanical displacement and (C) trace of strain in cross-section of the mechanical flapping mode. (D) Mechanical displacement and (E) trace of strain tensor in cross-section of the mechanical swelling mode. Cross-section profiles in (c) and (e) correspond to the in-plane position marked by the dashed line across (b) and (d). The black dot in (c) and (e) indicates the ideal location of an SiV, at the strain maximum of the respective mechanical modes. *Panels B and D: reprinted with permission from Burek, M.J. et al. 2016. Diamond optomechanical crystals. Optica 3, 1404–1411, © The Optical Society.*

10^5 at temperatures of 4 K (Chia et al., 2018). Much higher cooperativities can be reached by operating at milli-Kelvin temperatures, which decreases the qubit decoherence rate γ_s and improves mechanical quality factors due to reduced coupling of mechanical modes to thermal phonons (Chia et al., 2018).

With a high-cooperativity spin-phonon interface, coherent phenomena involving color center spins and single phonons can be realized. Cooling of a mechanical resonator using a color center spin (Delord et al., 2017; Giannelli et al., 2016; MacQuarrie et al., 2017) has been experimentally realized with a levitated nanodiamond (Delord et al., 2020), while proposals exist for phonon-mediated spin-state squeezing (Bennett et al., 2013; Rao et al., 2016), and phonon lasing using color center spins (Kepesidis et al., 2013). In further analogy to photonic systems, various phonon-mediated quantum network architectures have been proposed. Fig. 8A shows a proposed spin-phonon quantum network consisting of SiV centers embedded in a diamond waveguide, where single phonons created by a Raman process transfer quantum states between qubits (Lemonde et al., 2018). In another proposal, shown in Fig. 8B–D, diamond color center spins facilitate quantum state

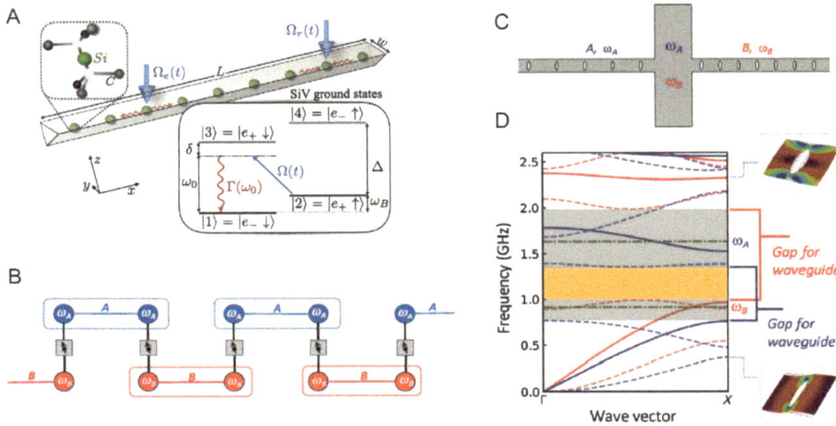

Fig. 8 (A) Proposed quantum network architecture using SiVs in a triangular cross-section diamond nanobeam; inset shows the level structure of the SiV ground-state manifold. (B) Proposed quantum network architecture consisting of emitters in cavities coupled to waveguides. Each waveguide alternates in resonant frequency between ω_a and ω_b. (C) Schematic for a proposed quantum network node consisting of a rectangular diamond resonator supporting mechanical modes with frequencies resonant with both ω_a and ω_b, and connected to phononic crystal waveguides A (B) to its left (right) resonant with ω_a (ω_b). (D) Phononic crystal bandstructure showing overlapping band gaps for the two waveguides A and B in (C). *Panel A: reprinted figures with permission from Lemonde, M.-A. et al. 2018. Phonon networks with silicon-vacancy centers in diamond waveguides. Phys. Rev. Lett. 120, 213603, Copyright 2018 by the American Physical Society. Panels B–D: reprinted from Kuzyk, M.C., Wang, H. 2018. Scaling phononic quantum networks of solid-state spins with closed mechanical subsystems. Phys. Rev. X, 8 licensed under the terms of the Creative Commons Attribution 4.0 International License.*

transfer in a one-dimensional chain of phononic cavities sandwiched between waveguides with alternating resonant frequencies (Kuzyk and Wang, 2018). State transfer proceeds via a triple swap process; first mapping the first color center spin state to the resonant frequency of the first mechanical resonator, then propagating the excitation through the waveguide to the second mechanical resonator, and finally mapping the phonon onto a spin in the second resonator.

3.4 Summary

The interfacing of color centers with mechanical degrees of freedom has great potential for quantum information processing and on-chip quantum networking applications. Although much progress is still necessary to fully exploit the capabilities of these systems, currently available systems already reach the strong coupling regime where proof-of-principle experiments are possible.

4. Optomechanical devices

In the absence of single phonon counters, operating mechanical resonators in the quantum regime can be challenging. Thus, it may be advantageous to couple them to other degrees of freedom which can be probed with higher efficiency. In the previous section, we discussed color center qubits with spin degrees of freedom as sensitive probes of such modes. Here we present the optomechanical approach where the motion of a mechanical resonator modulates an optical mode incident on the same resonator (Aspelmeyer et al., 2014). Properties of the mechanical mode can then be read out from itinerant photons.

4.1 Optomechanical crystals

One candidate for interfacing mechanical motion with optical modes in diamond is the optomechanical crystal (OMC) (Burek et al., 2016) which is a simultaneous photonic and phononic crystal (Fig. 9A and B). The OMC co-localizes an optical mode in the telecom wavelength range (~200 THz) with two few-GHz frequency scale mechanical resonances, termed flapping and swelling modes (Fig. 7B and D). The mechanical quality factors in current diamond OMC devices in ambient conditions, measured to be a few thousand, are comparable to those in similar OMCs fabricated from other materials such as silicon (Eichenfield et al., 2009), gallium arsenide(Balram et al., 2014), silicon nitride (Grutter et al., 2014), and gallium phosphide (Schneider et al., 2019). The mechanical quality factors in diamond OMCs improve to a few hundred thousand at 4 K (Chia et al., 2018). The wavelength-scale confinement of the optical and mechanical modes in all three directions results in single-photon optomechanical coupling rates g_o ~100 to 200 kHz in diamond (Burek et al., 2016). Using the optomechanical interaction, a blue-detuned pump laser can amplify the mechanical mode (Fig. 9C) to phonon lasing (Fig. 9E), and a red-detuned pump can damp the mechanical mode (Fig. 9D) (Aspelmeyer et al., 2014). The same interaction can also be used to optically probe the mode's phonon occupancy and coherently drive photon-phonon transfer (Fig. 9F) (Chan et al., 2011).

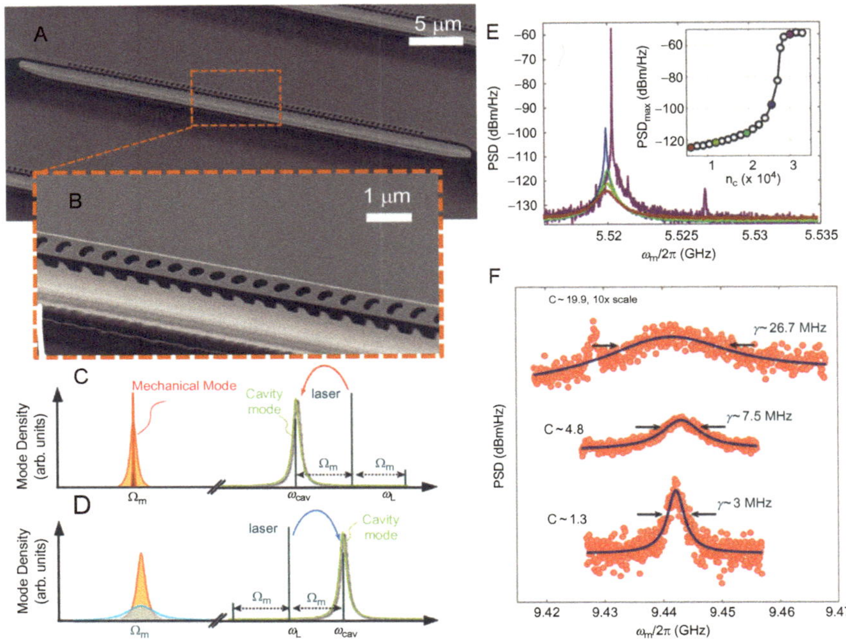

Fig. 9 (A) Scanning electron microscope (SEM) image of diamond OMC, taken at 60 degree tilt. (B) Close-up SEM of central region of diamond OMC. (C) Scattering picture of cavity optomechanics: for a blue-detuned pump at ω_l, optical cavity enhances lower motional sideband at $\omega_l - \Omega_m$, which scatters into optical cavity resonance at ω_{cav} by creating a phonon, leading to amplification of the mechanical mode at Ω_m. (D) For a red-detuned pump, the cavity enhances the upper motional sideband at $\omega_l + \Omega_m$; scattering into the optical cavity resonance is accompanied by annihilation of a phonon, leading to damping of the mechanical mode. (E) Optomechanically-induced amplification of the OMC flapping mode, where each RF spectrum corresponds to their respective colored points in the inset. Phonon lasing occurs when the threshold is crossed (purple spectrum) (F) Optomechanically-induced damping of the OMC swelling mode as optomechanical cooperativity is increased. *Panels A, B and E: reprinted with permission from Burek, M.J. et al. 2016. Diamond optomechanical crystals. Optica 3, 1404–1411, © The Optical Society. Panels C and D: reprinted with permission from Aspelmeyer, M., Kippenberg, T.J., Marquardt, F. 2014. Cavity optomechanics. Rev. Mod. Phys. 86, 1391–1452, Copyright 2014 by the American Physical Society.*

Coherent photon-phonon interactions can be achieved when the optomechanical cooperativity $C_{om} = 4n_c g_0^2/\kappa_o \gamma_m$ is greater than 1, where n_c is the intracavity photon number, κ_o is the total optical mode loss rate, and γ_m is the intrinsic mechanical mode damping rate. The interaction can be shown to be coherent by observing optomechanically induced transparency (OMIT), whereby tuning a strong laser pump into resonance with the red sideband and sweeping a weak probe across the optical resonance results in destructive interference between the probe photons and the pump photons scattered by the mechanical mode (Burek et al., 2016; Safavi-Naeini et al., 2011; Weis et al., 2010).

4.2 Microdisks

Cavity optomechanical effects in diamond have also been observed in microdisks (Fig. 10A) (Lake et al., 2018, 2019; Mitchell et al., 2016, 2019), which support multiple optical whispering gallery modes (WGMs). These WGMs couple to mechanical radial breathing modes with frequencies of ~2 GHz and mechanical quality factors of a few thousand (Mitchell et al., 2016). The microdisks exhibit optomechanical cooperativities greater than 1,

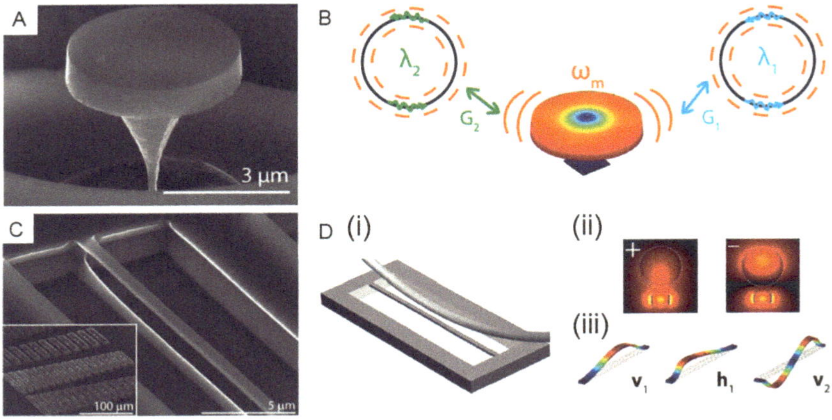

Fig. 10 (A) SEM of microdisk in Mitchell et al. (B) Scheme for wavelength conversion between two optical modes at $\lambda_1$and λ_2 via optomechanical couplings G_1 and G_2 to the mechanical radial breathing mode at ω_m. (C) SEM of diamond waveguides. (D) (i) Schematic of waveguide-optomechanical system composed of diamond waveguide and tapered fiber, (ii) Odd (−) and even (+) optical supermodes in waveguide-optomechanical system, (iii) Displacement profiles of relevant mechanical modes. *Panels A and B: reprinted with permission from Mitchell, M., Lake, D.P., Barclay, P.E. 2019. Optomechanically amplified wavelength conversion in diamond microcavities. Optica 6, 832, © The Optical Society. Panels C and D: adapted from Khanaliloo, B., Jayakumar, H., Hryciw, A.C., Lake, D.P., Kaviani, H., Barclay, P.E. 2015. Single-crystal diamond nanobeam waveguide optomechanics. Phys. Rev. X, 5, licensed under the terms of the Creative Commons Attribution 4.0 license.*

allowing coherent optomechanical interactions to be demonstrated via OMIT (Lake et al., 2018; Mitchell et al., 2019).

Since microdisks support multiple optical modes, they are a natural platform for multimode cavity optomechanics. One application is optomechanically-amplified wavelength conversion, which uses a pump that is blue-detuned from the first optical mode to create a phonon in the mechanical mode, and a pump red-detuned from the second optical mode to annihilate the phonon and scatter into the second optical resonance (Fig. 10B) (Mitchell et al., 2019). Another application is optical pulse storage and manipulation (Lake et al., 2019), where an optical signal is stored in the radial breathing mechanical mode, and the the signal can be stored for longer periods of time or be phase-shifted by coupling the mechanical mode to a second optical mode.

4.3 Waveguides

Optomechanical effects in single-crystal diamond nanostructures are not limited to optical cavities but can also be observed in waveguide-coupled geometries (Fig. 10C) (Khanaliloo et al., 2015b). In such waveguide-optomechanical systems, optomechanical coupling occurs between optical modes hybridized between the waveguide and an optical fiber, and mechanical modes from flexural motion (Fig. 10D). The mechanical mode displacement modulates the gap between the optical fiber taper and the diamond waveguide, which in turn modulates the mechanical mode damping rate. Phonon lasing can be observed in waveguide-optomechanical systems where the mechanical mode damping rate approaches zero.

4.4 Summary

The optomechanical systems described above can be used to couple between photonic and phononic degrees of freedom at disparate energy scales, typically hundreds of THz frequencies for photonic modes and MHz to GHz frequencies for phononic modes. It is also possible to couple these systems with strain-susceptible spin qubits defined in color centers, as discussed in Section 3, to form a hybrid quantum system with photonic, phononic, and spin degrees of freedom (Wallquist et al., 2009). Optical modes can be completely detuned from the optical transitions of color center qubits. Telecom frequencies are often used due to their low losses in optical fibers over long distances. Optical power scattered into the mechanical sidebands can be used to measure the phonon occupation of the mechanical mode

(Chan et al., 2011). This could be a useful measurement tool for experiments involving coherent interactions between color center spins and nanomechanical resonator modes. In another approach, the color center can be directly coupled to both the optical and mechanical modes. This offers a path towards transferring quantum information between qubits, photons and phonons, thus taking advantage of the strengths of each of these information carriers.

5. Conclusions and outlook

Diamond nanofabrication techniques have come a long way since the first demonstration of diamond nanowires (Babinec et al., 2010) 10 years ago. This progress has enabled realization of sophisticated quantum photonic devices (Burek et al., 2014) that provide high-fidelity spin-photon interfaces, culminating in recent demonstration of memory-enhanced quantum communication protocols which form the basis for further developments in quantum networking technology (Bhaskar et al., 2020). At the same time, these advances have inspired development of the nascent field of quantum phononics (quantum acousto-dynamics) that relies on phonons as on-chip information carriers. Coherent acoustic control of the SiV center electron spin has been recently demonstrated using surface acoustic waves (Maity et al., 2020). Efficient spin-phonon interfaces can be realized by fabricating complex nanoscale electromechanical (Machielse et al., 2019) and optomechanical (Burek et al., 2016) systems in diamond, and leveraging the large strain susceptibility of group IV diamond color centers (Meesala et al., 2018; Sohn et al., 2018).

Going forward, the next steps for diamond quantum photonic platform include engineering optically mediated entangling gates between spins and progressing towards long-distance device-independent quantum key distribution (DIQKD). One of the main challenges will be scaling the technology up. In order to realize quantum repeaters for example, it will be important to integrate a large number of quantum photonic devices into chip-scale systems (Wan et al., 2019), as well as to improve the yield and precision of techniques for the incorporation of stable spin qubits with minimal inhomogeneous broadening. Promising approaches include shallow implantation followed by regrowth (Rugar et al., 2020) and implantation and annealing at high temperatures (Iwasaki et al., 2017). We also expect that for experiments at milli-Kelvin temperatures, acoustic control of spin qubits may replace microwave and optical control approaches. Finally, to reduce the cost and complexity of the experiments and quantum repeater nodes, it will be important to move away from milli-Kelvin temperatures. This can be accomplished either by using the emerging color centers with larger strain susceptibilities like the SnV center (Iwasaki et al., 2017; Trusheim et al., 2020), or using either phononic density of state engineering or strain (Sohn et al., 2018) to control the incoherent interactions of the current workhorse—the SiV center. One of the near-term goals for the field of quantum phononics will be the experimental demonstration of high cooperativity between a single spin and a single phonon, followed by phonon-mediated interaction between two spin qubits, and finally realization of quantum gates using this approach.

Acknowledgments

The authors would like to thank the alumni of the Laboratory for Nanoscale Optics for their pioneering contributions on diamond nanophotonic, nanomechanical and optomechanical devices: Parag Deotare, Jennifer Choy, Birgit Hausmann, Michael Burek, Haig Atikian, I-Chun Huang, Pawel Latawiec, Young-Ik Sohn, Srujan Meesala, Linbo Shao, Irfan Bulu, Murray McCutcheon, Khadijeh Bayat, and Vivek Venkataraman. Diamond device fabrication was performed in part at the Center for Nanoscale Systems (CNS), a member of the National Nanotechnology Coordinated Infrastructure Network (NNCI), which is supported by the National Science Foundation under NSF award no. 1541959. CNS is part of Harvard University. The authors would also like to thank their collaborators: Mikhail Lukin, Amir Yacoby, Evelyn Hu, Hongkun Park, Ronald Walsworth (Harvard), Jelena Vučković (Stanford), Mete Atatüre (Cambridge), Ronald Hanson (TU Delft), Dirk Englund (MIT), Fedor Jelezko (Ulm), Patrick Maletinsky (Basel), Matthew Markham, Daniel J. Twitchen (Element Six), Edward Bielejec (Sandia National Laboratories). The authors acknowledge financial support from the following agencies: National Science Foundation (NSF), Defense Advanced Research Projects Agency (DARPA), Office of Naval Research (ONR), Army Research Office (ARO), Air Force Office of Scientific Research (AFOSR), and MITRE Corporation. C. C. acknowledges support from the Agency for Science, Technology and Research (A*STAR), Singapore. B. M. acknowledges support from the NSF GRFP. B. P. acknowledges support from the European Union's Horizon 2020 research and innovation programme through a Marie Sklodowska-Curie Action Individual Fellowship (COHESiV, grant agreement No 840968). M. C. acknowledges support from the NDSEG Fellowship.

References

Abobeih, M.H., et al., 2018. One-second coherence for a single electron spin coupled to a multi-qubit nuclear-spin environment. Nat. Commun. 9, 2552.

Abobeih, M.H., et al., 2019. Atomic-scale imaging of a 27-nuclear-spin cluster using a quantum sensor. Nature 576, 411–415.

Aharonovich, I., Englund, D., Toth, M., 2016. Solid-state single-photon emitters. Nat. Photonics 10, 631–641.

Arcizet, O., et al., 2011. A single nitrogen-vacancy defect coupled to a nanomechanical oscillator. Nat. Phys. 7, 879–883.
Aspelmeyer, M., Kippenberg, T.J., Marquardt, F., 2014. Cavity optomechanics. Rev. Mod. Phys. 86, 1391–1452.
Atatüre, M., Englund, D., Vamivakas, N., Lee, S.Y., Wrachtrup, J., 2018. Material platforms for spin-based photonic quantum technologies. Nat. Rev. Mater. 3, 38–51.
Atikian, H.A., et al., 2017. Freestanding nanostructures via reactive ion beam angled etching. APL Photonics 2, 051301.
Awschalom, D.D., Hanson, R., Wrachtrup, J., Zhou, B.B., 2018. Quantum technologies with optically interfaced solid-state spins. Nat. Photonics 12, 516–527.
Babinec, T.M., et al., 2010. A diamond nanowire single-photon source. Nat. Nanotechnol. 5, 195–199.
Babinec, T.M., Choy, J.T., Smith, K.J.M., Khan, M., Lončar, M., 2011. Design and focused ion beam fabrication of single crystal diamond nanobeam cavities. J. Vac. Sci. Technol., B 29, 010601.
Balasubramanian, G., et al., 2009. Ultralong spin coherence time in isotopically engineered diamond. Nat. Mater. 8, 383–387.
Balram, K.C., Davanço, M., Lim, J.Y., Song, J.D., Srinivasan, K., 2014. Moving boundary and photoelastic coupling in GaAs optomechanical resonators. Optica 1, 414–420.
Barfuss, A., Teissier, J., Neu, E., Nunnenkamp, A., Maletinsky, P., 2015. Strong mechanical driving of a single electron spin. Nat. Phys. 11, 820–824.
Bar-Gill, N., Pham, L.M., Jarmola, A., Budker, D., Walsworth, R.L., 2013. Solid-state electronic spin coherence time approaching one second. Nat. Commun. 4, 1743.
Barson, M.S.J., et al., 2017. Nanomechanical sensing using spins in diamond. Nano Lett. 17, 1496–1503.
Bassett, L.C., Heremans, F.J., Yale, C.G., Buckley, B.B., Awschalom, D.D., 2011. Electrical tuning of single nitrogen-vacancy center optical transitions enhanced by photoinduced fields. Phys. Rev. Lett. 107, 266403.
Batzer, M., et al., 2020. Single crystal diamond pyramids for applications in nanoscale quantum sensing. Opt. Mater. Express 10, 492–500.
Bayn, I., Meyler, B., Salzman, J., Kalish, R., 2011. Triangular nanobeam photonic cavities in single-crystal diamond. New J. Phys. 13, 025018.
Becker, J.N., et al., 2018. All-optical control of the silicon-vacancy spin in diamond at millikelvin temperatures. Phys. Rev. Lett. 120, 053603.
Bennett, S.D., et al., 2013. Phonon-induced spin-spin interactions in diamond nanostructures: application to spin squeezing. Phys. Rev. Lett. 110, 156402.
Bernien, H., et al., 2012. Two-photon quantum interference from separate nitrogen vacancy centers in diamond. Phys. Rev. Lett. 108, 43604.
Bernien, H., et al., 2013. Heralded entanglement between solid-state qubits separated by three metres. Nature 497, 86–90.
Bhaskar, M.K., et al., 2017. Quantum nonlinear optics with a germanium-vacancy color center in a nanoscale diamond waveguide. Phys. Rev. Lett. 118, 223603.
Bhaskar, M.K., et al., 2020. Experimental demonstration of memory-enhanced quantum communication. Nature 580, 60–64.
Bluvstein, D., Zhang, Z., Jayich, A.C.B., 2019. Identifying and mitigating charge instabilities in shallow diamond nitrogen-vacancy centers. Phys. Rev. Lett. 122, 076101.
Bradac, C., Gao, W., Forneris, J., Trusheim, M.E., Aharonovich, I., 2019. Quantum nanophotonics with group IV defects in diamond. Nat. Commun. 10, 5625.
Bradley, C.E., et al., 2019. A ten-qubit solid-state spin register with quantum memory up to one minute. Phys. Rev. X 9, 031045.
Burek, M.J., et al., 2012. Free-standing mechanical and photonic nanostructures in single-crystal diamond. Nano Lett. 12, 6084–6089.

Burek, M.J., et al., 2014. High quality-factor optical nanocavities in bulk single-crystal diamond. Nat. Commun. 5, 5718.

Burek, M.J., et al., 2016. Diamond optomechanical crystals. Optica 3, 1404–1411.

Burek, M.J., et al., 2017. Fiber-coupled diamond quantum nanophotonic interface. Phys. Rev. Appl. 8, 024026.

Capelli, M., et al., 2019. Increased nitrogen-vacancy centre creation yield in diamond through electron beam irradiation at high temperature. Carbon N. Y. 143, 714–719.

Chan, J., et al., 2011. Laser cooling of a nanomechanical oscillator into its quantum ground state. Nature 478, 89–92.

Chen, Y.C., et al., 2017. Laser writing of coherent colour centres in diamond. Nat. Photonics 11, 77–80.

Chen, Y.-C., et al., 2019. Laser writing of individual nitrogen-vacancy defects in diamond with near-unity yield. Optica 6, 662.

Chia, C., et al., 2018. Diamond optomechanical crystals at cryogenic temperatures. In: Conference on Lasers and Electro-Optics. https://doi.org/10.1364/cleo_si.2018.sf2j.2.

Childress, L., et al., 2006. Coherent dynamics of coupled electron and nuclear spin qubits in diamond. Science 314, 281–285.

Choy, J.T., et al., 2011. Enhanced single-photon emission from a diamond–silver aperture. Nat. Photonics 5, 738–743.

Choy, J.T., et al., 2013. Spontaneous emission and collection efficiency enhancement of single emitters in diamond via plasmonic cavities and gratings. Appl. Phys. Lett. 103, 161101.

Chu, Y., et al., 2014. Coherent optical transitions in implanted nitrogen vacancy centers. Nano Lett. 14, 1982–1986.

Cui, S., et al., 2015. Reduced plasma-induced damage to near-surface nitrogen-vacancy centers in diamond. Nano Lett. 15, 2887–2891.

Davies, G., Hamer, M.F., Price, W.C., 1976. Optical studies of the 1.945 eV vibronic band in diamond. Proc. R. Soc. Lond. A Math. Phys. Sci. 348, 285–298.

Delord, T., Nicolas, L., Chassagneux, Y., Hétet, G., 2017. Strong coupling between a single nitrogen-vacancy spin and the rotational mode of diamonds levitating in an ion trap. Phys. Rev. A 96, 063810.

Delord, T., Huillery, P., Nicolas, L., Hétet, G., 2020. Spin-cooling of the motion of a trapped diamond. Nature 580, 56–59.

Dietrich, C.P., Fiore, A., Thompson, M.G., Kamp, M., Höfling, S., 2016. GaAs integrated quantum photonics: towards compact and multi-functional quantum photonic integrated circuits. Laser Photonics Rev. 10, 870–894.

Doherty, M.W., et al., 2012. Theory of the ground-state spin of the NV^- center in diamond. Phys. Rev. B Condens. Matter 85, 205203.

Dory, C., et al., 2019. Inverse-designed diamond photonics. Nat. Commun. 10, 3309.

Dow, A.B.A., Popov, C., Schmid, U., Kherani, N.P., 2013. Super-high-frequency SAW transducer utilizing AlN/ultrananocrystalline diamond architectures. IEEE Trans. Ultrason. Ferroelectr. Freq. Control 60, 1581–1586.

Duan, L.M., Lukin, M.D., Cirac, J.I., Zoller, P., 2001. Long-distance quantum communication with atomic ensembles and linear optics. Nature 414, 413–418.

Eichenfield, M., Chan, J., Camacho, R.M., Vahala, K.J., Painter, O., 2009. Optomechanical crystals. Nature 462, 78–82.

Evans, R.E., Sipahigil, A., Sukachev, D.D., Zibrov, A.S., Lukin, M.D., 2016. Narrow-linewidth homogeneous optical emitters in diamond nanostructures via silicon ion implantation. Phys. Rev. Appl. 5, 044010.

Evans, R.E., et al., 2018. Photon-mediated interactions between quantum emitters in a diamond nanocavity. Science 362, 662–665.

Faraon, A., Santori, C., Huang, Z., Acosta, V.M., Beausoleil, R.G., 2012. Coupling of nitrogen-vacancy centers to photonic crystal cavities in monocrystalline diamond. Phys. Rev. Lett. 109, 033604.

Faraon, A., et al., 2013. Quantum photonic devices in single-crystal diamond. New J. Phys. 15, 025010.

Gaidarzhy, A., Imboden, M., Mohanty, P., Rankin, J., Sheldon, B.W., 2007. High quality factor gigahertz frequencies in nanomechanical diamond resonators. Appl. Phys. Lett. 91, 203503.

Giannelli, L., Betzholz, R., Kreiner, L., Bienert, M., Morigi, G., 2016. Laser and cavity cooling of a mechanical resonator with a nitrogen-vacancy center in diamond. Phys. Rev. A 94, 053835.

Gimeno-Segovia, M., Rudolph, T., Economou, S.E., 2019. Deterministic generation of large-scale entangled photonic cluster state from interacting solid state emitters. Phys. Rev. Lett. 123, 070501.

Golter, D.A., Oo, T., Amezcua, M., Stewart, K.A., Wang, H., 2016. Optomechanical quantum control of a nitrogen-vacancy center in diamond. Phys. Rev. Lett. 116, 143602.

Goss, J.P., Jones, R., Breuer, S.J., Briddon, P.R., Öberg, S., 1996. The twelve-line 1.682 eV luminescence center in diamond and the vacancy-silicon complex. Phys. Rev. Lett. 77, 3041–3044.

Grutter, K.E., Davanço, M., Srinivasan, K., 2014. Si_3N_4 nanobeam optomechanical crystals. IEEE J. Sel. Top. Quantum Electron. 21, 61–71.

Hausmann, B.J.M., et al., 2011. Single-color centers implanted in diamond nanostructures. New J. Phys. 13, 045004.

Hausmann, B.J.M., et al., 2012. Integrated diamond networks for quantum nanophotonics. Nano Lett. 12, 1578–1582.

Hausmann, B.J.M., et al., 2013. Coupling of NV centers to photonic crystal nanobeams in diamond. Nano Lett. 13, 5791–5796.

Hensen, B., et al., 2015. Loophole-free bell inequality violation using electron spins separated by 1.3 kilometres. Nature 526, 682–686.

Hong, S., et al., 2012. Coherent, mechanical control of a single electronic spin. Nano Lett. 12, 3920–3924.

Huang, T.-Y., et al., 2019. A monolithic immersion metalens for imaging solid-state quantum emitters. Nat. Commun. 10, 2392.

Hughes, A.E., Runciman, W.A., 1967. Uniaxial stress splitting of doubly degenerate states of tetragonal and trigonal centres in cubic crystals. Proc. Phys. Soc. Lond. 90, 827–838.

Humphreys, P.C., et al., 2018. Deterministic delivery of remote entanglement on a quantum network. Nature 558, 268–273.

Iwasaki, T., et al., 2017. Tin-vacancy quantum emitters in diamond. Phys. Rev. Lett. 119, 253601.

Jelezko, F., et al., 2004. Observation of coherent oscillation of a single nuclear spin and realization of a two-qubit conditional quantum gate. Phys. Rev. Lett. 93, 130501.

Kalb, N., et al., 2017. Entanglement distillation between solid-state quantum network nodes. Science 356, 928–932.

Kepesidis, K.V., Bennett, S.D., Portolan, S., Lukin, M.D., Rabl, P., 2013. Phonon cooling and lasing with nitrogen-vacancy centers in diamond. Phys. Rev. B 88, 064105.

Kepesidis, K.V., Lemonde, M.-A., Norambuena, A., Maze, J.R., Rabl, P., 2016. Cooling phonons with phonons: acoustic reservoir engineering with silicon-vacancy centers in diamond. Phys. Rev. B Condens. Matter 94, 214115.

Khanaliloo, B., Jayakumar, H., Hryciw, A.C., Lake, D.P., Kaviani, H., Barclay, P.E., 2015. Single-crystal diamond nanobeam waveguide optomechanics. Phys. Rev. X 5, 041051.

Khanaliloo, B., Mitchell, M., Hryciw, A.C., Barclay, P.E., 2015a. High-Q/V monolithic diamond microdisks fabricated with quasi-isotropic etching. Nano Lett. 15, 5131–5136.

Kimble, H.J., 2008. The quantum internet. Nature 453, 1023–1030.

Kirsch, P., Assouar, M.B., Elmazria, O., Mortet, V., Alnot, P., 2006. 5GHz surface acoustic wave devices based on aluminum nitride/diamond layered structure realized using electron beam lithography. Appl. Phys. Lett. 88, 223504.

Kolkowitz, S., et al., 2012. Coherent sensing of a mechanical resonator with a single-spin qubit. Science 335, 1603–1606.

Kuzyk, M.C., Wang, H., 2018. Scaling phononic quantum networks of solid-state spins with closed mechanical subsystems. Phys. Rev. X 8, 041027.

Lake, D.P., Mitchell, M., Kamaliddin, Y., Barclay, P.E., 2018. Optomechanically induced transparency and cooling in thermally stable diamond microcavities. ACS Photonics 5, 782–787.

Lake, D.P., Mitchell, M., Barclay, P.E., 2019. Processing telecom wavelength light with an optically tunable memory. arXiv physics.optics, 1912.06118.

Lee, C.L., Gu, E., Dawson, M.D., 2007. Micro-cylindrical and micro-ring lenses in CVD diamond. Diamond Relat. Mater. 16, 944–948.

Lee, J.C., Aharonovich, I., Magyar, A.P., Rol, F., Hu, E.L., 2012. Coupling of silicon-vacancy centers to a single crystal diamond cavity. Opt. Express 20, 8891–8897.

Lee, K.W., et al., 2016. Strain coupling of a mechanical resonator to a single quantum emitter in diamond. Phys. Rev. Appl. 6, 034005.

Lee, D., Lee, K.W., Cady, J.V., Ovartchaiyapong, P., Bleszynski Jayich, A.C., 2017. Topical review: spins and mechanics in diamond. J. Opt. 19, 033001.

Lemonde, M.-A., et al., 2018. Phonon networks with silicon-vacancy centers in diamond waveguides. Phys. Rev. Lett. 120, 213603.

Lemonde, M.-A., Peano, V., Rabl, P., Angelakis, D.G., 2019. Quantum state transfer via acoustic edge states in a 2D optomechanical array. New J. Phys. 21, 113030.

Lesik, M., et al., 2014. Perfect preferential orientation of nitrogen-vacancy defects in a synthetic diamond sample. Appl. Phys. Lett. 104, 113107.

Li, P.-B., Li, X.-X., Nori, F., 2019. Band-gap-engineered spin-phonon, and spin-spin interactions with defect centers in diamond coupled to phononic crystals. arXiv quant-ph, 1901.04650.

Lipson, M.G., 2005. Modulating, and emitting light on silicon-challenges and opportunities. J. Lightwave Technol. 23, 4222.

Machielse, B., et al., 2019. Quantum interference of electromechanically stabilized emitters in nanophotonic devices. Phys. Rev. X 9, 031022.

MacQuarrie, E.R., Gosavi, T.A., Jungwirth, N.R., Bhave, S.A., Fuchs, G.D., 2013. Mechanical spin control of nitrogen-vacancy centers in diamond. Phys. Rev. Lett. 111, 227602.

MacQuarrie, E.R., et al., 2015a. Coherent control of a nitrogen-vacancy center spin ensemble with a diamond mechanical resonator. Optica 2, 233.

MacQuarrie, E.R., Gosavi, T.A., Bhave, S.A., Fuchs, G.D., 2015b. Continuous dynamical decoupling of a single diamond nitrogen-vacancy center spin with a mechanical resonator. Phys. Rev. B Condens. Matter 92, 224419.

MacQuarrie, E.R., Otten, M., Gray, S.K., Fuchs, G.D., 2017. Cooling a mechanical resonator with nitrogen-vacancy centres using a room temperature excited state spin–strain interaction. Nat. Commun. 8, 14358.

Maity, S., et al., 2018. Spectral alignment of single-photon emitters in diamond using strain gradient. Phys. Rev. Appl. 10, 024050.

Maity, S., et al., 2020. Coherent acoustic control of a single silicon vacancy spin in diamond. Nat. Commun. 11, 193.

Maletinsky, P., et al., 2012. A robust scanning diamond sensor for nanoscale imaging with single nitrogen-vacancy centres. Nat. Nanotechnol. 7, 320–324.
Marseglia, L., et al., 2011. Nanofabricated solid immersion lenses registered to single emitters in diamond. Appl. Phys. Lett. 98, 133107.
Maze, J.R., et al., 2011. Properties of nitrogen-vacancy centers in diamond: the group theoretic approach. New J. Phys. 13, 025025.
Meesala, S., et al., 2016. Enhanced strain coupling of nitrogen-vacancy spins to nanoscale diamond cantilevers. Phys. Rev. Appl. 5, 034010.
Meesala, S., et al., 2018. Strain engineering of the silicon-vacancy center in diamond. Phys. Rev. B: Condens. Matter Mater. Phys. 97, 205444.
Michl, J., et al., 2014. Perfect alignment and preferential orientation of nitrogen-vacancy centers during chemical vapor deposition diamond growth on (111) surfaces. Appl. Phys. Lett. 104, 102407.
Mitchell, M., et al., 2016. Single-crystal diamond low-dissipation cavity optomechanics. Optica 3, 963–970.
Mitchell, M., Lake, D.P., Barclay, P.E., 2019. Optomechanically amplified wavelength conversion in diamond microcavities. Optica 6, 832.
Mouradian, S., Wan, N.H., Schröder, T., Englund, D., 2017. Rectangular photonic crystal nanobeam cavities in bulk diamond. Appl. Phys. Lett. 111, 021103.
Najafi, F., et al., 2015. On-chip detection of non-classical light by scalable integration of single-photon detectors. Nat. Commun. 6, 5873.
Naydenov, B., et al., 2010. Increasing the coherence time of single electron spins in diamond by high temperature annealing. Appl. Phys. Lett. 97, 242511.
Neuman, T., et al., 2020. A phononic bus for coherent interfaces between a superconducting quantum processor, spin memory, and photonic quantum networks. arXiv quant-ph, 2003.08383.
Neumann, P., et al., 2010. Single-shot readout of a single nuclear spin. Science 329, 542–544.
Nguyen, C.T., et al., 2019. Quantum network nodes based on diamond qubits with an efficient nanophotonic interface. Phys. Rev. Lett. 123, 183602.
Nielsen, M.A., 2004. Optical quantum computation using cluster states. Phys. Rev. Lett. 93, 040503.
Ohno, K., et al., 2012. Engineering shallow spins in diamond with nitrogen delta-doping. Appl. Phys. Lett. 101, 082413.
Ohno, K., et al., 2014. Three-dimensional localization of spins in diamond using 12C implantation. Appl. Phys. Lett. 105, 052406.
Osterkamp, C., et al., 2015. Stabilizing shallow color centers in diamond created by nitrogen delta-doping using SF6 plasma treatment. Appl. Phys. Lett. 106, 113109.
Ovartchaiyapong, P., Lee, K.W., Myers, B.A., Jayich, A.C.B., Budker, D., 2014. Dynamic strain-mediated coupling of a single diamond spin to a mechanical resonator. Nat. Commun. 5, 4429.
Pezzagna, S., Naydenov, B., Jelezko, F., Wrachtrup, J., Meijer, J., 2010. Creation efficiency of nitrogen-vacancy centres in diamond. New J. Phys. 12, 065017.
Pfaff, W., et al., 2014. Unconditional quantum teleportation between distant solid-state quantum bits. Science 345, 532–535.
Pigeau, B., et al., 2015. Observation of a phononic Mollow triplet in a multimode hybrid spin-nanomechanical system. Nat. Commun. 6, 8603.
Pingault, B., et al., 2014. All-optical formation of coherent dark states of silicon-vacancy spins in diamond. Phys. Rev. Lett. 113, 263601.
Pingault, B., et al., 2017. Coherent control of the silicon-vacancy spin in diamond. Nat. Commun. 8, 15579.
Piracha, A.H., et al., 2016. Scalable fabrication of integrated nanophotonic circuits on arrays of thin single crystal diamond membrane windows. Nano Lett. 16, 3341–3347.

Rabl, P., et al., 2009. Strong magnetic coupling between an electronic spin qubit and a mechanical resonator. Phys. Rev. B Condens. Matter 79, 041302.

Rabl, P., et al., 2010. A quantum spin transducer based on nanoelectromechanical resonator arrays. Nat. Phys. 6, 602–608.

Radtke, M., Render, L., Nelz, R., Neu, E., 2019. Plasma treatments and photonic nanostructures for shallow nitrogen vacancy centers in diamond. Opt. Mater. Express 9, 4716.

Rao, D.D.B., Momenzadeh, S.A., Wrachtrup, J., 2016. Heralded control of mechanical motion by single spins. Phys. Rev. Lett. 117, 077203.

Raussendorf, R., Browne, D.E., Briegel, H.J., 2003. Measurement-based quantum computation on cluster states. Phys. Rev. A 68, 022312.

Reiserer, A., Rempe, G., 2015. Cavity-based quantum networks with single atoms and optical photons. Rev. Mod. Phys. 87, 1379–1418.

Riedrich-Möller, J., et al., 2011. One- and two-dimensional photonic crystal microcavities in single crystal diamond. Nat. Nanotechnol. 7, 69–74.

Robledo, L., et al., 2011. High-fidelity projective read-out of a solid-state spin quantum register. Nature 477, 574–578.

Rodriguez-Madrid, J.G., et al., 2012. Super-high-frequency SAW resonators on AlN/diamond. IEEE Electron Device Lett. 33, 495–497.

Rogers, L.J., Jahnke, K.D., Doherty, M.W., Dietrich, A., 2014a. Electronic structure of the negatively charged silicon-vacancy center in diamond. Phys. Rev. B: Condens. Matter Mater. Phys. 89, 235101.

Rogers, L.J., et al., 2014b. All-optical initialization, readout, and coherent preparation of single silicon-vacancy spins in diamond. Phys. Rev. Lett. 113, 1–5.

Ruf, M., et al., 2019. Optically coherent nitrogen-vacancy centers in micrometer-thin etched diamond membranes. Nano Lett. 19, 3987–3992.

Rugar, D., Budakian, R., Mamin, H.J., Chui, B.W., 2004. Single spin detection by magnetic resonance force microscopy. Nature 430, 329–332.

Rugar, A.E., et al., 2020. Generation of tin-vacancy centers in diamond via shallow ion implantation and subsequent diamond overgrowth. Nano Lett. 20, 1614–1619.

Safavi-Naeini, A.H., et al., 2011. Electromagnetically induced transparency and slow light with optomechanics. Nature 472, 69–73.

Sangtawesin, S., et al., 2019. Origins of diamond surface noise probed by correlating single-spin measurements with surface spectroscopy. Phys. Rev. X 9, 031052.

Santori, C., et al., 2010. Nanophotonics for quantum optics using nitrogen-vacancy centers in diamond. Nanotechnology 21, 274008.

Schneider, K., Baumgartner, Y., Hönl, S., Welter, P., Hahn, H., 2019. Optomechanics with one-dimensional gallium phosphide photonic crystal cavities. Optica 6, 577.

Schreck, M., Gsell, S., Brescia, R., Fischer, M., 2017. Ion bombardment induced buried lateral growth: the key mechanism for the synthesis of single crystal diamond wafers. Sci. Rep. 7, 44462.

Schröder, T., et al., 2017. Scalable focused ion beam creation of nearly lifetime-limited single quantum emitters in diamond nanostructures. Nat. Commun. 8, 15376.

Schuetz, M.J.A., et al., 2015. Universal quantum transducers based on surface acoustic waves. Phys. Rev. X 5, 031031.

Sipahigil, A., et al., 2012. Quantum interference of single photons from remote nitrogen-vacancy centers in diamond. Phys. Rev. Lett. 108, 143601.

Sipahigil, A., et al., 2014. Indistinguishable photons from separated silicon-vacancy centers in diamond. Phys. Rev. Lett. 113, 113602.

Siyushev, P., Metsch, M.H., Ijaz, A., Binder, J.M., Bhaskar, M.K., 2017. Optical and microwave control of germanium-vacancy center spins in diamond. Phys. Rev. B: Condens. Matter Mater. Phys. 96, 081201.

Sohn, Y.I., et al., 2018. Controlling the coherence of a diamond spin qubit through its strain environment. Nat. Commun. 9, 1–6.
Sternschulte, H., Thonke, K., Sauer, R., Münzinger, P.C., Michler, P., 1994. 1.681-eV luminescence center in chemical-vapor-deposited homoepitaxial diamond films. Phys. Rev. B Condens. Matter 50, 14554–14560.
Sukachev, D.D., et al., 2017. Silicon-vacancy spin qubit in diamond: a quantum memory exceeding 10 ms with single-shot state readout. Phys. Rev. Lett. 119, 223602.
Tamarat, P., et al., 2006. Stark shift control of single optical centers in diamond. Phys. Rev. Lett. 97, 083002.
Tao, Y., Boss, J.M., Moores, B.A., Degen, C.L., 2014. Single-crystal diamond nanomechanical resonators with quality factors exceeding one million. Nat. Commun. 5, 3638.
Tchernij, S.D., et al., 2017. Single-photon-emitting optical centers in diamond fabricated upon Sn implantation. ACS Photonics 4, 2580–2586.
Teissier, J., Barfuss, A., Appel, P., Neu, E., Maletinsky, P., 2014. Strain coupling of a nitrogen-vacancy center spin to a diamond mechanical oscillator. Phys. Rev. Lett. 113, 020503.
Teraji, T., et al., 2015. Homoepitaxial diamond film growth: high purity, high crystalline quality, isotopic enrichment, and single color center formation. Phys. Status Solidi A 212, 2365–2384.
Togan, E., et al., 2010. Quantum entanglement between an optical photon and a solid-state spin qubit. Nature 466, 730–U4.
Trusheim, M.E., et al., 2020. Transform-limited photons from a coherent tin-vacancy spin in diamond. Phys. Rev. Lett. 124, 023602.
van Dam, S.B., et al., 2019. Optical coherence of diamond nitrogen-vacancy centers formed by ion implantation and annealing. Phys. Rev. B 99, 161203.
Wallquist, M., Hammerer, K., Rabl, P., Lukin, M., Zoller, P., 2009. Hybrid quantum devices and quantum engineering. Phys. Scr. 2009, 014001.
Wan, N.H., Mouradian, S., Englund, D., 2018. Two-dimensional photonic crystal slab nanocavities on bulk single-crystal diamond. Appl. Phys. Lett. 112, 141102.
Wan, N.H., Lu, T.-J., Chen, K.C., et al., 2019. Large-scale integration of near-indistinguishable artificial atoms in hybrid photonic circuits. Nature 583, 226–231.
Wehner, S., Elkouss, D., Hanson, R., 2018. Quantum internet: a vision for the road ahead. Science 362, eaam9288.
Weis, S., et al., 2010. Optomechanically induced transparency. Science 330, 1520–1523.
Xie, L., Zhou, T.X., Stöhr, R.J., Yacoby, A., 2018. Crystallographic orientation dependent reactive ion etching in single crystal diamond. Adv. Mater. 30, 1705501.
Yale, C.G., et al., 2013. All-optical control of a solid-state spin using coherent dark states. Proc. Natl. Acad. Sci. U. S. A. 110, 7595–7600.
Yamamoto, T., et al., 2013. Extending spin coherence times of diamond qubits by high-temperature annealing. Phys. Rev. B 88, 075206.
Ziegler, J.F., Ziegler, M.D., Biersack, J.P., 2010. SRIM—the stopping and range of ions in matter (2010). Nucl. Inst. Methods Phys. Res. B 268, 1818–1823.

Index

Note: Page numbers followed by "*f*" indicate figures and "*t*" indicate tables.

A

Active photonic elements, 161–163, 164*f*
Angle etching technique, 153–154
Avalanche photodiodes (APDs), 106, 163–165

B

Band diagram, p-i-n diamond membrane, 206, 206*f*
Binary detection logic, 11–12, 11*t*
  probability density function (PDF), 12*f*
Boron, 181–182
Boron-doped diamond (BDD), 56–58
  electrodeposition, 87–88, 88*f*
  grain boundaries, junctions and dislocations, 70
  optical image, 199–200, 199*f*
  for supercapacitors, 89

C

Cathodoluminescence (CL) spectroscopy, 78–82, 81*f*
Cavity, photonic crystal, 154–155, 156*f*
Charge-assisted engineering, color centers, 21–25, 23*f*
Chemical vapor deposition (CVD) diamond, 32–33, 150
  dislocation in, 72, 73*f*
  for electronics, sensors and microelectromechanical systems, 85
  grain boundaries, 69–70
  laser-processing, 88
  properties, 177*t*
  thermal resistance measurements, 82–83, 84*f*
  transmission electron microscopy (TEM)
    heteroepitaxy (*see* Heteroepitaxial CVD diamond synthesis)
    homoepitaxy (*see* Homoepitaxial CVD diamond synthesis)
Color centers, 66
  charge-assisted engineering, 21–25, 23*f*
  fabrication of, 185–186
  NV (*see* Nitrogen-vacancy (NV) centers)
  SiV (*see* Silicon vacancy (SiV) centers)
Compressed graphite, 67–69
Conductive diamond, 87–88
Continuous-wave PDMR (CW-PDMR), 113, 115–118, 117*f*
Crystal defects, 65
  defect complexes, 73–74
  dislocations, 72
  grain boundaries, 69–72
  inclusions and voidites, 66–69
  platelets, 69–72
  point defects, 73–74
  stacking faults, 69–72
CTRIM program, 3–6, 5*f*
Cutting edges, CVD-coated ultra-sharp diamond, 89

D

Defect complexes, 73–74
Defective single-crystal diamond films, 52
Depletion region, 194
Detonation nanodiamond (DND), 58–59
  boron stability in, 65
  nitrogen-vacancy centers, 65
  size-dependent reactivity, 60–61
  structure, defects and n impurities, 62, 63–64*f*
  transmission electron microscopy, 60–61, 61*f*
Device independent quantum key distribution (DIQKD), 243–244
Diamond
  crystal structure, 186, 186*f*
  defect complexes, 74
  defects in (*see* Crystal defects)
  dielectric function, 77–78
  dipole orientation, 187, 187*f*
  dislocations in, 72
  electroluminescence, 210–211
  electroluminescence of, 191–197

Diamond (*Continued*)
electron- or proton-irradiation treatments, 128
nitrogen-vacancy (NV) centers, 107–108, 108*f*
overgrowth, 181, 182*f*
point defects, 73–74, 107
ring resonators fabrication, 188, 189*f*
single photon emission, 174–175
spin-qubit interactions, 222, 222*f*
Diamond devices, 176–178
Diamond-like carbon, 59
Diamond membrane
devices, 175
electrical characteristics, 208, 209*t*
fabrication of, 178–181, 180*f*
flip and thin, 182–184, 183*f*
optical resonators, 186–191
p-type, 181–182
induced strain, 180–181, 181*f*
optical characterization, 184–185
performance of, 210–211, 211*t*
Raman spectroscopy, 183–184, 184*f*
single crystal, 179–180, 180*f*, 197–209
Diamond nanoneedles
elastic deformation, 75
in situ TEM deformation testing, 75, 76*f*
Diamond pillars
plastic deformation, 75–77
SiV centers, 80
Dielectric function, of diamond, 77–78
Dirac-delta function, 195–196
Direct laser writing (DLW), 153–154
Dislocations, in diamond, 72
DiVincenzo criteria, 2

E

Elastic deformation, diamond nanoneedles, 75
Electrically triggered photon, 196–197, 197*f*
Electrochemical etching, 179–180
Electroluminescence, of diamond, 191–197, 210–211
Electro-mechanical phase shifter, 162, 163*f*
Electron-beam induced etching (EBIE), 153–154
Electron beam lithography, 188
photonic crystal cavities, 190, 191*f*
Electron energy loss spectroscopy (EELS), 34–35, 78
Electron loss near-edge fine structure (ELNES), 62
Energy-dispersive X-ray spectroscopy (XEDS), 34–35
Epitaxial lateral overgrowth (ELO), 54

F

Face centered cubic (FCC) unit, 186
5keV nanoimplanter, 7, 8*f*
Fluorescent nanodiamond (FND)–gold hybrid nanoparticles, 65
Focused ion beam (FIB) milling, 153–154, 190, 192*f*
nanopores, 12
Frenkel defect, 19

G

Gallium nitride (GaN), 177–178, 177*t*, 209*t*
properties, 177*t*
Gallium phosphide (GaP), 159
GaP-on-diamond devices, 159, 160*f*
Gauss's law, 195–196
Germanium vacancy (GeV) centers, 159
Ground state level anticrossing (GSLAC), 138–139, 139–140*f*

H

Heteroepitaxial CVD diamond synthesis, 36, 37*f*
boron segregation at dislocations, 72
on Ir (001) substrate
early stages, 49–52, 50*f*
layer growth, 52–53
microstructure phenomena, 51*f*
on Ir (111) substrates, 53–54
wafer scale single-crystal diamond, 54–55
Hexagonal diamond. *See* Lonsdaleite
Highly-oriented diamond (HOD) films, 36–38, 37*f*
High-overtone bulk acoustic resonator (HBAR), 234*f*
High-pressure high-temperature (HPHT) synthesis, 35
nanodiamond, 58–59
polycrystalline boron-doped superconducting diamond, 67

High-resolution high-angle annular dark-field (HAADF)-STEM
CVD diamond, 73*f*
diamond nanoneedle, 76*f*
natural diamond, 71*f*
polycrystalline diamond, 68–69*f*
single diamond particle, 81*f*
High-resolution (HR)TEM, 33–34
detonation nanodiamond (DND), 62, 63–64*f*
Homoepitaxial CVD diamond synthesis, 36, 55–58, 57*f*
B-doping induced dislocation formation, 56–58
optimization strategies, 55
4H-silicon carbide (4H-SiC), 177*t*
H-treated nanodiamond, 61–62, 61*f*
100kV ion implanter, 6–7, 6*f*
Hybrid diamond-based photonic architectures, 159–161, 160*f*
Hydrofluoric acid (HF), 12–14
Hydrogen, 21
Hydrogen silsesquioxane (HSQ), 188

I

Identical-location STEM, 87–88, 88*f*
Implanted diamonds, 21–22
Inclusions
in natural diamond, 67
polycrystalline boron-doped superconducting diamond, 67
Indium gallium nitride (InGaN), 177–178
Inductively coupled plasma—reactive ion etching (ICP-RIE), 182–183, 183*f*
In situ TEM deformation, diamond nanopillars, 75–77, 76*f*
Integrated photonic circuits. *See* Photonic integrated circuits (PICs), diamond-based
Interdigital transducers (IDTs), 235
Ion beam induced charge (IBIC) technique, 10
Ion channeling, 3–6
Ion implantation, 1–2. *See also* Single ion implantation
CTRIM simulation, 3–6, 5*f*
100kV implanter, 6–7, 6*f*
nanobeam setup, 7–9, 8*f*
SRIM simulation, 3–6, 4*f*
Ion-implanted diamond, defects in, 69
Ion track etching, 12–14

L

Laser-nucleated cleaving (LNC), 178–179
Laser writing, 88
Liquid metal ion source (LIMS), 14
Lonsdaleite, 70

M

Mach Zehnder interferometer, 162
Magnetometry
microwave-free, 138–140
PDMR NV, 134–138
Man-made platelets, 71
Mechanical devices, 230–231
frequency tuning, 231–233
phonons as on-chip quantum information carriers, 235–238
qubit control, 233–235
Metal–oxide–semiconductor capacitors (MOSCAPs), 86
MgV centers, temperature evolution, 24, 24*f*
Microdisks, 241–242, 241*f*
Microwave-free magnetometry, 138–140
Microwave plasma-enhanced CVD (MPCVD), 35, 176–177, 179
homoepitaxial CVD diamond, 56
Molybdenum di-Sulfide (MoS_2), 177–178, 207, 209*t*
Multiply twinned diamond particles (MTP), 54
Muscovite, 14, 15*f*

N

Nanoapertures, 12–17, 16*f*
Nanobeam cavity, 157–158, 158*f*
Nanocrystalline diamond (NCD), 59
boron-doped, 78
thermal conductivity analysis, 83
Nanodiamond, 32. *See also* Detonation nanodiamond (DND)
applications, 59
core–shell model, 62
fluorescent ND-Au hybrid nanoparticles, 65
HPHT synthesis, 58–59
positive zeta potential, 62
properties of, 60

Nano electromechanical systems (NEMS), 175
Nanoneedles
elastic deformation, 75
in situ TEM deformation testing, 75, 76*f*
Nanophotonic component fabrication, 152–154
Nanophotonic crystal devices, 227, 228–229*f*
Natural diamond
high-resolution high-angle annular dark-field (HAADF)-STEM, 71*f*
lonsdaleite, 70
nitrogen inclusions, 67
platelet structure, 71, 71*f*
properties, 177*t*
voidites, 66–67
Natural platelets, 71
n-doped semiconductor
band bending, 194, 195*f*
energy band, 193–194, 194*f*
ND1 vacancy, 19
Negative Cs imaging (NCSI), 74
Neutral nitrogen-vacancy (NV0), 80, 81*f*
Niobium titanium nitride (NbTiN), 165
Nitrogen, 66
Nitrogen inclusions, 67
Nitrogen-vacancy (NV) centers, 66, 107–108, 108*f*, 220–221
creation efficiency, 21–22
creation of, 17–19
detonation nanodiamond (DND), 65
photon bunching from CL, 81
quantum computer and, 25–27, 26*f*
temperature evolution, 24, 24*f*
for waveguide integrated single photon, 158–159
Nuclear spin readout, PDMR, 132–134, 133*f*

O

1D crystal defects, 65
1D photonic crystal cavities, 227
Optical detection of magnetic resonances (ODMR), 106, 141
vs. PDMR performance, 137, 137*t*
Optical resonators, 154–155
Optomechanical cavity, 236–237, 237*f*
Optomechanical crystals (OMC), 239–241, 240*f*
Opto-mechanical phase shifter, 162, 163*f*

P

Passive photonic elements, 154–155
Patterned nucleation growth (PNG), 54
PDMR. *See* Photoelectric detection of magnetic resonances (PDMR)
p-doped semiconductor
band bending, 194, 195*f*
energy band, 193–194, 194*f*
Phonons, as on-chip quantum information carriers, 235–238
Photocurrent density, 114–115
Photoelectric detection of magnetic resonances (PDMR), 106, 108*f*
continuous-wave, 115–118, 117*f*
contrast limitation, 124
diamond material properties on performance
background photocurrent, 121–125
photoelectric detection rate, 125–129, 126*f*
experimental implementation, 113–121
for magnetic field sensing
microwave-free magnetometry, 138–140
PDMR NV magnetometry, 134–138
nuclear spin readout, 132–134, 133*f*
NV ionization mechanisms, 108–112, 109*f*
vs. ODMR performance, 137, 137*t*
photocurrent detection, 114–115
pulsed, 118–121, 120*f*
spin dependence of NV center ionization, 112–113
Photoelectric detection techniques, 106–108
Photoelectric quantum effect, 106
Photoelectric readout, of NV spin
continuous-wave and pulsed PDMR, 113–121
single NV spin, 129–132, 130–131*f*
Photoluminescence (PL), 78–82
Photoluminescence excitation (PLE), 175–176
Photonic crystal cavities, 188–191

electron beam lithography, 190, 191f
FIB milling, 190, 192f
patterned silicon mask, 190, 193f
Photonic devices
freestanding, 225–230
surface-etched devices, 223–225, 223f
Photonic integrated circuits (PICs), diamond-based, 149–150
active components, 161–163, 164f
nanophotonic component fabrication, 152–154
passive photonic elements, 154–155
single photon emitters, 155
coupling to diamond photonic structures, 157–159
hybrid architectures, 159–161, 160f
waveguide integrated single-photon detectors, 163–166
Photon-phonon interaction, 222, 222f
Photons, 106
electrically triggered, 196–197, 197f
p-i-n diamond membrane, 203–209, 204f, 206f
band diagram, 206, 206f
optical image, 203–205, 204–205f
p-i-n ZnO, 206–207
Planar defects, 71
Planar photonic crystal cavity, 157–158, 158f
Plastic deformation, diamond pillars, 75–77
Platelets, in natural diamond, 71, 71f
p-n diamond junctions
characterization of, 200–203
depletion region, thickness of, 202–203
engineering of, 201, 202f
fabrication of, 200
Point defects, 73–74
Polycrystalline diamond (PCD), 152, 175
defects and sp^2 hybridization, 70–71
grain junctions and boundaries, 67, 68–69f
thermal conductivity analysis, 83
thermal properties measurement, 82–83
Porous diamond, for supercapacitors, 89
Printed circuit board (PCB), 113–114
p-type diamond membranes
characterization of, 199–200, 199f
fabrication of, 181–182
Pulsed PDMR, 118–121, 120f
Purcell effect, 175
Purcell factor, 157–158, 187, 187f

Q

Quality factor (Q-factor), 154–155
Quantum computer, concept of, 25–27, 26f
Quantum network architecture, 237–238, 238f
Quantum supremacy, 174
Quantum technology, 174, 220
Qubit, 9, 21–22

R

Rabi oscillations, 130–131f, 132–134
Racetrack resonator, 154–155, 156f
Raman spectroscopy, 183–184, 184f
Ramsey sequences, 134–135
Reactive ion etching (RIE), nanopillar fabrication, 223–224
Refractive index, 150
Ring resonators fabrication, diamond, 188, 189f
$R_{PDMR/ODMR}$, 121–123, 122f
$R_{Quadratic/Linear}$, 121–123, 122f

S

Scanning electron microscope (SEM)
diamond cantilever, 231, 232f
diamond optomechanical cavity, 236–237, 237f
microdisk, 241–242, 241f
nanopores, 14, 15f
optomechanical crystal (OMC), 239, 240f
suspended diamond waveguide, 225, 226f
Scanning TEM (STEM), 33–34
Schottky power devices, 85–86
Secondary ion mass spectrometry (SIMS), 200, 201f
Silicon carbide (SiC), 177–178, 209t
Silicon nitride (SiN), 159–160
Silicon nitride–diamond hybrid devices, 159, 160f
Silicon (Si), properties of, 177t
Silicon vacancy (SiV) centers, 174–175, 220–221
in diamond crystallites, 80
in diamond pillars, 80

Silicon vacancy (SiV) centers (*Continued*)
quantum network architecture using, 237–238, 238*f*
stability and fluorescence properties, 79–80
for waveguide integrated single photon, 158–159
Single crystal diamond membranes, 179–180, 180*f*
for nanoelectronics, 197–209
p-i-n junctions, 203–209, 204*f*, 206*f*
p-n junctions
characterization of, 200–203
fabrication of, 200
p-type diamond membranes, 199–200, 199*f*
Single crystalline diamond (SCD), 152
Single diamond particle, CL spectroscopy analysis, 80, 81*f*
Single ion implantation, 9–17, 9*f*
concept for, 13*f*
ion detection, 13*f*
nanoapertures, 12–17
Single NV spin photoelectric detection, 129–132, 130–131*f*
Single-photon detectors, waveguide integrated, 163–166
Single photon emission, 174
Single photon emitters, 155, 174
coupling to diamond photonic structures, 157–159
hybrid diamond-based photonic architectures, 159–161, 160*f*
Solid immersion lenses (SILs), 223*f*, 224
Source of negatively charged ions (SNICS), 6–7, 6*f*
Space charge region (SCR), 194, 195*f*
Spherical nanodiamond, 58–59
Spin-phonon interaction, 222, 222*f*
Spin-phonon quantum network, 237–238, 238*f*
Spin-photon interaction, 222, 222*f*
Spin-to-charge conversion (SCC), 112
SRIM program, 3, 4*f*
Stark shift, 231
Step-free diamond surfaces, 56
Stopping range of ions in matter (SRIM), 179–180
Straggling, 3–6
Superconducting nanowire single photon detectors (SNSPDs), 151
waveguide integrated, 163–165, 166*f*
Surface acoustic wave (SAW), 234*f*, 235
Surface-etched devices, 223–225, 223*f*
Suspended photonic devices, 222

T

TEM. *See* Transmission electron microscopy (TEM)
Thin down approach dry etching, 153–154
Thin film bonding diamond techniques, 228–229*f*
Tin vacancy (SnV) centers, temperature evolution, 24, 24*f*
Transition-metal dichalcogenides (TMDs), 207
Transmission electron microscopy (TEM), 33–34
for CVD diamond synthesis, 35–36
heteroepitaxy on Ir (100) and on Ir (111), 49–55
heteroepitaxy on Si and SiC, 36–49
homoepitaxial crystal growth, 55–58
electronic properties measurement, 77–78
H-treated nanodiamond, 61–62, 61*f*
in situ deformation testing, diamond nanopillars, 75–77, 76*f*
for process and device technology, 83–89, 88*f*
thermal properties measurement, 82–83, 84*f*
Transparency zone, 15
Tungsten di-selenide (WSe_2), 177–178, 207, 209*t*
Two-color PDMR, 125
2D air-suspended PhC-cavities, 192*f*
2D chalcogenide materials, 207
2D crystal defects, 65
3D crystal defects, 65
Type Ib diamond, 20
Type IIa diamond, 20
Type IIb diamond, 20

U

Ultrafast TEM (UTEM), 82
Ultra-nanocrystalline diamond (UNCD), 59, 77
Undercut etching, 153–154

V

Vacancies, in diamond
  creation of, 17–19, 18*f*
  implantation-induced, 22
  radiation-induced damage, 19–20
Valence EELS (VEELS), 77–78
Voidites, 66–67

W

Wafer scale single-crystal diamond heteroepitaxy, 54–55
Waveguide, 154
  couplers, 154
  SNSPDs, 163–165, 166*f*
Waveguide-coupled microdisk resonators, 159, 160*f*
Waveguide integrated single-photon detectors, 163–166
Waveguide-optomechanical systems, 241*f*, 242
Whispering gallery modes (WGMs), 241–242

X

XEDS. *See* Energy-dispersive X-ray spectroscopy (XEDS)

Z

Zeeman splitting, 134–135
Zeeman tuning, 236–237
Zero-dimensional defects, 65

FSC
www.fsc.org
MIX
Papier | Fördert
gute Waldnutzung
FSC® C083411

Kolibri 360 GmbH
Ferdinand-Jühlke-Straße 7
99095 Erfurt, Deutschland
produktsicherheit@kolibri360.de